Studies in Systems, Decision and Control

Volume 360

The series "Studies in Systems, Decision and Control" (SSDC) covers both new developments and advances, as well as the state of the art, in the various areas of broadly perceived systems, decision making and control–quickly, up to date and with a high quality. The intent is to cover the theory, applications, and perspectives on the state of the art and future developments relevant to systems, decision making, control, complex processes and related areas, as embedded in the fields of engineering, computer science, physics, economics, social and life sciences, as well as the paradigms and methodologies behind them. The series contains monographs, textbooks, lecture notes and edited volumes in systems, decision making and control spanning the areas of Cyber-Physical Systems, Autonomous Systems, Sensor Networks, Control Systems, Energy Systems, Automotive Systems, Biological Systems, Vehicular Networking and Connected Vehicles, Aerospace Systems, Automation, Manufacturing, Smart Grids, Nonlinear Systems, Power Systems, Robotics, Social Systems, Economic Systems and other. Of particular value to both the contributors and the readership are the short publication timeframe and the world-wide distribution and exposure which enable both a wide and rapid dissemination of research output.

Indexed by SCOPUS, DBLP, WTI Frankfurt eG, zbMATH, SCImago.

All books published in the series are submitted for consideration in Web of Science.

More information about this series at http://www.springer.com/series/13304

Vitaliy P. Babak · Serhii V. Babak ·
Volodymyr S. Eremenko · Yurii V. Kuts ·
Mykhailo V. Myslovych · Leonid M. Scherbak ·
Artur O. Zaporozhets

Models and Measures in Measurements and Monitoring

Springer

Vitaliy P. Babak
Institute of Engineering Thermophysics
National Academy of Sciences of Ukraine
Kyiv, Ukraine

Volodymyr S. Eremenko
Igor Sikorsky Kyiv Polytechnic Institute
National Technical University of Ukraine
Kyiv, Ukraine

Mykhailo V. Myslovych
Institute of Electrodynamics
National Academy of Sciences of Ukraine
Kyiv, Ukraine

Artur O. Zaporozhets
Institute of Engineering Thermophysics
National Academy of Sciences of Ukraine
Kyiv, Ukraine

Serhii V. Babak
Verkhovna Rada of Ukraine
Kyiv, Ukraine

Yurii V. Kuts
Igor Sikorsky Kyiv Polytechnic Institute
National Technical University of Ukraine
Kyiv, Ukraine

Leonid M. Scherbak
Kyiv International University
Kyiv, Ukraine

ISSN 2198-4182 ISSN 2198-4190 (electronic)
Studies in Systems, Decision and Control
ISBN 978-3-030-70782-8 ISBN 978-3-030-70783-5 (eBook)
https://doi.org/10.1007/978-3-030-70783-5

This Springer imprint is published by the registered company Springer Nature Switzerland AG
The registered company address is: Gewerbestrasse 11, 6330 Cham, Switzerland

Introduction

Nowadays, measurements are an integral component in all spheres of the life of modern society—from everyday life to high-tech industries, energy, aviation, astronautics, various branches of the scientific and industrial complex. Ensuring comfortable living conditions for a person at home ("smart home"), production of electric and thermal energy, management of complex technological processes, monitoring the dynamics in space and time of climatic and ecological environmental conditions, control of transport objects on earth, water, air and space diagnostics of technical and other objects of varying complexity, product quality control and management—all these processes require high-precision measurements of a significant number of physical quantities, which develop in space and time, and processing of constantly growing volumes of measurement information.

Measurement is almost the only source of obtaining objective quantitative information about the surrounding material world, the level and quality of society. Expanding the dynamic range of measurements in studies of the micro- and macroworld, increasing the requirements for accuracy and reliability of measurement results, increasing the range of measured electrical and non-electrical physical quantities—all this requires an in-depth understanding of the synthesis and analysis of measurement procedures and operations, the development of an appropriate theoretical basis, information and hardware providing.

The cornerstone of measurement theory and practice has always been and remains the fundamental concepts of model and measure. It is known that models and measures have always reflected and served as a kind of indicator of the achievements of science and technology at a certain historical stage of their development. On their basis, measuring instruments were developed, the use of which in practice contributed to the acquisition of new knowledge. In practice, the hypothesis of the famous English physicist A. Michelson (middle of nineteenth century) about new discoveries in physics, when the relative measurement accuracy reaches the level of 10^{-n} ($n \geq 6$), was confirmed.

It is precisely the probability models of the measured quantities, processes, signals as information carriers, as well as physical and probabilistic measures that determine the measurement result, provide it with the properties of objectivity and reliability. Therefore, the issues of improving and developing models and measures in the measurement methodology play an increasingly important role in finding ways to achieve high accuracy of measurements and expanding their areas of application. This monograph is devoted to questions of generalization and research of features and the application of constructive models and measures in the modern measurement methodology.

The monograph consists of 8 chapters.

Chapter 1 discusses general issues of measurement problems, including the main provisions and definitions of measurements, proposes a system of postulates on which the theory and practice of measurements are based, and the concept of a mathematical model of measurement uncertainty, and briefly describes measurements as the inverse problem of the theory of signals and systems. The concept of measure and its development from a philosophical category to objects of study and use in mathematics and metrology are studied in detail. The authors proposed and reviewed the concept of matching physical and probabilistic measures in measurements and proposed a classification of physical and probabilistic measures for measurements, and the use of various measures in the structure of information-measuring systems.

Chapter 2 systematizes mathematical models of measuring signals and fields, considers the main spatiotemporal models of quasi-determined signals and provides the necessary information about the spaces of signals, in particular linear, metric, normalized and Hilbert. Separately, the use of orthogonal bases in the study of deterministic signals is considered, and theoretical information about the model of random signals and fields is presented.

Chapter 3 discusses models and measures for measuring random angular quantities and provides basic concepts, terms, definitions and characteristics that are used in the statistical analysis of angular data. The most characteristic probability distributions of random angles (von Mises and wrapped normal distribution) are analyzed. A comparative analysis of the numerical characteristics of random variables and random angles is performed. The basic information about the models developed by the authors and measures of random phase shifts of cyclic signals for phase measurements is presented.

Chapter 4 gives an example of the construction and use of models and measures in the diagnosis of electric power facilities, describes in detail the processes of generating diagnostic signals and models of the formation of training sets (measures) for diagnosing the state of electric power equipment and offers the option of constructing diagnostic spaces based on the measurement of information signals, as well as constructing decisive rules for the diagnosis and classification of certain types of defects in the nodes of electro-energy objects.

Chapter 5 shows examples of the use of models and measures on a circle for solving problems of precision ultrasonic thickness measurement of products from materials with significant attenuation, processing the results of bag scale-free phase measurements based on numerical systems of residual classes in phase range finders and direction finders. The features of statistical data processing in environmental monitoring systems based on unmanned aerial vehicles (UAVs) are considered.

Chapter 6 discusses the features of using models and measures for standardless measurements of the characteristics of composite materials and substantiates the possibility of using neural networks in computerized diagnostic systems to classify defects and build appropriate naming scales.

Chapter 7 discusses air pollution monitoring using UAVs. The specifics and content of the information support of the measuring system using UAVs are described. A model of the vector random field is proposed for the case of the formation of a local pollution field, the characteristics of which are estimated in the framework of the correlation theory. The structure of the multifunctional measuring system using UAVs for remote monitoring of air pollution is analyzed, and the results of experimental studies of monitoring air pollution by radionuclides near the Chernobyl Nuclear Power Plant are presented.

Chapter 8 considers the features of monitoring atmospheric air pollution and proposes a promising direction for the development of an extensive network of monitoring systems. A method for localizing the source of pollution according to monitoring data in the polar coordinate system is proposed. The methods of statistical processing of monitoring data within the framework of the correlation theory are substantiated. The structure of a spatially branched air pollution monitoring network using modern information technologies is described. The measuring modules of the network are developed, and the results of their experimental studies are presented. A method for assessing the level of air pollution based on monitoring results and statistical characteristics of local atmospheric pollution fields is proposed.

A number of formulated provisions and statements are debatable. The authors will be grateful to all readers who will send feedback, comments and suggestions on the material presented in the monograph.

Vitaliy P. Babak
Serhii V. Babak
Volodymyr S. Eremenko
Yurii V. Kuts
Mykhailo V. Myslovych
Leonid M. Scherbak
Artur O. Zaporozhets

Contents

Abbreviations

ADC	Analog-to-digital signal conversion
AFCS	Automatic flight control system
APCS	Automatic process control system
DC	Direct current
ECM	Electronic countermeasure
EDR	Exposure dose rate
EE	Electric power equipment
EM	Electrical machine
EPF	Electric power facility
FID	Filter-injection device
FPM	Frequency–pulse modulation
GPS	Global Positioning System
GCE	Ground control equipment
IMS	Information-measuring system
IRPU	Information receiving and processing unit
ISO	International Organization for Standardization
LRP	Linear random process
MC	Magnetic circuit
MPC	Maximum permissible concentration
NPP	Nuclear power plant
OCE	Onboard control equipment
PCRP	Periodically correlated random process
PDF	Probability density function
PWM	Pulse width modulation
RMS	Root mean square
RO	Research object
RP	Radio pulse
SAR	Synthetic-aperture radar
SCA	Sample circular average
SD	Standard deviation
SM	Steering mechanism
SRC	System of residual class

SRL	Selective resulting length
SRNS	Satellite radio navigation system
TPP	Thermal power plant
UAS	Unmanned aerial system
UAV	Unmanned aerial vehicle
UTM	Ultrasonic thickness measurement
WEU	Wind electric unit

Chapter 1
Problems and Features of Measurements

Abstract The general problems of measurements and monitoring are considered, including the issues of displaying the dynamics of a physical quantity in space and time, the stability of unit's measures of a physical quantity, spatial coordinates and time, data protection and measurement results. The system of postulates on which the theory of measurements is based is given. The role of models and measures for the formation of the result of measuring physical quantities is shown. The concept of a mathematical model of measurement uncertainty is considered. The concept of "measure" and its development from a philosophical category to objects of study and use in mathematics and metrology are studied in detail. The charge is considered as an object of mathematics for use in measurements of alternating physical quantities. The features of measures for measuring physical quantities distributed on a straight line and on a circle are analyzed. The concept of matching physical and probabilistic measures in measurements is proposed. It is concluded that obtaining a meaningful measurement result is based on the use of a consistent inextricably linked set of physical and probabilistic measures. The classification of physical and probabilistic measures for measurements is given. The use of various measures (physical, mathematical, probabilistic) in the structure of information-measuring systems is considered. Measurements are briefly characterized as the inverse problem of the theory of signals and systems.

Keywords Measurement postulates · Physical measures · Mathematical measures · Probabilistic measures · Model of measurement uncertainty · Measures on the line · Measures on the circle

1.1 Main Provisions

Measurements are considered one of the main directions of cognition of the world. Currently, measurement operations are carried out on such a large scale worldwide that it has become an integral part of our lives. With a deeper formulation of measurement problems, it is precisely the absence of their results makes it impossible to solve a number of urgent and important problems of our time. For example, climatic and

V. P. Babak et al., *Models and Measures in Measurements and Monitoring*,
Studies in Systems, Decision and Control 360,
https://doi.org/10.1007/978-3-030-70783-5_1

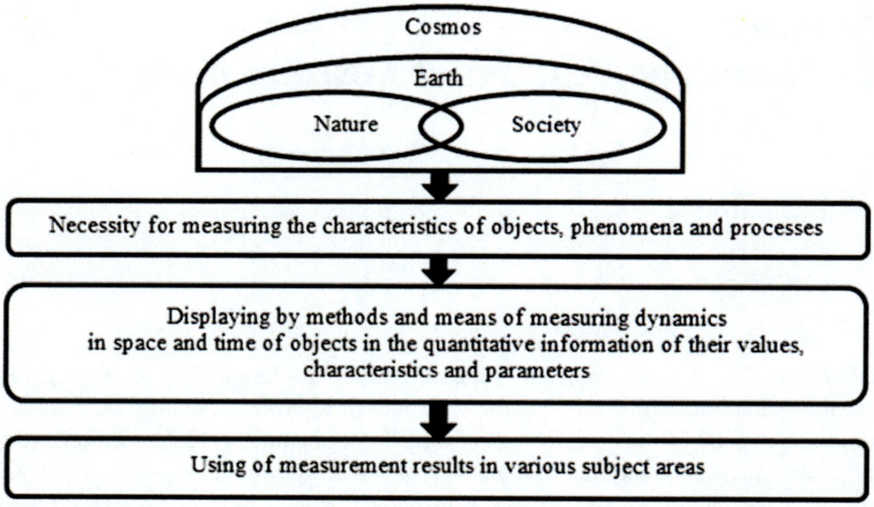

Fig. 1.1 Schematic illustration of the measurements using

environmental problems on our planet are not being fully resolved due to the lack of the necessary databases for measuring dynamics in time and space of environmental characteristics. This is primarily due to insufficiently substantiated mathematical models, spatially undeveloped hardware and relevant information support for environmental studies [1–3].

For revealing the content and essence of measurements, it briefly presents the main provisions of the measurements and considers the use of models and measures as the main objects in a wide range of studies [4, 5]. The general approach to the use of measurements can be represented by a schematic illustration (Fig. 1.1).

In metrology, as the science of measurements and their practical application, there are three main areas of research:

- measurement theory, including the theoretical foundations of creating physical and mathematical models, methods, measures, information support and hardware for measurement processes;
- applied metrology, including the development of hardware and software tools and measurement systems, a reference base of physical quantities, etc.;
- legal metrology, which includes legislative acts, standards and recommendations, the implementation of which are regulated and monitored by the state to ensure the uniformity of measurements in the country and abroad.

Measurement definition. Given the wide range of measurements using, it is advisable to give 3 options for its definitions.

Philosophy. Measurement is a creation and using at each stage of the development of society of effective measuring instruments as the main resource for understanding the world.

Metrology. Measurement is a reflection of the dynamics in space and time of the values and characteristics of various physical quantities into a quantitative result based on interaction with the object of study of special measuring instruments.

In a broad sense. Measurement is a obtaining and using the results of experimental measuring procedures for monitoring and controlling the modes of functioning of objects and systems in various fields of science, technology and economy, monitoring the level and quality of life of a society.

Each science becomes more structured if an appropriate system of laws, axioms, or postulates is formed for it. Examples are: Euclidean axioms in geometry; Kolmogorov axioms in probability theory; Newton's laws in mechanics; periodic table in chemistry; fundamental laws of thermodynamics and electrodynamics [6–10].

For measurement, as a combination of theory and practice, such a system of postulates can be proposed:

Postulate 1. The dynamics of changes in the properties, values and characteristics of the physical quantities of objects of nature and civilization occur and manifest in space and time.

Postulate 2. The quantitative result of measuring a physical quantity is formed by the interaction of the object of study with the measuring tool, based on the using of comparative operations with the measurement of a unit of physical quantity and processing of the obtained measurement data.

Postulate 3. The theory and practice of each measurement are united and reflected by a mathematical model of dynamics in space and time of a physical quantity as a model of uncertainty and the corresponding operators of measuring instruments for its transformation to evaluate the measurement results.

The main problems of measurement include the following:

Display problem. Substantiation and mapping of dynamics in space and time of a physical quantity into a model of measurement uncertainty and the corresponding operators of its transformation in links and modules of measuring instruments.

The problem of stability measures. Ensuring spatial homogeneity and temporal stationarity of measures and standards in different places and at different points in time in order to ensure the uniformity of measurements and achieve specified results and their accuracy indicators.

The problem of data protection and measurement results. Implementation of a set of measures and the use of protective equipment during generating, transmitting, processing and presenting data and measurement results in accordance with information security requirements in order to minimize the impact of both natural and deliberate interference and unauthorized access to measurement information.

Systems of units of physical quantities. From the history of the development of measurements, it is known that the first means of measurement were the human senses. The measurements themselves began from the moment when a certain material object or process (for example, a finger of a certain length, a hand of a tribal leader, a bag of grain, a time interval that corresponds to a certain phase of the Sun, Moon in the sky, etc.) was set in correspondence (display) an abstract unit—as a primary numerical measure, the size of a physical quantity. The emergence of a number system unit became the starting point for the development of geometry, mathematics, and other sciences. Subsequently, various systems of units of physical quantities were formed and used. The facts of the formation and transition to another system of units testified to the corresponding stages of the development of terrestrial civilization, that is, they served as peculiar indicators of their activity, and the potential of science and technology for the corresponding time period was used to implement units of physical quantities.

An example is the process of forming a modern system of SI units, on the basis of which the results of measuring the values of the macrocosm are obtained on the basis of the values of the microcosm. So, for example, the unit of time second (atomic second) of the SI system is defined as follows: second (s) is equal to 9,192,631,770 periods of radiation corresponding to the transition between two ultrafine levels of the ground state of the cesium-133 atom. Today, with using the SI system, measurements are made of up to 100 electrical and magnetic quantities and more than 4000 quantities of a different physical nature.

The measurement results serve as primary useful information for further operations of monitoring, diagnostics, management, forecasting, etc. A concept or term "monitoring" is also used that is directly related to the measurement, primarily with numerical information about the object under study or a system that develops in space and time.

Monitoring is a research process in the space and time of functioning and the state of various objects and systems to obtain the dynamics of their current values and characteristics according to measurement data for solving monitoring, diagnostics, control and forecast tasks.

Based on the measurement data in various subject areas, we can give examples of approximate monitoring time intervals (Fig. 1.2), which largely determine

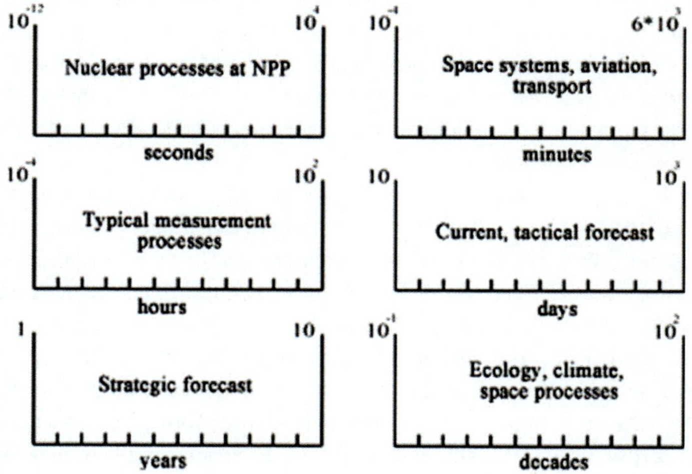

Fig. 1.2 Examples of time intervals (scales) for monitoring in various subject areas

the methodology for conducting the corresponding measurement processes of both macrocosm objects and microcosm objects.

The measurement process and the main stages of its implementation. Measuring operations have characteristic features that describe:

- interaction of the research object with the primary measuring transducer (sensor) of the measuring instrument;
- the conversion of the measurement information of the study object—information of a physical quantity into a measuring signal, which is its physical carrier;
- comparison of the signal level with the size of the measurement unit and the formation of numerical measurement data;
- assessing and presenting quantitative results and characteristics of measurement uncertainty.

Performing such operations makes it possible to reflect the specifics of the measurement process, classify hardware and software (systems), and describe the next steps [11, 12].

Organizational and preparatory phase. At this stage, the technical task for conducting the measurement experiment is agreed upon by the customer and the contractor, the deadlines for completion, allocation of necessary funds and other resources, including financial ones, are determined (Fig. 1.3).

The stage of creating information support and hardware. At this stage, the contractor develops and substantiates information support (models, measures, algorithms, programs) and hardware (tools, systems, equipment) for conducting the measured experiment; a program and methodology for conducting such an experiment are created and agreed upon (Fig. 1.4).

Final stage. At this stage, the performer conducts a full-scale measuring experiment, the measurement data are processed and the result and indicators of measurement accuracy are obtained. These results are documented and transmitted to the customer for further use (Fig. 1.5).

Integrally, the problem of measuring physical quantities with distinguished models and measures is illustrated by the following schematic representation (Fig. 1.6) [13–15].

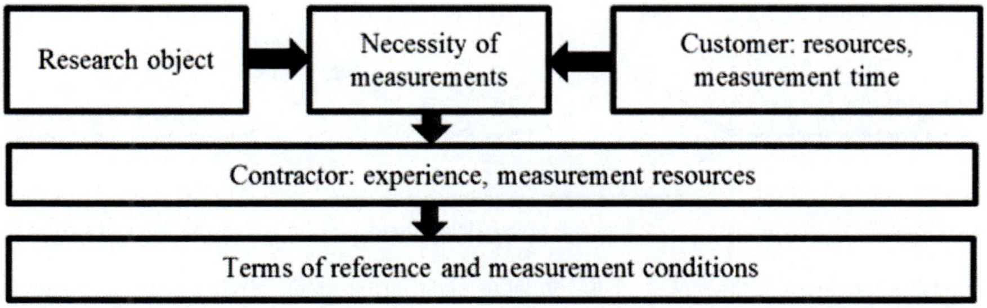

Fig. 1.3 The structural diagram of the organizational and preparatory phase of the measurement process

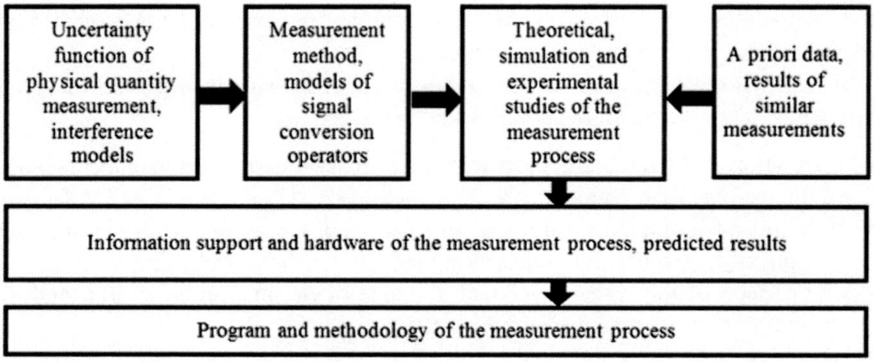

Fig. 1.4 Block diagram of the stage of creating information support and hardware for the measurement process

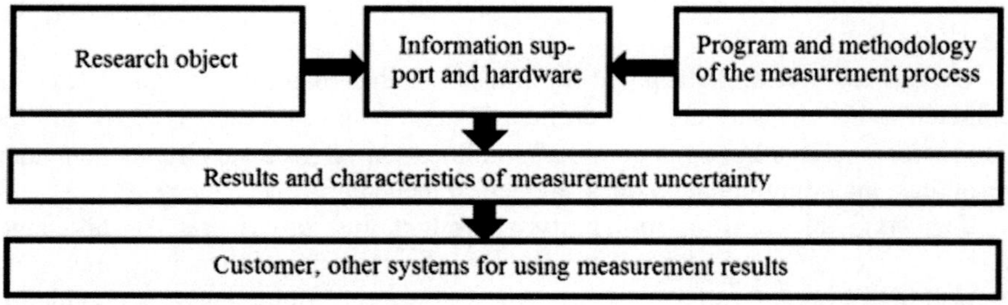

Fig. 1.5 Block diagram of the final stage of the measurement process

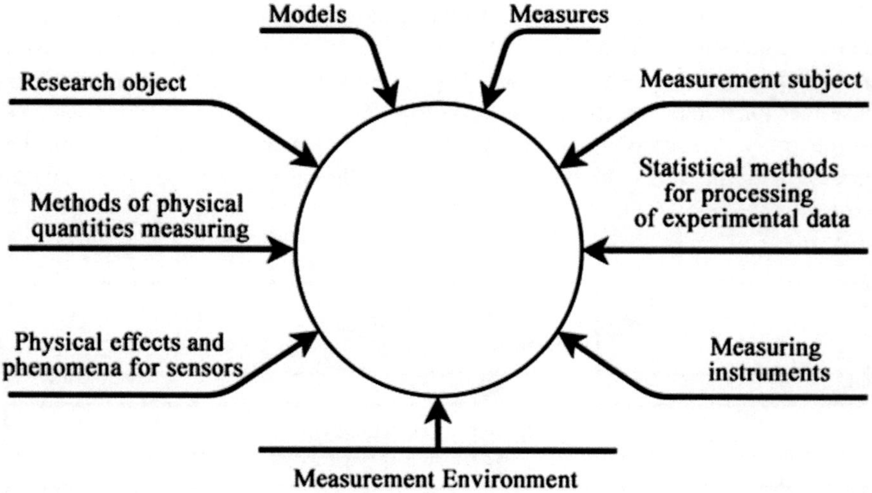

Fig. 1.6 Schematic representation of the formation of the measurement result of physical quantities

Next, it move on to a more detailed consideration of models and measures in measurement problems as the main subjects of research in this work.

1.2 Mathematical Model of Measurement Uncertainty

Solutions to a wide range of measurement problems are combined based on the use of physical and mathematical models of the studied objects, processes and phenomena. Despite the fact that such models are secondary, that is, copies, reflections that do not fully correspond to real objects, their role in the measurement methodology is very significant and has a fundamental character. Each of the models, physical and mathematical, complement each other, which allows to increase the efficiency of the measurement process, for example, justify ways to improve the accuracy of measurement results [16–20].

In order to take into account the specificity and characteristic features of measurement, it is proposed to distinguish a class of mathematical models that describe the primary information on measuring physical quantities of various objects of research on a wide range of models. The class of such models will be called the class of measurement uncertainty models [21–23].

The mathematical model of measurement uncertainty is a one-dimensional or multidimensional Hilbert, determinate or random function, or their combinations, the value and numerical characteristics of which are estimated by the results and indicators of measurement accuracy [24].

A mathematical model of the uncertainty of measuring a physical quantity is created on the basis of a combination of knowledge, hypotheses, initial and boundary conditions generated from a priori research data of a quantity recorded using mathematical objects, terms and symbols in the form of a logically consistent, consistent structure that reflects the dynamics in space and time, properties, value and characteristics of the studied value, and their quantitative assessment must be carried out by processing the measurement data experiment.

It is known that the energy characteristics of the second order (for example, energy, power, dispersion) of Hilbert functions are finite, that is, such functions are physically realizable, and are not a mathematical idealization (such as, for example, a continuous random process of white noise that has infinite dispersion) [25].

The above definition of the uncertainty model uses the most typical functions which used in theory and practice of measurement, such as the implementation of such a model. These functions are called measurement uncertainty functions.

Measurement uncertainty model is:

- necessary component of the statement of each measurement task;
- hypothetical functional dependence in the absence of the necessary a priori data for constructing the functional dependence, which is typical at the initial stages of the study of new physical quantities;

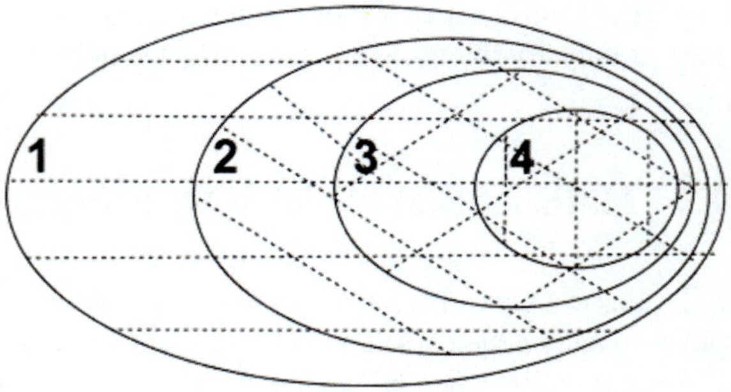

Fig. 1.7 Schematic representation of the evolution of the measurement uncertainty model in the form of an Euler-Venn diagram

- theoretical tools for constructing a mathematical model of the studied value according to measurements with achieved accuracy indicators.

The evolution of measurement uncertainty models can be illustrated by the following diagram in the form of an Euler-Venn diagram (Fig. 1.7).

Figure 1.7 indicates: 1—a common set of models; 2—class of primary models of measurement uncertainty; 3—model of the studied value using quantitative results and achieved indicators of measurement accuracy; 4—an idealized measurement model used in a computer measurement experiment and in other cases, taking into account the quantitative measurement result.

In the process of creating, justifying and using such models, it is necessary to take into account the characteristic features of the measurement, which we dwell on in more detail [26–28].

The most complex and important tasks of the measurement process, ensuring the success of the measurement experiment, is to perform the following measurement operations:

- using hardware: the formation of measurement information; converting the existing information by the primary transducer (sensor) into a measuring signal; comparing the signal with the degree of physical quantity and the formation of numerical information (measurement data);
- using information support: justification and creation of a model of a physical quantity; creation of a model for the conversion of measurement information; substantiation of the measuring signal model.

The above measurement operations are schematically illustrated in Fig. 1.8.

In the general case, 3 options can be distinguished for the formation of the primary measurement information of the research object (RO) (Fig. 1.9).

Given into account the diversity of measurement processes in a wide range of subject sectors of the economy, science and technology, the classification of measurement uncertainty models is given (Fig. 1.10). For the signs of classification, subject areas of use, the type and nature of models, and combinations of models were selected.

Any classification is to some extent conditional. Therefore, the classification given in Fig. 1.10 may be supplemented or modified in accordance with the introduction of new features and structures.

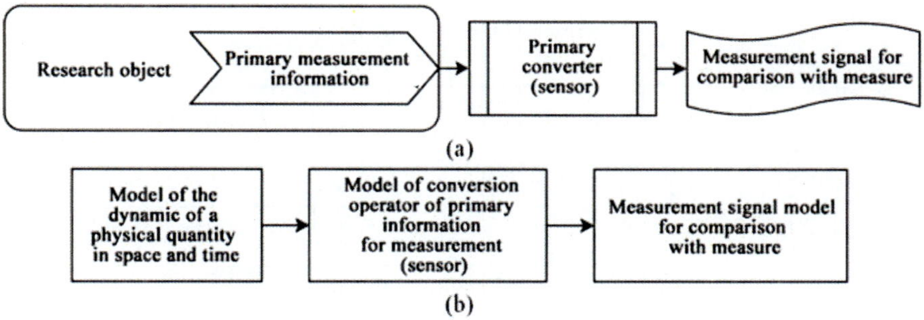

Fig. 1.8 Schematic illustration of the first operations of a measurement experiment based on the use of: **a** hardware; **b** information support

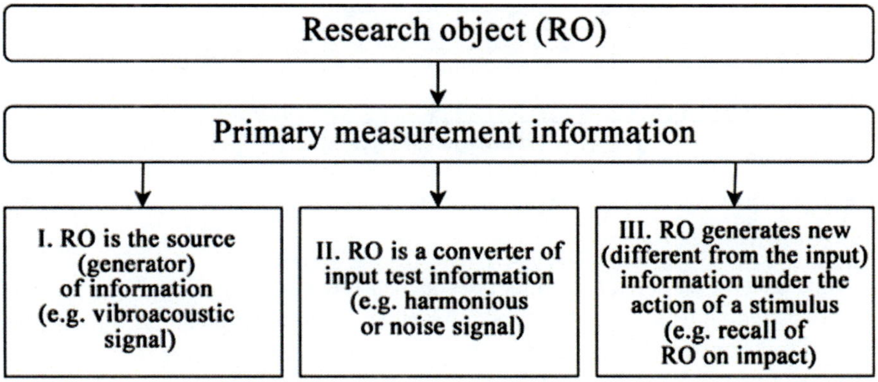

Fig. 1.9 Schematic illustration of the formation of variants of primary information (signals) of measurement of the RO

1.3 Measures, Their Properties and Application in Measurements

The growing necessity for measurements, the constant complication of procedures and means of measurement, the penetration of measurements into new areas of the material world required at each stage of human development a corresponding logical and methodological understanding of the concept of measure. In accordance with the needs of society from ancient times to the present, there is a process of development of the philosophical category of measure—from an ethical-moral concept to an abstract mathematical theory of measure. For a deeper understanding of its content and role in the theory of cognition of the world as a whole, and in metrology in particular, it briefly consider the emergence, development and use of this concept in philosophical, mathematical and metrological aspects [29–31].

The following is a series of facts from the history of the development of the measure.

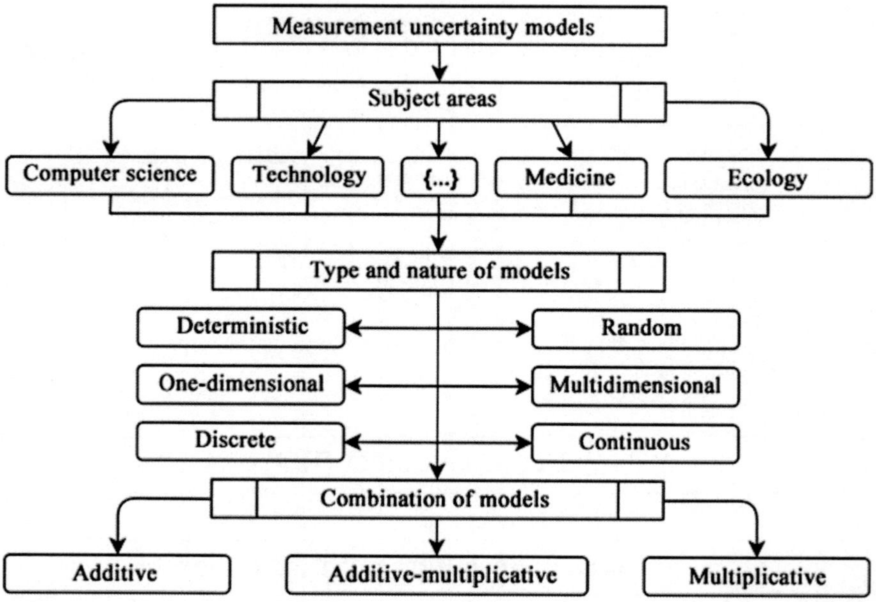

Fig. 1.10 Classification of measurement uncertainty models

In ancient mythology, the symbol of the measure was the goddess Nemesis, who was depicted with scales in her hands. An ethical and moral meaning was invested in this concept, which was understood as a comparison of moderation, balance in actions and words, as opposed to redundancy and negative actions of people.

The famous ancient Greek scientist Pythagoras and his students (6–4 centuries BC) tried to quantify the measure using numbers and proportions. For example, they considered a sign of justice equivalence, equality, and tried to express them by numbers. So, justice was defined as the first square numbers, that is, as 4 or 9. Discussing the problem of proportionality, they called the quantities falling under the same measure one-dimensional, and those that did not fall—non-one-dimensional.

Later, Aristotle (445–385 BC), one of the greatest ancient Greek philosophers and scientists, reflected as follows: "Measure (from the Greek "*metron*") is what quantity is known; and the quantity is known either as a single or a number, and any number is a singular, therefore any quantity is such that it is known…". So the only one, being the beginning of a number and expressed in some indivisible degree—the scale of measurement, has a numerical form. "The measure of the number is accurate: after all, the unit is perceived as something indivisible in all respects." In all other cases, the measure acts as a model of measurement, for example, liquid or granular, which has a weight and size. Recounting various types of measures as units of measurement, Aristotle expressed the opinion, that one can't limit oneself to only one measure and several measures should be distinguished that are presented in various numerical ratios or proportions, in addition, a measure as a unit of measurement should be uniform with the measured one.

Another ancient Greek philosopher Protagoras (490–420 BC), one of the most famous sophists, proclaimed: "Man is the measure of all things: for the existing - that they exist, for the nonexistent - that they do not exist." This statement already contains the idea of a different quality of things and a transition from one quality to another. But only in the eighteenth century the outstanding German philosopher Hegel (1770–1831) developed this idea. Analyzing the change in quality in categories of quantity, he came to the philosophical category of measure. It was a study of the inextricable link between the categories of quality and quantity that allowed Hegel to convincingly show that changes in existence are not only a transition of one quantity to another, but also a transition of qualitative into quantitative and vice versa. According to Hegel, a measure is a unity of quality and quantity, which are determined by the very nature of the object. Such a definition reflects the inextricable relationship between the qualitative and quantitative side of each object or phenomenon of the material world. Measure means that any specific quality is associated with a certain quantity. Each qualitatively peculiar object or phenomenon has certain quantitative characteristics. The last have the property of variability and mobility.

According to Hegel, the transition from quantitative to qualitative changes means the change of one measure to another. If quantitative changes occur within a measure characteristic of this phenomenon, their quality remains unchanged. If quantitative changes lead the phenomenon beyond its measure, a change in quality will occur. For example, the temperature of the water does not affect its liquid state until it reaches the boiling point, when the water passes into steam. The boundaries of such transitions are actually a measure and play the role of peculiar standards (internal) with respect to objects. In this sense, a measure is an internal property of an object, based on the unity of its quantitative and qualitative characteristics.

The transformation of quantity into quality has a reverse process—the transition of quality into quantity. The newly formed quality corresponds to a new measure, that is, a new concrete unity of quality and quantity, which makes it possible to implement further quantitative changes of the new quality and the next transition of quantity into quality. The law of mutual transition of quantitative and qualitative changes, discovered by Hegel, describes the mechanism of self-development of the phenomena of the material world.

The concept of measures began to be actively used and filled with new content in various areas of society.

In the use of measures should highlight:

- a measure in the broad sense, mainly as a qualitative category;
- a measure in the narrow sense, mainly as a quantitative category.

Characterizing the integral role of a measure, the following should be noted. With the development of industry and production, expansion and strengthening of trade relations between states, attempts were made to introduce various measures as uniform rules of interstate relations. These attempts were both successful and not quite. And this can be judged in our time [32].

In this work, it will be consider measures that are used in metrology in accordance with the Concept of Uncertainty [22].

The success of the use of unified metrology measures by the international community can be considered one of the unique facts in the history of terrestrial civilization. Metrology has become an international communication tool for scientists, researchers, production specialists and others in all countries of the world.

The role of measures in assessing measurement uncertainty. Such a statement, first of all, has a philosophical basis, as mentioned earlier. In the process of creating measuring instruments, a measure is a fundamental element on the basis of which quality is evaluated quantitatively. It forms a fundamentally new quantitative information, reduces the uncertainty of the studied objects, phenomena and processes. This fact makes it possible to adjust the well-known triad of research methodologies "model → algorithm → program" to the more reasoned "model → measure → algorithm → program".

The effectiveness of the practical implementation and use of measures is determined by their uncertainty values when reproducing units of physical quantities. The smaller the value of the uncertainty, the higher the class of the standard or measuring instrument when implementing the appropriate measure. It is known that each country has national standards for units of physical quantities. The nomenclature of standards and their characteristics of uncertainty are estimated at the level of gross product and is considered the national treasury of each country. The most developed countries in the world have the largest range of national standards for units of physical quantities.

Math measures. Measure theory is one of the areas of mathematics important for practical application, within the framework of which methods for creating various types of measures are substantiated and determined.

The material, which will be given below, is adapted to the problems of measurement. A measure, as a mathematical object, has found practical application for creating measures of various physical quantities—mass, length, area, volume, and the others.

In the general case, a measure is a countably additive function of sets; it takes only non-negative values, including infinity. In mathematics, a number of measures are studied, for example: Jordan, Lebesgue, Lebesgue-Stieltjes measures, stochastic measure, etc. So, the definition of the degree of Jordan is close to the definition of area and volume in space. The Lebesgue-Stieltjes measure is used in probability theory. A stochastic measure is a random countably additive function of sets. Moreover, to prove the condition of countably additivity of the stochastic measure, various types of convergence are used, namely: probability convergence; RMS; with probability 1. But the most widely used for theory and practice of measurement are numerical measures, that is, measures that take numerical values.

Nowadays, there is an interesting fact of generalizing measures related to measurement. It is known that earlier the measure took only non-negative numerical values. For practical measurement problems, the question arose, how to determine the degree of a physical quantity with a negative electric charge? A simple and natural generalization of a degree is charge. The term "charge" is borrowed from physics. In physics, a charge differs from a mass in that the mass is always non-negative, and the charge can be either positive or negative. The same difference between the measure and the charge as mathematical objects. This fact significantly expands the boundaries of the use of methods of measure theory in metrology.

As examples, it gives the definition of a probability measure and charge.

Probability measure. The mathematical model of a physical quantity is a random variable $\xi(\omega)$, $\omega \in \Omega$, which is widely used in measurement theory. With unknown characteristics, $\xi(\omega)$ is the subject of measurement and is defined as a function of the measurement uncertainty.

It is known that $\xi(\omega)$ as a random function has the next features:

- the domain of definition is given by the probability space (Ω, F, P), where $\xi(\omega)$ is the space of elementary events $\omega \in \Omega$, F is the algebra (σ-algebra) of the subset Ω, P is the probability of elementary events: $P(\varnothing) = 0$, $P(\Omega) = 1$ and $P \in [0, 1]$;
- the range of values is given by the probability space $\left(X, \mathrm{B}, P_\xi\right)$, where $X \subseteq R$ is the numerical subset of the number line R, B is the algebra (σ-algebra) of the subset X, and P_ξ is the probability of the values $\xi(\omega)$.

The random function $\xi(\omega)$ is completely determined by the distribution function:

$$F(x) = P\{\omega \in \Omega : \xi(\omega) < x\},\ x \in R.$$

The probability measure $\xi(\omega)$ as a function of sets cannot be directly written in general form, for it the generating function is $F(x)$.

The probabilistic measure $P_\mu(x_1, x_2)$ determines the probability that a random continuous function $\xi(\omega)$ takes values from a continuous numerical interval:

$$[x_1, x_2], x_1, x_2 \in R, x_1 > x_2,$$

thus,

$$P_\mu[x_1, x_2] = P\{\omega \in \Omega : x_1 \le \xi(\omega) \le x_2\} = F(x_2) - F(x_1).$$

Thus, the one-dimensional probability measure $P_\mu[x_1, x_2] \geq 0$ is integral, normalized, and dimensionless with a range of values.

For more complex random functions, for example, a two-dimensional random vector $\Xi_2(\omega) = (\xi_1(\omega), \xi_2(\omega))$, where $\{\xi_i(\omega_i), i = 1, 2\}$ are one-dimensional random variables, the probability measure of the vector is also more difficult to determine using the two-dimensional distribution function of the vector $F(x, y)$.

So, for example, a probabilistic measure of finding the values of a random vector $P_\mu\left[x \pm h_x, y \pm h_y\right]$ in the framework of a two-dimensional number space of the plane (x, y)

$$\left((x - h_x) < \xi_1(\omega) < (x + h_x), (y - h_y) \le \xi_2(\omega) < (y + h_y)\right)$$

is determined by the expression

$$P_\mu\left[x \pm h_x, y \pm h_y\right] = F\left(x + h_x, y + h_y\right) - F\left(x - h_x, y + h_y\right) - F\left(x + h_x, y - h_y\right) + F\left(x - h_x, y - h_y\right), \quad x, y \in R, \ h_x, h_y > 0.$$

Consider the following example of a charge as a measure using topological space.

Charge as a measure. By definition, charge is also a function of sets. To reveal the essence of the charge, it can be use a measurable topological space as the corresponding formalization of the numerical measurement data obtained with using the charge as a measure. In the future, the use of a topological space will make it possible to evaluate the uncertainty of a measurement result of type B.

The construction of a dimensional topological space with a charge as an object of the theory of measure is performed in stages with comments on the interpretation of the measurement process.

Due to the fact that the construction of a topological space is a theory, and obtaining real measurement data is practice, that is, the implementation of the theory, the stages of the theory are denoted by a single number, and comments by practice are denoted by a double number.

1. The numerical set of measurement data X with elements of $x \in X$ is specified.

   1.1 The set X is a discrete sequence of measurement data that can be obtained using various measures, including the charge.

2. Each element of $x \in X$ is associated with a vicinity of $U(x)$ as a system of open sets.

   2.1 For each value of the sequence of measurement data, there is a vicinity whose size is determined integrally by the uncertainties of the physical quantity and the means of measurement.

3. The system of open sets $I, U(x) \in I, x \in X$ is created, which is called a topology.

   3.1 The topology can be used to predict the values of further measurement data.

4. The pair (X, I) is called a topological space if the following two conditions are satisfied:

   (a) for $\forall x \in X$, it takes place

$$\exists U(x) \in I; \tag{1.1}$$

   (b) for $\forall x \in X$ and $(U(x), V(x))$, it take places $\exists W(x) \in I$, where

$$W(x) \subseteq U(x) \cap V(x); \tag{1.2}$$

4.1 Basically, the fulfillment of the above conditions is the formation of the algebra (∃) of the subsets X and I to formalize the construction of a topological space.

5. The topological space (X, I) is called measurable if, in addition to satisfying conditions (1.1) and (1.2), an additional class of subsets I of it is a σ-algebra.

   5.1 The fulfillment of the additional condition, including the countable (infinity) case, is important for the correct formalization in order to take into account all possible cases of the class of I subsets.

6. The topological space (X, I) is called Hausdorff if, in addition to satisfying conditions (1.1) and (1.2), the following condition is additionally fulfilled:

   (c) for $\forall x, y \in X$ it take places $\exists[U(x), U(y)] \in I : U(x) \cap U(y) = \varnothing, x \neq y$.

   6.1 A property of a Hausdorff topological space is each pair of points of such a space has vicinities that do not intersect according to the axiom of the Hausdorff branch, and has fundamental importance in obtaining measurement data. Thanks to it, the two values of the measurement data can be considered different using the resolution of the measuring tool.

7. The real function H is given in the space A on the n-algebra of the subsets B such that:

$$q(A) \in \Sigma \subseteq R, A \in I, \tag{1.3}$$

for which the following conditions are true:

   (d) $q(\varnothing) = 0$;
   (e) function q is countably additive, i.e., for

$$A = \sum_{i=0}^{\infty} A_i, A_i \cap A_j = \varnothing, i \neq j, \tag{1.4}$$

   it get

$$q\left(\sum_{i=0}^{\infty} A_i\right) = \sum_{i=0}^{\infty} q(A_i). \tag{1.5}$$

   7.1 From conditions (c) and (d) it follows that the charge as a measure corresponds to all its conditions, with the exception of the condition of non-negativity. The charge can be both negative and positive. This is important during implementing units of physical quantities that may have different signs.

8. The collection (X, I, q) is called a measurable topological space with a charge. This set is a mathematical model of a more general nature in comparison with a probabilistic measure for the implementation of metrology measures as material objects for measuring means (systems). The above measures relate to quantities given on the number line. At the same time, a significant part of the measurement problems is associated with the study of random angles. The last, as mathematical objects, have a certain specificity, which leads to the necessity of setting probabilistic measures on the circle. These issues will be discussed in more detail in Chap. 3.
Further it will be considering the implementation of physical measures in metrology.

Physical measures. The concept of measure is also determining for metrology. The modern interpretation of the concept of measure for applied technical applications has been forming since the seventeenth century, when measurement procedures became methodologically significant and very important for the development of material production, and the range of measured physical quantities began to increase rapidly. Today, the following definition is generally accepted in metrology: a measure is a measuring device that implements reproduction and (or) preservation of a physical quantity of a given size. There is another definition: a measure (the degree of a physical quantity, a measure of a quantity) is a means of measurement in the form of a body, substance or device, designed to reproduce and preserve the physical quantity of one or more specified sizes, the values of which are expressed in established units and are known from necessary accuracy. In these definitions, the following initial concepts are used as quantity and physical quantity. In a broad sense, quantity is a generalization of specific concepts, for example, length, flat angle, area, mass, and the like; physical quantity is understood as a property that is common in a qualitative sense to many physical objects (their states), and individual for each of them in quantitative terms. Usually physical quantities have corresponding dimensions. Quantities as separate independent objects do not exist, but are certain characteristics of a group of objects of the material world. Therefore, a measure, in the above definitions, is understood as a material carrier of a physical quantity of a given size.

The given definitions of the measure are focused on the practical implementation of measurements, and do not reflect its physical and mathematical essence. In this sense, a measure of a physical quantity is a certain numerical function $\mu(\cdot)$, which assigns to each subset A of the set of values of a physical quantity X some inalienable (generally called) number $N = \mu(A)$. For the measure of a physical quantity, the axiomatics of the measure of sets is preserved. A certain difference is associated with the need to measure named and negative physical quantities. The latter feature is determined by the choice of zero of the measurement scale and can be formally taken into account by artificial introduction, according to a certain logical condition, of the "$-$" sign, or by taking into account a certain constant C: $\mu(A) = \mu(A) - C$. For example, in the case of a transition from absolute temperature to temperature of the Celsius scale $C = 273.15$ K.

A characteristic property of a quantity is the possibility of their measurement, that is, $\left[x_j, x_{j+1}\right)$ comparison with a certain quantity of the same kind and of a certain size, which, by agreement, is taken as the unit of measurement, that is, as measure. Methodologically, the use of set measures for measurements can be justified as follows. The size of a physical quantity, as a result of the combination of various aggravating factors, cannot be determined with arbitrarily high accuracy. Therefore, it is advisable to divide the entire measurement range, for example, of a scalar physical quantity X, into half-intervals of the form X_j, $j = 1, 2, 3, \ldots$. Each of these half-intervals do not intersect and represent a certain set of sizes of the physical quantity. Based on the totality of such sets, a numerical function can be constructed; it has the properties of a measure.

As a result of measurements, an unnamed number N is obtained, which expresses the ratio of the size x of the measured quantity X to the measure $\mu(1_x)$, that determined for a given unit set 1_x of values X. The choice of 1_x and $\mu(1_x)$ is carried out by agreement and justified within the framework of the adopted system of units of physical quantities. The main purpose of the measurement is to find the value x of the physical quantity X as a definition of a quantitative assessment of its size:

$$Y = N \cdot \mu(1_x), \tag{1.6}$$

it's implemented on the basis of the reproduction of a physical measure of the size of a physical quantity. In Eq. (1.6), the errors in measuring and reproducing the size of a physical quantity are not taken into account. The process of measuring the scalar quantity X in case $\mu(\cdot)$ is given by a step function and shown schematically in Fig. 1.11.

From Fig. 1.11a it can be seen that any size of a physical quantity from the set of its values within each of the half-intervals is displayed by one named number. Unambiguous measures of physical quantities reproduce one of its size, ambiguous—several sizes. In general, measures of sets and physical quantities are external to the subject of measurement.

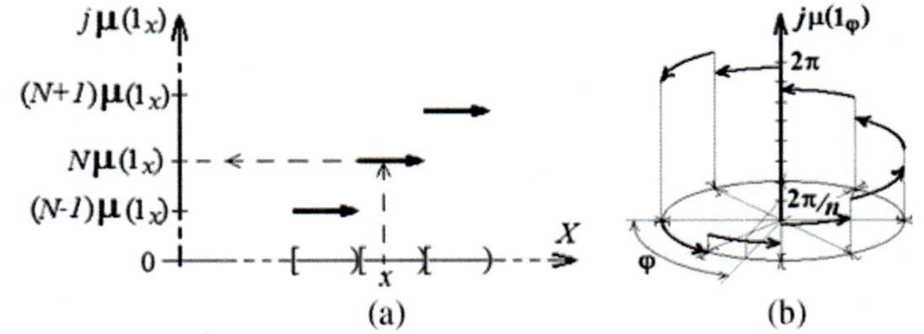

Fig. 1.11 Graphic illustration of the process of measuring a physical quantity distributed on a straight line (**a**) and on a circle (**b**)

The above information on physical measures refers to quantities distributed on a numerical line. Further it will be considering the features of measures of angular quantities, which have a certain specificity.

It is known that for an arbitrary circle with a finite radius, the value of the central angle is defined as the ratio of the length of the arc on which this angle is based to the length of the radius of the circle. The total center angle is 2π. Naturally, the basis for determining the measure of the angle is the number π. The measure of the angle (the number π or its part) can be reproduced either physically using dials (which takes place, for example, in quadrants, protractor, sextants, etc.), or mathematically based on special calculations. The last makes angular measurements (which include phase measurements of cyclic signals) especially attractive from the point of view of ensuring the uniformity of measurements and the achievement of high accuracy indices [33–38].

An illustration of the process of measuring a flat angle if its measure $\mu(1_x)$ is given by a step function and is shown in Fig. 1.11b. In general, the formation of the value of the measured angle occurs similarly to (1.6) with the difference that the step function is specified in a cylindrical coordinate system. An important feature of the formation of measures for angular measurements is that the set of values of the angle is determined through a circle—a closed curve. Angles 0 and 2 π correspond to one point of the circle at which the measure of the angle changes stepwise by 2π.

1.4 Concept of Harmonization of Physical and Probabilistic Measures in Measurements

In the structure of empirical methods of cognition of the world of measurements, due to objectivity and information content, occupy a special place. Metrology, as the science of measurements, in its development has passed a difficult path from a deterministic to a probabilistic approach. Even in recent decades, the concept and content of measurements has significantly transformed. To confirm this fact, it suffices to give the following definitions of measurements. The following is given: "Measurement is a reflection of the properties of an object that manifest themselves in relation to equivalence, order, and additivity by a limited number of called natural numbers". In the international dictionary of metrology VIM-3 (2003), the following broader definition is given, which emphasizes the probabilistic nature of the measurement result: "Measurement is the process of obtaining one or more values of a quantity that can be reasonably attributed to a quantity". In the same dictionary, the measurement result is explained as a set of values attributed to the measured quantity together with any other available and relevant information (for example, it can be the law of probability distribution, standard or extended uncertainty, coverage probability, etc.). The stochastic nature of measurement as a process is emphasized by another characteristic—coverage probability, that is, the probability that the set of true values of the measured value is in the specified range of coverage. Most often, the measurement

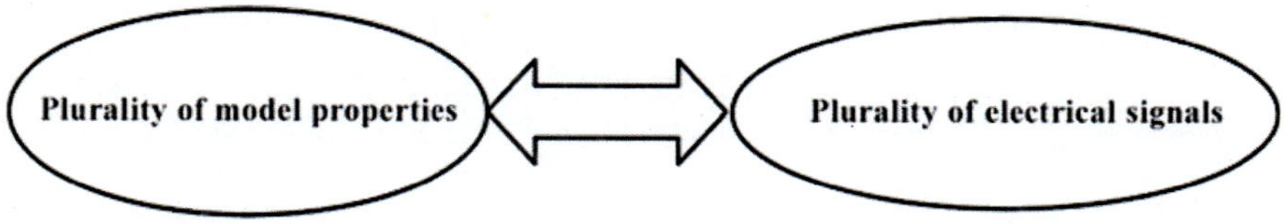

Fig. 1.12 Graphic illustration of the principle of isomorphism

result is presented in the form of a coverage interval that is specified by its limits and corresponding coverage probability [39].

The basis of all known definitions of measurement is the displaying the set of values of the measured quantity, the quality of the objects of study into a number—named or unnamed. Such displaying is the most convenient and allows to formalize the measurement result and apply mathematical methods for their processing. Let briefly dwell on the logical and mathematical concepts of isomorphism and homomorphism, and their use in measurement theory. These concepts express uniformity (isomorphism) or similarity (homomorphism) of the structure of systems (sets, processes, constructions, etc.).

Two systems, considered abstractly from the nature of their constituent elements, are isomorphic to each other, if each element in one system corresponds to one element of the second system, and each connection in one system corresponds to one connection of the second system, and vice versa (Fig. 1.12).

Such a one-to-one correspondence is called an isomorphism. A complete isomorphism takes place only between abstract, idealized objects (for example, the correspondence between a geometric figure and its analytical expression in the form of a formula; a straight line segment as many points and many real numbers). Most often, isomorphism is not associated with all, but only with some properties fixed in the cognitive aspect and relations of the compared objects, which may differ in other respects. It is the property of isomorphism that allows to study the relationships existing in the RO according to their models.

In the case when the action of physical laws is studied under certain conditions, a theoretical analysis of mathematical formulas, equations of measurements in the macrocosm is performed, then the displaying of plurality of quality values of the studied quantities or processes and plurality of numbers can be considered as isomorphic.

The proof or justification of isomorphism for measurement problems plays a fundamental role. This is due to the following. In the general case, the measurement problem is the inverse problem of the theory of signals and systems. In the case of proof of isomorphism, a solution to such a problem exists.

Unlike isomorphism, "homomorphism" is an unambiguous correspondence of objects (systems) in one direction only. Therefore, a homoform image is an incomplete, approximate reflection of the structure of the original. Homomorphism relationships are more general (and weaker). Therefore, any homomorphism is a homomorphism, but not vice versa.

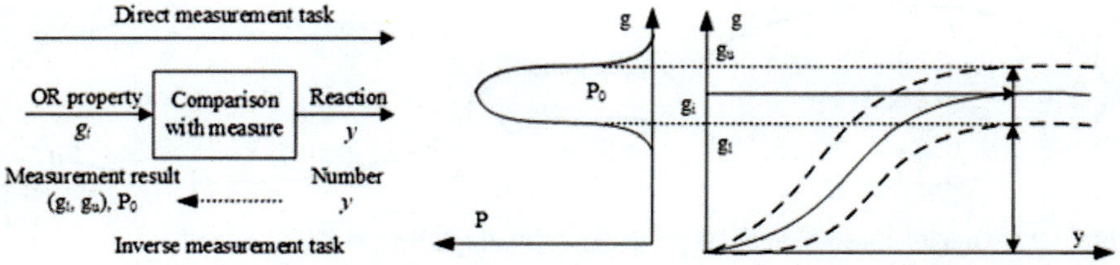

Fig. 1.13 Schematic illustration of the formation of the measurement result (g_l, g_u—lower and upper bounds of uncertainty)

In a measuring experiment, the set M values of the research object properties is homomorphic to the set M' parameters and characteristics of the received information signals. In general, non-ambiguous reflection, inherent in homomorphism, allows solving direct problems of measurement theory (modeling) and complicates solving inverse problems (actually obtaining measurement results). The objective of the direct task is to find the consequences of known or given causes (information signals as a function of certain properties of RO), that is, along the cause-effect relationships (Fig. 1.13).

The inverse tasks are to find the causes (properties of the AU by known signals), that is, in the direction "against" cause-effect relationships. Direct tasks arise at the design stage, analysis of the measuring instrument, and inverse problems arise during measurement (control).

The need for joint use of physical and probabilistic measures to form the measurement result is shown in Fig. 1.14. A physical measure gives only one number (or a vector in the case of multidimensional measurements). The measurement result, as a vector formed by a certain set of values prescribed to the measured value, together with other relevant information, is formed using a probabilistic measure.

In the structure in Fig. 1.14 it has:

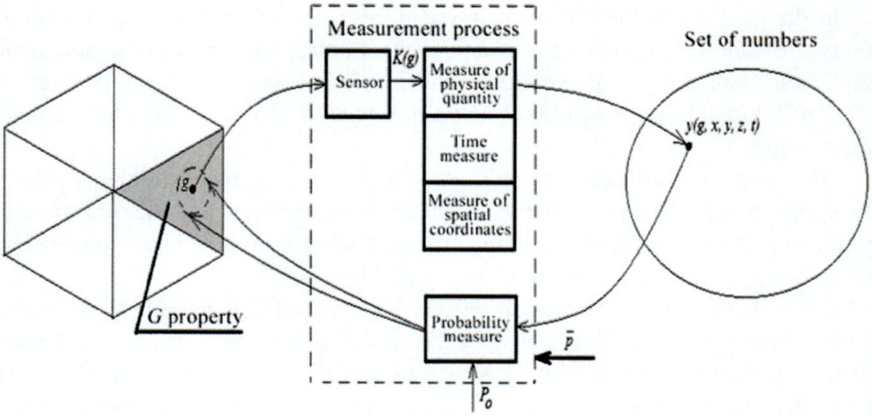

Fig. 1.14 Sharing physical and probabilistic measures in measurements

1. The sensor can perform the conversion of the physical quantity g into another—$K(g)$, for which it is much easier to create a measure $K_0(g)$.
2. The application of the measure of physical magnitude $K_0(g)$ to the size $K(g)$, in the general case of measuring the field at a point in space with Cartesian coordinates (x, y, z) at time t, gives the number.

$$y(g, x, y, z, t) = [K(g, x, y, z, t)/K_o(g)]^+,$$

where $[\cdot]^+$ is the designation of the operation of extracting the integer part of the number.
3. Due to the action of the vector $\overline{p}$ of unaccounted for influencing factors, the inverse transformation $y \rightarrow g$ gives not one value, but a certain vicinity.
4. The measurement result is obtained not only as a result of technical means of comparing the size of the measured physical quantity with the value taken as the unit of its measurement, but also as a result of the application of a probability measure, a certain mathematical apparatus and mathematical statistics.
5. The conversion of various physical quantities into an angular quantity (flat angle, phase shift of signals) is a convenient type of conversion, since the unit of measurement of the last (radian or π) is reproduced by computer technology with almost unlimited accuracy regardless of the place and time of the measurement.
6. The stochastic approach in measurement theory is of particular importance in the case of measurements of physical quantities having a pronounced probabilistic nature, for example, in the case of nano-measurements, the study of quantum effects, and the like.

Thus, obtaining a meaningful measurement result is based on the use of a consistent, inextricably linked set of physical and probabilistic measures, the classification of which is shown in Fig. 1.15.

Example of the use of measures in information-measuring systems (IMS). The solution to the problem of ensuring the uniformity and stationarity of a measure in the IMS theory in each specific measurement case has its own specifics and characteristic features. At the same time, general properties of use and types of measures can be distinguished:

- M1—measure of a unit of value;
- M2—measure of the unit of spatial coordinates of the measurement place;
- M3—measure of the unit of measurements time;
- M4—probabilistically normalized measure of the result and accuracy of measurements in the statistical processing of measurement data;
- M5—degree of protection of information of a particular measurement process.

Figure 1.16 is a schematic illustration of the use of these measures during converting measurement information by the functional modules of the IMS.

A special case of information protection is to increase the noise immunity of measurements; it is used for almost all measurement processes. In the structure of

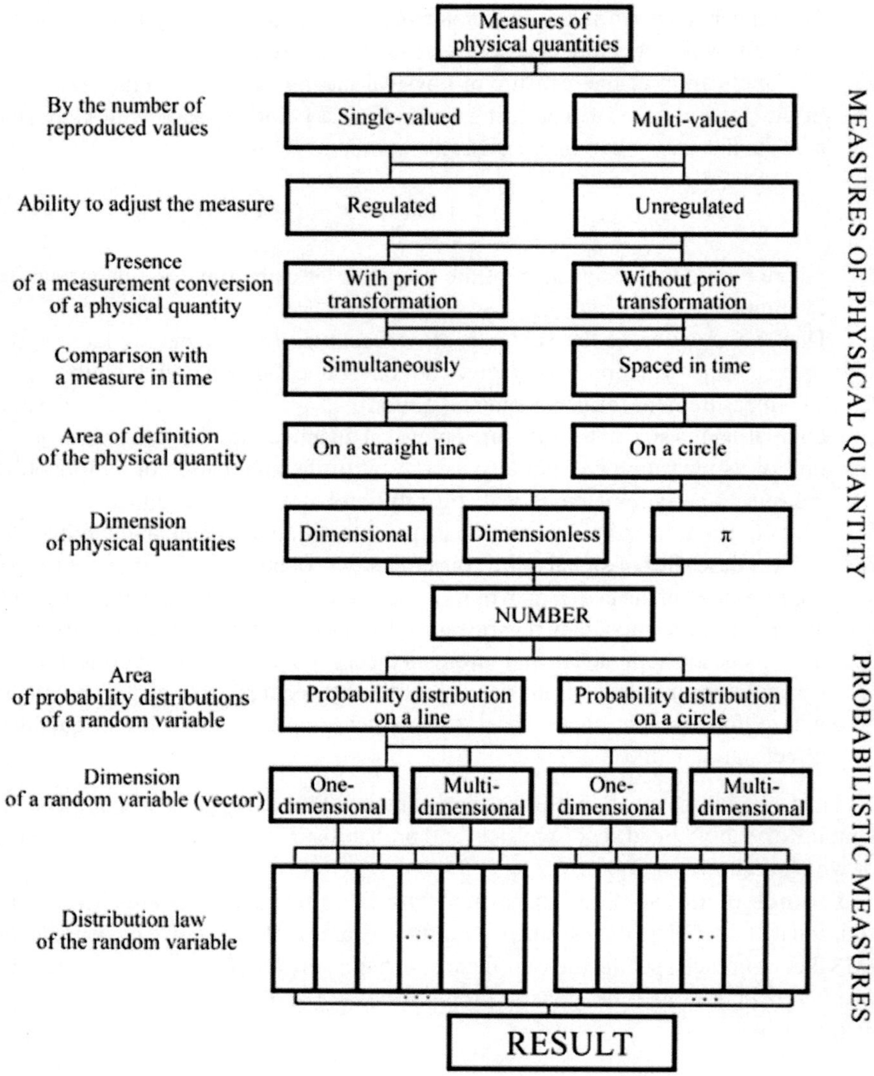

Fig. 1.15 Classification of physical and probabilistic measures for measurements

the IMS, as a rule, there is a signal filtering module that provides an increase in the signal/noise ratio at its output.

In the case of measurements of n ($n > 1$) quantities, there are n measures of units of the studied quantities. The first three degrees are physical measures (in the general case, it has $n + 2$ measures), which are implemented by the corresponding standards of measures with a given accuracy. The uniformity and stationarity of such measures are determined by the characteristics of technical devices and systems

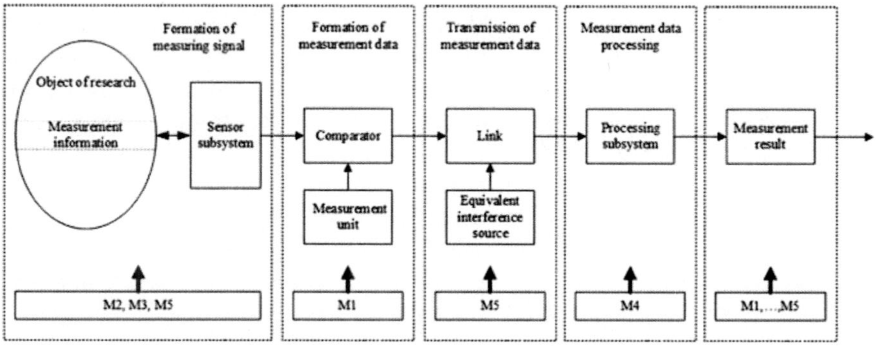

Fig. 1.16 Schematic representation of the use of different measures in the structure of IMS

for their formation. This makes it possible to a certain extent to control the provision of uniformity and stationarity of physical measures during the measurement process in various places of space and at different times, while ensuring a comparison of the measurement results and thus fulfill the requirements of the uniformity of measurements.

A probabilistic standardized measure is a non-physical measure, but a measure of the totality of the action of various random factors on the value and characteristics of the data and the measurement result during they are carried out. Using a probabilistic measure in the statistical processing of measurement data makes it possible to increase the accuracy of the measurement result in comparison with the accuracy of the measurement data.

The degree of protection of information during measurements is complex. The measure is formed by a large number of factors, the effect of most of which is random. This makes it possible to determine such a measure as probabilistic, which can be applied both for individual operations (for example, transferring measurement data via communication channels, recording the measurement result), and for the entire measurement process as a whole.

The accuracy value of these five measures is transformed, respectively, into the accuracy value of the data and measurement results in the corresponding modules of the IMS.

1.5 Measurement as the Inverse Problem of the Theory of Signals and Systems

It will justify the use of a number of well-known methods and the formulation of a number of statements of measurement problems that reflect certain facets of the measurement problems diversity and emphasize the main role of models and measures for their solution. Naturally, such problem statements and methods for solving them use the results of the latest achievements of a number of natural and technical sciences, including mathematical analysis, mathematical physics, quantum mechanics, computational mathematics, probability theory and mathematical statistics, theory of signals and systems, theory of automatic control, measurement and digital signal processing, etc. [40–44].

Measurement task as the basic task of reducing the uncertainty of research results. The term or concept of uncertainty has a wide range of interpretations and interpretations by analogy with the concepts of information and plurality. In a broad sense, the following meaning is embedded in the concept of uncertainty: this is the degree of ignorance about the RO, which can be reduced by performing certain operations to obtain quantitative information, and this is confirmed by real facts.

Thus, the famous French scientist in the field of physics and information theory, professor of the Sorbonne (France), Harvard and Columbia University (USA) L. Brillouin in his works "Science and Information Theory" (1962), "Scientific Uncertainty and Information" (1964) noted following. The successes of science in such fields as thermodynamics, classical and quantum mechanics, statistics and information were obtained due to the information uncertainty existing at that time in studies of the corresponding phenomena and objects. It was also noted that the given examples of achievements obtained with certain indicators of accuracy. An interesting conclusion of the author about the methodology of scientific research, and, accordingly, the methodology of research of metrology as an empirical science: "…all these problems, as well as science itself, will never be completed. Full completion would mean death for any research".

The terminology of uncertainty began to appear in scientific publications in the first half of the twentieth century, and it applied only to measurement problems. The well-known uncertainty principle of quantum mechanics (the principle of the German nuclear physicist Heisenberg) defines the boundaries of the accuracy characteristics (RMS values) of the simultaneous measurement of the spatial coordinate of a quantum particle x with momentum energy e in the form:

$$\sigma_x \cdot \sigma_e \geq h/2,$$

where $h = 6.62607015 \times 10^{-34}$ [J s] is Planck constant, σ_x and σ_e are standard deviations of the measurement data.

Further studies have confirmed the fact that the principle of measurement uncertainty is used not only in quantum mechanics (microcosm), but also in classical mechanics, electrodynamics (macrocosm). These are simultaneous measurements of current and voltage, characteristics of electric and magnetic fields, time and frequency characteristics of a radar signal. At the same time, the above uncertainty principle applies to two measured quantities connected at the same time functionally, but for each of them individually does not impose any restrictions on the measurements accuracy.

As the theory and practice of measuring physical quantities has shown, uncertainty exists and manifests itself in every measurement experiment.

Given the broad interpretation of the concept of uncertainty and with a view to its more specific use, the following classification is proposed:

- uncertainty in the broad sense as a qualitative characteristic;
- uncertainty in the narrow sense as a quantitative characteristic, mainly during assessing a quantitative measurement result.

1993 was a year of revolutionary change in justifying the use of the concept of uncertainty in evaluating measurement results. The working group of the International Metrological Community, including: International Bureau of Weights and Measures (BIMP), the International Electric Commission (IEC), the International Organization for Standardization (ISO), etc., in 1993 published the Guide to the Expression of Uncertainty in Measurement—GUM [26].

This publication has officially introduced the concept of measurement uncertainty internationally. For example, according to the normative document ISO 14064-1-2007, the definition of uncertainty is given in the next edition.

Uncertainty is a parameter, associated with the result of a measurement, that characterizes the dispersion of the values that could reasonably be attributed to the measurand.

In general, measurement uncertainty involves many components. Some of them can be estimated using the statistical distribution of the sequence of measurement data and be characterized by standard deviation (SD). Estimates of other components can only be based on previous experience or other information.

Today, international metrology, including Ukraine, in assessing the quality of measurement results has moved from the classical concept (Classical Approach, CA) to a new concept based on measurement uncertainty (Uncertainty Approach, UA). According to the new UA, instead of the fundamental concept of "measurement error", which uses the known or hypothetical true value of a physical quantity, the measurement uncertainty is applied, which corresponds to the range (interval) of location values of the quantity true value.

Thus, the main indicator of measurement quality is the uncertainty of the measurement result. A decrease in uncertainty is a sign of increased efficiency, that is, an increase in measurement accuracy. Thus, the uncertainty can be selected as a criterion for assessing the quality of various measuring operations, primarily the functioning of measurement tools and systems.

Further it will be formulated practical recommendations for using the concept of uncertainty.

So, for example, if it adapts the uncertainty to the measurement data of a physical quantity, the mathematical model of which is described by a random variable with unknown mathematical expectation and variance—the measurement uncertainty function, then the uncertainty of the true value of the studied quantity can be estimated with a given probability by a confidence interval with the corresponding statistical estimates of the mathematical expectation and dispersion, or other methods that essentially form the same information.

For the one-dimensional case, the measurement result is presented in the form:

$$y = \overline{x} \pm \Delta x[x], \tag{1.7}$$

where y is a measurement result; $\overline{x}$ is a reference value of the measurement result (in most cases, this is the arithmetic mean of multiple measurements or the estimate of the mathematical expectation of a random variable); Δx is a expanded uncertainty of the measurement result; $[x]$ is a measuring unit.

Given the heterogeneity and diversity of measurement tasks, the value of the measurement uncertainty is determined depending on the setting of a specific task in different ways, namely:

- standard uncertainty of measurement (type A);
- standard uncertainty of measurement (type B);
- total standard measurement uncertainty;
- extended measurement uncertainty.

It is known that the result of measuring the physical quantity Y is represented by the named number y. For example, $y = (10.2 \pm 0.1)$m is a record of the result of measuring the length in units of the SI system.

According to the concept of uncertainty, the generalized measurement equation has the form:

$$Y = f(X_1, X_2, \ldots, X_m). \tag{1.8}$$

In this equation, Y is the measurement result, which is represented by the multidimensional function $f\left(\left\{X_j, j = \overline{1, m}\right\}\right)$, the range of which is uniquely determined by the numerical values obtained from the measurement data of the set of arguments $\left\{X_j, j = \overline{1, m}\right\}$.

For each specific case of the measurement process, a measurement equation of the form (1.8) takes an appropriate functional dependence, for the solution of which one or another measurement method is used with its subsequent implementation by hardware-software means (systems).

Methods of analysis and synthesis in measurement problems. It will be considering methods of analysis and synthesis of the theory of signals and systems adapted to measurement problems. To describe the use of such methods, the following one-dimensional version of the measurement problem will be considered.

Statement of the measurement problem. A signal is received at the input of the measuring instrument, which is described by the uncertainty function in the form of an additive mixture of a determinate function of time $s(t)$ and interference—a stationary Gaussian random process $\xi(\omega, t)$ with zero mathematical expectation at a fixed value of t, that is:

$$\eta(\omega, t) = s(t) + \xi(\omega, t), \omega \in \Omega, t \in T. \tag{1.9}$$

The measuring mean, as a hardware-software tool, is described by the conversion operator of the input signal $Z[\cdot]$. At the output of the mean, it needs to get feedback—the measurement result in the form

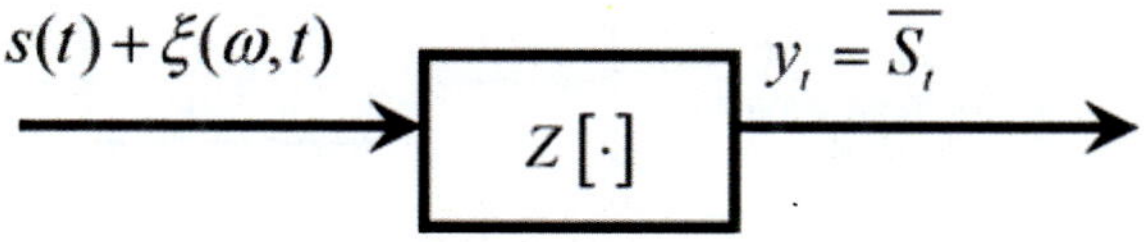

Fig. 1.17 Schematic illustration of a measurement task using a measuring mean

$$y_t = \overline{s}_t + \Delta s, \tag{1.10}$$

where $\overline{s}$ is the reference value of the signal $s(t)$ at a fixed t, Δs is the corresponding value of its uncertainty.

Schematically such a statement of the measurement problem can be represented in Fig. 1.17.

Further it will be considering the following methods of analysis and synthesis of transformations of information signals by means and measurement systems. Despite the fact that the methods of analysis and synthesis are ideally different, their relationship during using complement each other to solve measurement problems.

Synthesis method. This method is used to solve the well-known optimization problems of determining the transformation operator of a measuring instrument $Z[\cdot]$ according to a given optimality criterion. The only optimality criterion for measurement tasks is to minimize the value of the measurement uncertainty of physical quantities (type A or B, total standard or extended).

Thus, on the class of uncertainty functions for measuring physical quantities, the optimal transformation operator $Z[\cdot]$ of the measuring instrument minimizes the uncertainty of the measurement result in accordance with the synthesis method that is used.

Analysis methods. The use of this method, in comparison with the synthesis method, has a large range of tasks, namely:

Direct task. Given input action is the uncertainty function for measuring physical quantities $s(t)$ and the transformation operator $Z[\cdot]$. It is necessary to find the response characteristics, that is, $y(t) = Z[s(t)], t \in T$.

Inverse task. Given feedback $y(t)$ is the result and measurement uncertainty, the transformation operator $Z[\cdot]$ of the measuring instrument. It is necessary to determine the characteristics and the measurement uncertainty function $x(t)$ of the physical quantity or their combination.

Identification task. This task is called the black box task. It is necessary to determine the transformation operator $Z[\cdot]$ of the measuring instrument for the given functions $s(t)$ and $y(t)$.

The results of using analysis and synthesis methods have been published in a significant number of scientific and technical works on theories of signals and systems, automatic control, optimal systems, etc. In metrology and, first of all, in the theory of information-measuring systems, the inverse problem and the identification problem are typical problems measurements.

Figure 1.18 is an illustrative diagram of an inverse measurement task. To obtain the necessary information on the measurement results, it is necessary to solve the inverse task:

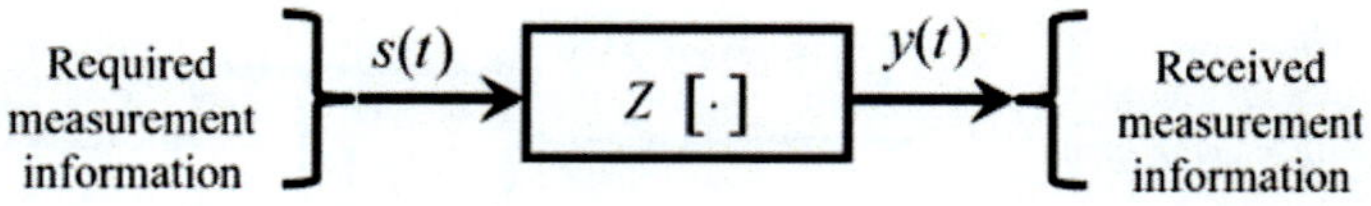

Fig. 1.18 Inverse measurement task diagram

$$s(t) = Z^{-1}[y(t)],$$

where $Z^{-1}[\cdot]$ is a inverse operator for the given transformation operator $Z[\cdot]$ of the measuring instrument.

In the general case, inverse problems are relevant not only for measurements, but also for mathematical physics, thermodynamics, geophysics, geodesy, astronomy, medical displaying, including computed tomography, remote sensing of the Earth and others. Inverse problems in most practical cases are incorrectly posed and have considerable difficulty in solving them [45, 46]. Methods and results of solving inverse problems are published in a significant number of scientific and technical publications. So, for example, the following publications are known: "Methods for solving uncorrelated problems" by A. M. Tikhonov, V. Ya. Arseninin (1974); "Inverse problems of heat conduction" in two volumes by Yu. M. Matsevytyi (2002).

In practice, the tasks of identifying measuring instruments are always solved in order to confirm their metrological characteristics. A series of full-scale measuring experiments of each measuring instrument (system) is carried out and the real metrological characteristics of its transformation operator $Z[\cdot]$ are determined. It should be noted that identifying the transformation operator $Z[\cdot]$ for each measuring instrument is a difficult task and this complexity is caused by the following. The transformation operator $Z[\cdot]$ implements the measurement Eq. (1.8) and integrally describes the sequence of such measurement operations:

- interaction of the RO with the primary measuring transducer (sensor, sensor) and the formation at its output of the measuring signal as a physical medium of measurement information;
- generation of measurement data of the numerical values of the generated signal based on a comparative operation with the unit (scale) of the investigated physical quantity;
- transmission of measurement data over the channel for transmitting information and accounting for the effects of interference;
- processing measurement data, received during influence of interference, and determining the result and measurement uncertainty.

Thus, the transformation operator $Z[\cdot]$ of the measuring instrument (system) is a composite operator, which is formed by a sequence of operators of the measuring instrument modules.

In the previous paragraph 4, it was considered the general conditions for solving the inverse measurement task. The idealized case of an isomorphic transformation of the information signal $s(t)$ by the corresponding operator $Z[\cdot]$ of the measuring instrument does not exist in practice. Therefore, the solution to this problem is obtained through the use of the concept of a combination of physical and probabilistic measures.

References

1. Forbes, A.B.: Measurement uncertainty and optimized conformance assessment. Measurement **39**(9), 808–814 (2006). https://doi.org/10.1016/j.measurement.2006.04.007
2. Acciaio, B., Penner, I.: Dynamic risk measures. In: Di Nunno, G., Øksendal, B. (eds.) Advanced Mathematical Methods for Finance, pp. 1–34. Springer, Berlin, Heidelberg (2011). https://doi.org/10.1007/978-3-642-18412-3_1
3. Zhu, X., Kuo, W.: Importance measures in reliability and mathematical programming. Ann. Oper. Res. **212**, 241–267 (2014). https://doi.org/10.1007/s10479-012-1127-0
4. Martyniuk, G., Onykiienko, Y., Scherbak, L.: Analysis of the pseudorandom number generators by the metrological characteristics. Eastern-Euro. J. Enterp. Technol. **1**, 9(79), 25–30 (2106). https://doi.org/10.15587/1729-4061.2016.60608
5. Petrov, A., Chernyakov, Y., Steblyanko, P., Demichev, K., Haydurov, V.: Development of the method with enhanced accuracy for solving problems from the theory of thermo-pseudoelastic-plasticity. Eastern-Euro. J. Enterp. Technol. **4**, 7(94), 25–33 (2018). https://doi.org/10.15587/1729-4061.2018.131644
6. Cabrelli, C.A., Molter, U.M.: The Kantorovich metric for probability measures on the circle. J. Comput. Appl. Math. **57**(3), 345–361 (1995). https://doi.org/10.1016/0377-0427(93)E0213-6
7. Contreras, G., Lopes, A.O., Thieullen, Ph.: Lyapunov minimizing measures for expanding maps of the circle. Ergodic Theory Dyn. Syst. **21**(5), 1379–1409 (2001). https://doi.org/10.1017/S0143385701001663
8. Pesin, Y., Weiss, H.: The multifractal analysis of Gibbs measures: motivation, mathematical foundation, and examples. Chaos **7**, 89 (1997). https://doi.org/10.1063/1.166242
9. Bayart, F., Grivaux, S.: Invariant Gaussian measures for operators on Banach spaces and linear dynamics. Proc. Lond. Math. Soc. **94**(1), 181–210 (2007). https://doi.org/10.1112/plms/pdl013
10. Stadek, J., Gaska, A.: Evaluation of coordinate measurement uncertainty with use of virtual machine model based on Monte Carlo method. Measurement **45**(6), 1564–1575 (2012). https://doi.org/10.1016/j.measurement.2012.02.020
11. Parthasarathy, K.R.: Probability Measures on Metric Spaces, 288 p. Academic Press, Cambridge (1967)
12. McMaster, R.B.: A statistical analysis of mathematical measures for linear simplification. Am. Cartographer **13**(2), 103–116 (1986). https://doi.org/10.1559/152304086783900059
13. Marcelo, V., Jiagang, Y.: Physical measures and absolute continuity for one-dimensional center direction. Annales de l'I.H.P. Analyse non linéaire **30**(5), 845–877 (2013). https://doi.org/10.1016/j.anihpc.2012.11.002
14. Johnson, A., Rudolph, D.J.: Convergence under ×q of ×p invariant measures on the circle. Adv. Math. **115**(1), 117–140 (1995). https://doi.org/10.1006/aima.1995.1052
15. Arnoldi, J.-F., Loreau, M., Haegeman, B.: Resilience, reactivity and variability: a mathematical comparison of ecological stability measures. J. Theor. Biol. **389**, 47–59 (2016). https://doi.org/10.1016/j.jtbi.2015.10.012
16. Dieck, R.H.: Measurement uncertainty models. ISA Trans. **36**(1), 29–35 (1997). https://doi.org/10.1016/S0019-0578(97)00004-9
17. Fenton, N., Krause, P., Neil, M.: Software measurement: uncertainty and causal modeling. IEEE Softw. **19**(4), 116–122 (2002). https://doi.org/10.1109/MS.2002.1020298
18. Rowe, W.D.: Understanding uncertainty. Risk Anal. **14**(5), 743–750 (1994). https://doi.org/10.1111/j.1539-6924.1994.tb00284.x

19. Giordani, A., Mari, L.: Measurement, models, and uncertainty. IEEE Trans. Instrum. Meas. **61**(8), 2144–2152 (2012). https://doi.org/10.1109/TIM.2012.2193695
20. Jakubiec, W., Plowucha, W., Starczak, M.: Analytical estimation of coordinate measurement uncertainty. Measurement **45**(10), 2299–2308 (2012). https://doi.org/10.1016/j.measurement.2011.09.027
21. Cox, M.G., Harris, P.M.: Measurement Uncertainty and Traceability **17**(3), 533 (2006). https://doi.org/10.1088/0957-0233/17/3/S13
22. Guide to the Expression of Uncertainty in Measurement, 1st edn, 112 p. ISO, Switzerland (1993)
23. Wilhelm, R.G., Hochen, R., Schwenke, H.: Task specific uncertainty in coordinate measurement. CIRP Ann. **50**(2), 553–563 (2001). https://doi.org/10.1016/S0007-8506(07)62995-3
24. Babak, V., Eremenko, V., Zaporozhets, A.: Research of diagnostic parameters of composite materials using Johnson distribution. Int. J. Comput. **18**(4), 483–494 (2019)
25. Eremenko, V., Zaporozhets, A., Babak, V., Isaienko, V., Babikova, K.: Using hilbert transform in diagnostic of composite materials by impedance method. Period. Polytech. Electr. Eng. Comput. Sci. **64**(4), 334–342 (2020). https://doi.org/10.3311/PPee.15066
26. Kessel, W.: Measurement uncertainty according to ISO/BIPM-GUM. Thermochim. Acta **382**(1–2), 1–16 (2002). https://doi.org/10.1016/S0040-6031(01)00729-8
27. Possolo, A., Toman, B.: Assessment of measurement uncertainty via observation equations. Metrologia **44**(6), 464 (2007). https://doi.org/10.1088/0026-1394/44/6/005
28. Philips, S.D., Borchardt, B., Estler, W.T., Buttress, J.: The estimation of measurement uncertainty of small circular features measured by coordinate measuring machines. Precis. Eng. **22**(2), 87–97 (1998). https://doi.org/10.1016/S0141-6359(98)00006-3
29. Zvaritch, V.N., Myslovych, M.V.: White noise in some simulation problems of information signals. Elektron. Model **40**(2), 17–26 (2018). https://doi.org/10.15407/emodel.40.02.017
30. Breuer, H.-P.: Separability criteria and bounds for entanglement measures. J. Phys. A: Math. Gen. **39**(38), 11847 (2006). https://doi.org/10.1088/0305-4470/39/38/010
31. Babak, V., Zaporozhets, A., Kuts, Y., Scherbak, L., Eremenko, V.: Application of material measure in measurements: theoretical aspects. In: Zaporozhets, A., Artemchuk, V. (eds.) Systems, Decision and Control in Energy II. Studies in Systems, Decision and Control. Springer, Cham, pp. 261–269 (2021). https://doi.org/10.1007/978-3-030-69189-9_15
32. Adesso, G., Bromley, T.R., Cianiaruso, M.: Measures and applications of quantum correlations. J. Phys. A: Math. Theor. **49**, 473001 (2016). https://doi.org/10.1088/1751-8113/49/47/473001
33. Keane, M., Murray, R., Young, L.-S.: Computing invariant measures for expanding circle maps. Nonlinearity **11**(1), 27 (1998). https://doi.org/10.1088/0951-7715/11/1/004
34. Herz, C.S.: Spectral synthesis for the circle. Ann. Math. **68**(3), 709–712 (1958). https://doi.org/10.2307/1970163
35. Jenkinson, O.: Geometric barycentres of invariant measures for circle maps. Ergodic Theory Dyn. Syst. **21**(2), 511–532 (2001). https://doi.org/10.1017/S0143385701001250
36. Dzhavilov, A.A.: Thermodynamic formalism and singular invariant measures for critical circle maps. Theor. Math. Phys. pp. 166–180 (2003). https://doi.org/10.1023/A:1022271903129
37. Graczyk, J., Swiatek, G.: Singular measures in circle dynamics. Commun. Math. Phys. **157**, 213–230 (1993). https://doi.org/10.1007/BF02099758
38. Di Roberto, C., Dempster, A.: Circularity measures based on mathematical morphology. Electron. Lett. **36**(20), 1691–1693 (2000). https://doi.org/10.1049/el:20001191
39. Babak, V.P., Babak, S.V., Myslovych, M.V., Zaporozhets, A.O., Zvaritch, V.M.: Methods and models for information data analysis. In: Diagnostic Systems For Energy Equipments. Studies in Systems, Decision and Control, vol. 281, pp. 23–70. Springer, Cham (2020). https://doi.org/10.1007/978-3-030-44443-3_2
40. Strichartz, R.S., Taylor, A., Zhang, T.: Densities of self-similar measures on the line. Exp. Math. **4**(2), 101–128 (1995). https://doi.org/10.1080/10586458.1995.10504313
41. Kuts, Y., Scherbak, L., Sokolovska, G.: Methods of processing broadband and narrowband radar signals. In: 2011 Microwaves, Radar and Remote Sensing Symposium, Ukraine, pp. 374–377 (2011). https://doi.org/10.1109/MRRS.2011.6053678

42. Fryz, M., Scherbak, L.: Conditional linear random process as a mathematical model of radar noise. In: 2011 Microwaves, Radar and Remote Sensing Symposium, Ukraine, pp. 367–370 (2011). https://doi.org/10.1109/MRRS.2011.6053675
43. Dergunov, A.V., Kuts, Y.V., Scherbak, L.N.: Comparative analysis of modern time-series analysis methods. In: 2011 Microwaves, Radar and Remote Sensing Symposium, Ukraine, pp. 378–381 (2011). https://doi.org/10.1109/MRRS.2011.6053679
44. Kuts, Y.V., Shengur, S.V., Scherbak, L.N.: Circular measurement data modeling and statistical processing in LabView. In: 2011 Microwaves, Radar and Remote Sensing Symposium, Ukraine, pp. 317–320 (2011). https://doi.org/10.1109/MRRS.2011.6053664
45. Zaporozhets, A.O., Khaidurov, V.V.: Mathematical models of inverse problems for finding the main characteristics of air pollution sources. Water Air Soil Pollut. **231**, 563 (2020). https://doi.org/10.1007/s11270-020-04933-z
46. Zaporozhets, A., Khaidurov, V., Tsiupii, T.: Optimization models of industrial furnaces and methods for obtaining their numerical solution. In: Zaporozhets, A., Artemchuk, V. (eds.) Systems, Decision and Control in Energy II. Studies in Systems, Decision and Control. Springer, Cham, pp. 121–139 (2021). https://doi.org/10.1007/978-3-030-69189-9_7

Chapter 2
Models of Measuring Signals and Fields

Abstract The methodological aspects of the analysis of multidimensional signals and fields are highlighted. Mathematical models of measuring signals and fields are systematized, the main spatio-temporal models of quasi-determined signals are considered—continuous and discrete, complex-valued, periodic. Some facts of the theory of orthogonal signals and orthogonal bases are considered, the possibilities of their use for measurements are analyzed. Signal models that are described by random processes are considered. The definitions of a random process stationary in the broad and narrow sense of random processes, a linear random process, a harmonized random process, a periodically correlated random process are given. Models of multidimensional signals and spatio-temporal random fields are considered. Characteristics of signals and spatio-temporal random fields are given taking into account their structure for research by means of measuring equipment. It is concluded that the creation of hardware and information support for measurements and monitoring requires the coordination of physical and probabilistic measures to assess the characteristics of multidimensional signals and fields in accordance with the requirements of the concept of measurement uncertainty.

Keywords Signals' mathematical models · One-dimensional signal · Multi-dimensional signal · Continuous signals · Discrete signals · Orthogonal bases · Random signal models · Field models

2.1 Mathematical Models of Signals and Their Classification

For performing the measurement tasks, 4 options are used for the formation of primary information in the interaction of various research objects (ROs) with the measurement system [1–3]:

V. P. Babak et al., *Models and Measures in Measurements and Monitoring*, Studies in Systems, Decision and Control 360,
https://doi.org/10.1007/978-3-030-70783-5_2

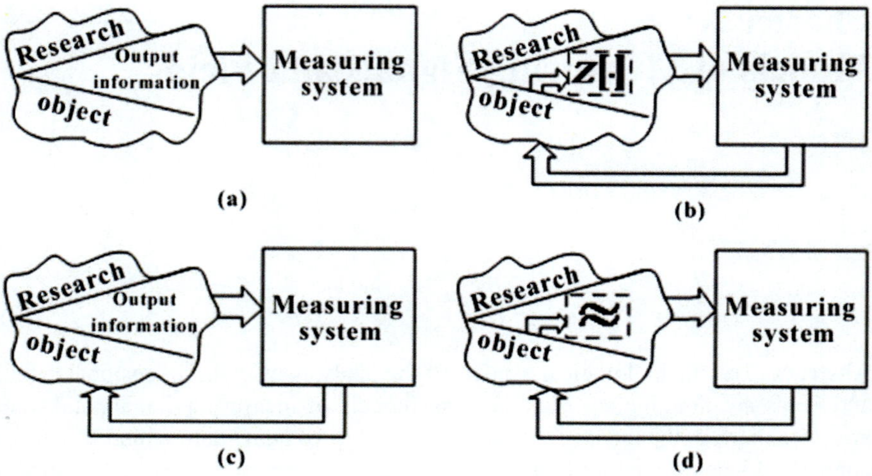

Fig. 2.1 Variants of interaction between RO and the measurement system

(a) the RO is the source (generator) of measurement information;
(b) Ro is a linear or nonlinear, inertial or inertialess, one-dimensional or multi-dimensional transducer of the impact (signal) acting on RO, and which is conventionally called test, and RO is described by a specific transformation operator $Z[\cdot]$;
(c) the RO is a source that forms its own, different from the input, reaction (response), and the input signal is considered as stimulus;
(d) the nature of the dynamics (changes over time) of the studied properties, states and characteristics of the RO is determined by the measurement system, that is, the measurement and control system.

Options for this interaction are shown schematically in Fig. 2.1.

For the first option (Fig. 2.1a), the basic mathematical model of the signal is a random process or a combination of deterministic and random processes whose characteristics reflect the properties of the RO. Options in Fig. 2.1b–d provide the formation of test signals that are set by the measurement system, and their model is justified based on the characteristics of RO, the conditions and tasks of measurement. These patterns may be deterministic, random, or their combinations.

Input signals are exposed of RO, as a result of which their characteristics change, and the signals themselves become physical carriers of information. Such signals are called measuring.

The possibilities of modern science and technology allow to implement three areas of research of the measurement process [4–8]:

- mathematical modeling based on the development of information support, the use of computer technology and the implementation of a computational (computer) measuring experiment;

- physical modeling based on the use of physical RO models, both homogeneous in physical nature and other physical nature, which includes a reasonable choice of measuring instruments and conducting a simulation measurement experiment;
- experimental or full-scale study of a real unit based on the use of measuring instruments and conducting a full-scale measuring experiment.

The basis for such studies are mathematical models of signals, the arguments of which in the general case are the variables of space $(x, y, z) = \boldsymbol{r} \in G \subseteq R^3$ and time $t \in T \subseteq R$.

Each time, the question arises before the researcher: how to choose a fairly simple model, which at the same time would fully reflect the basic and essential properties of the physical process? The process of creating and researching a mathematical model of a signal consists of the following steps.

At the *first stage*, the analysis of the results of studies of the basic physical laws, the influence of various factors and other information during the formation of the signal in space and time are carried out. On this basis, using mathematical objects, terms and symbols, a new displaying is created—a mathematical model of the signal, usually in an analytical form.

At the *second stage*, a wide range of theoretical and applied measurement problems is solved, in which the signal model is used.

At the *third stage*, the consistency of the results of theoretical and practical studies of the signal based on the selected criteria is evaluated. In some scientific works, the evaluation of the results of comparative analysis uses the term *adequate model*. It should be noted that in this case it is not talking about a comparative analysis of two objects of different physical nature (mathematical and material), but rather by comparing the characteristics, parameters, patterns, etc., that were obtained during an agreed study of the model and the physical phenomenon.

The *fourth stage* is for building a more advanced or new model in the case when the results of the analysis of the new data, facts about RO do not correspond to the proposed model. In some cases, for example, for harmonious signals, the need to complete such a stage in full may not arise—only certain parameters or characteristics of the model can be adjusted.

The study of signal models as mathematical objects in measurement problems is carried out using methods of functional analysis, measure theory, theory of random processes and mathematical statistics, theory of signals and systems, mathematical physics and measurement theory [9–12].

The one-dimensional signal is described by the function of one argument $\{u(x), x \in X\}$. If the argument of the function $u(x)$ is time, that is, $u(t)$, the components of the sequence of one-dimensional signals $\{u_i(t), t \in T\}, i = \overline{1, n}$ are functions of time. Examples of such signals are the voltage at the output of harmonic and pulse signal generators, acoustic transducer systems, etc.

The term *multidimensional signal* requires some specificity.

An ordered set of n one-dimensional signals $\left\{u_i(x),\ i = \overline{1, n}\right\}$ creates the so-called multi-channel signal, which is described by the vector $\boldsymbol{u}_n(t) =$

Fig. 2.2 Signal field implementation example

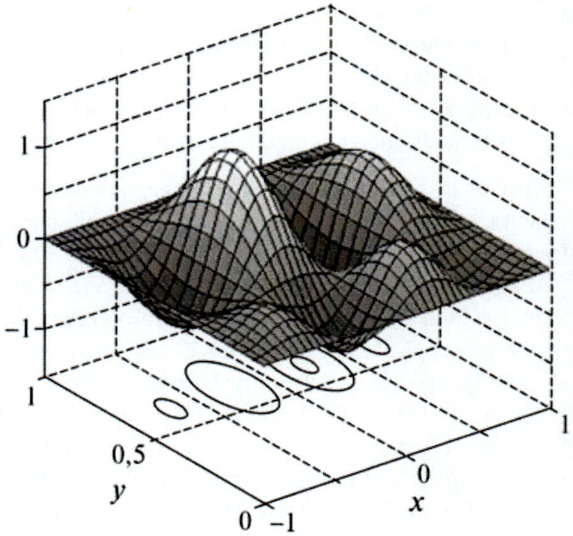

$\{u_1(t), \ldots, u_n(t)\}, t \in T$. Number n is the dimension of the vector $\boldsymbol{u}_n(t)$ is determined by the number of output channels of the research system.

The signal field is described by the function of several (at least two) variables (arguments) $u(\boldsymbol{x}) = u(x_1, x_2, \ldots, x_n)$. In the case when one of the variables is understood as time and the other three as space variables, that is, $u(\boldsymbol{r}; t) = u(x, y, z; t)$, or $u(x, y; t)$, or $u(x; t)$, then signal field models are called spatio-temporal. One of the implementations of such a field is shown only in Fig. 2.2.

An example of complex mathematical models is a vector signal field of the form

$$\boldsymbol{u}_n(\boldsymbol{r}; t) = \{u_1(\boldsymbol{r}, t), \ldots; u_n(\boldsymbol{r}, t)\}, \boldsymbol{r} \in G, t \in T.$$

Signals can be divided into continuous and discrete, but it is necessary to indicate the nature of their areas of definition and values.

The term "digital signals" is used to describe discrete signals both in time and in value, followed by their encoding. To do this, choose one of the known numerical systems: binary, octal, hexadecimal, etc. In digital signal processing, priority is given to the binary system, therefore, digital signals in most cases are numerical sequences of binary characters with a finite number of bits, for example 11001110.

Pulse signals are described by finite functions, that is, functions that acquire nonzero values only at finite time intervals. It should be noted that there are other definitions of pulse signals, for example, a sequence of signals that do not intersect on the time axis.

During describing multidimensional signals, the number of options for continuity and discreteness increases. For example, a signal field can be discrete in time, continuous in spatial arguments, and discrete in values.

Table 2.1 Signs of continuous and discrete signals

$t \in T$, area of definition	$s(t)$, set of values	
	Continuous	Discrete
Continuous	*Continuous (analog)*	*Quantized*
Discrete	*Discrete in time*	*Digital*

The knowledge of mathematical models of signals allows to carry out their comparative analysis, determine their numerical characteristics (parameters), evaluate the measurement results, establish the identity and difference of the signals and classify them. As a rule, the name of the measuring signal is determined by the type of its mathematical model, for example, deterministic signals are described by deterministic functions, random ones by random functions.

Problem of signal classification. The solution of this problem for signal models (hereinafter simply signals) is relevant and important for measurement. The values, parameters and characteristics of the signals are the subjects of research of various measurement tasks, therefore, the number of types or classes of measuring signals is significant. For example, if signals are classified by types of functions in functional analysis, then their number can reach thousands. If classified according to the laws of distribution of random signals, then their number will exceed three hundred. In the case of classification of signals from more general positions, namely, under the condition of substantiation of function spaces, their number will exceed several tens. To such spaces belong vector, linear, metric, normalized, measurable with measure, probabilistic, Hilbert etc. [13–18].

Any classification is conditional and depends on its purpose and features. In further materials of this work, examples of typical signals will be given, and they are widely used in solving problems of the theory and practice of measurement.

Continuous and discrete signals. Depending on the nature of the areas of definition and the values of signal $\{s(t) \in R, \quad t \in T\}$, it will be distinguished 4 types of signals, the signs of which are given in Table 2.1:

- continuous (analog)—signals that acquire arbitrary values and are defined for any time t;
- discrete in time with a sampling interval Δt—signals that are arbitrary in magnitude, but are not specified on the entire time axis, but only at certain times $t_i = (i - 1)\Delta t$, where $i = 1, 2, \ldots$;
- quantized with a quantization interval Δs—signals that are continuous functions of $s(t)$ in time acquire only discrete values of $s_j = (j - 1)\Delta s$ at $j = 1, 2, \ldots$;
- digital—signals that are discrete in level and time, as well as encoded, that is, each reference is assigned a number with a finite number of digits.

Complex signals. All previously considered signal models belong to the class of real ones. The mathematical model of complex signals is supplied in general by the formula:

$$\dot{z}(t) = x(t) + iy(t), \tag{2.1}$$

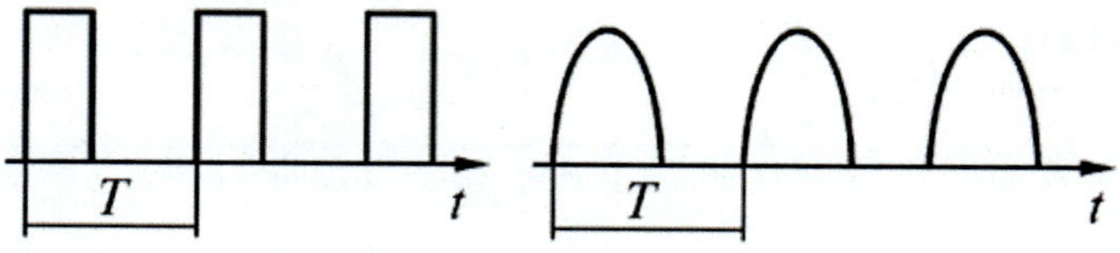

Fig. 2.3 Graphs of the implementation of periodic signals: **a** rectangular; **b** cut cosines

where $x(t)$ and $y(t)$ are real signals; $i = \sqrt{-1}$ is the imaginary unit.

A special case of (2.1) is the so-called complex-exponential signal, has the form:

$$\dot{z}(t) = Ae^{i(\omega_0 t + \phi_0)},$$

where A, ω_0, ϕ_0 are the amplitude, circular frequency, and initial phase of the signal respectively.

Using the Euler formula, this signal can be represented as:

$$\dot{z}(t) = A\cos(\omega_0 t + \phi_0) + iA\sin(\omega_0 t + \phi_0).$$

Periodic signals. Frequency can also be a classification feature of signals. On this basis distinguish: periodic and aperiodic (single) signals.

The general form of the mathematical model of a periodic signal has the form:

$$s(t) = s(t + nT_0), n = \pm 1, \pm 2, \ldots, \quad (2.2)$$

where T_0 is a period of signal. Examples of the implementation graphs of periodic signals are shown in Fig. 2.3.

The aperiodic (pulse or single) signal $s(t)$ is a special case of a periodic signal when the period T_0 tends to infinity, i.e. $s(t) = \lim_{T_0 \to \infty} s(t + nT_0)$.

Harmonious and modulated signals. The mathematical model of a harmonic signal is described by the expression

$$u(t) = A_0 \cos(2\pi f_0 t + \phi_0) = A_0 \cos(\omega_0 t + \phi_0), t \in T, \quad (2.3)$$

where A_0—amplitude, f_0—frequency, $2\pi f_0$—circular frequency, $T_0 = 1/f_0$—period.

The value $2\pi f_0 t + \phi_0$ for fixed t is called the signal phase, and the value of the phase with $t = 0$, that is, parameter ϕ_0, is called the initial phase.

A wide range of applications of the harmonic signal model is due to the following facts:

- a significant number of natural phenomena, processes, signals generated by technical systems, which are characterized by the property of cyclicality, regularity, rhythm, i.e. repeatability in time or in space, can be described by an idealized mathematical model (2.3);

- the use of signal (2.3) has deep historical roots—this is the study of the Italian physicist and astronomer Galileo Galilei (1564–1642) of oscillatory movements, the development of pendulum and spring clocks as mechanical oscillatory systems of the Dutch physicist Christian Huygens (1629–1695), the works of the English physicist Robert Hooke (1635–1703) in the theory of oscillations and research results of the famous French mathematician Jean Baptiste Joseph Fourier (1768–1830), as the founder of harmonic analysis in mathematics and physics;
- a whole branch of the national economy of developed countries—instrumentation in the middle and second half of the twentieth century produced a significant number of measuring instruments and signal parameter measurement systems (2.3), including voltmeters, ammeters, frequency meters and phase meters;
- model (2.3) is an integral component of other more complex information signals.

Subject to changes in the parameters of the harmonic signal, namely, amplitude, frequency or phase, a series of modulated signals with a harmonious carrier signal is formed. Such modulated signals are: amplitude-modulated; frequency-modulated and phase-modulated. The last two types of harmonic signal modulation are generally considered as signals with angular modulation. The plots of realizations of the amplitude-modulated signal with different modulation coefficients k_A are shown in Fig. 2.4.

In more detail, the results of studies of modulated signals with a harmonious carrier signal are considered in [19].

Pulse and modulated pulse signals. To represent pulsed signals, a Heaviside unit function of the form

$$\Phi(t) = \begin{cases} 1, \; t \geq 0, \\ 0, \; t < 0. \end{cases} \tag{2.4}$$

A periodic sequence of rectangular pulses using (2.4) is described by the expression

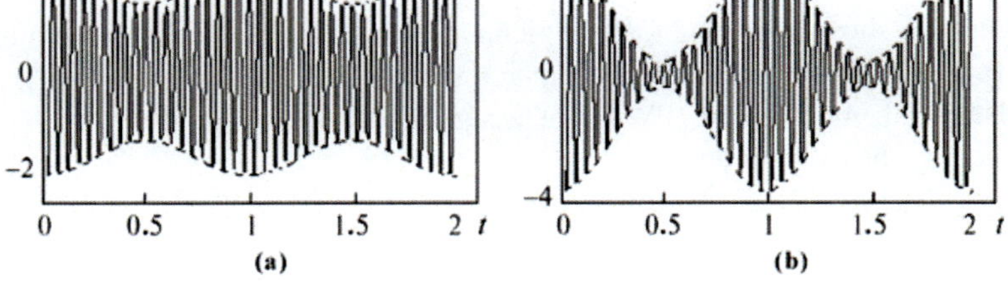

Fig. 2.4 Graphs of the implementation of amplitude-modulated signals with different values of k_A: **a** $k_a = 0.2$; **b** $k_a = 0.8$

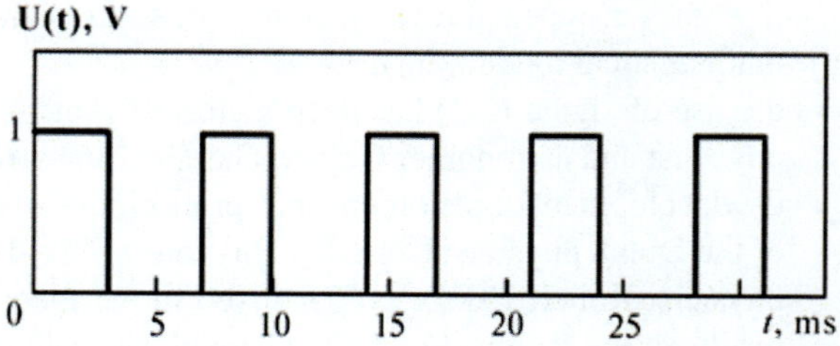

Fig. 2.5 Schedule for the implementation of a periodic sequence of rectangular pulses

$$u(t) = \sum_{k=0}^{n} A_0[\Phi(\tau_0 + kT_0 - t) - \Phi(kT_0 - t)], \tag{2.5}$$

where T_0—repetition period, τ_0—duration, $A_0 > 0$—amplitude of the pulses are the parameters of the sequence of rectangular pulses.

The implementation schedule of a periodic sequence of pulses for specific n, T_0, τ_0, and A_0 is shown in Fig. 2.5.

Provided that the parameters of the periodic sequence of rectangular pulses change in time, namely, the amplitude, duration, and repetition period, a series of modulated pulse signals is formed by analogy with modulated signals with a harmonious carrier signal.

Further it will be considering some types of modulated pulse signals.

An amplitude-modulated pulse signal or a signal with amplitude-pulse modulation (APM) using (2.5) is described by the expression

$$U_{\mathrm{APM}}(t) = \sum_{k=0}^{n} A(t)[\Phi(\tau_0 + kT_0 - t) - \Phi(kT_0 - t)], \tag{2.6}$$

where, in accordance, the function $A(t)$ is the result of modulation by the information signal of the parameter A_0 of the pulse set.

The measurements also use a sequence of radio pulses (RPs), which is described by the expression

$$u_{RP}(t) = \sum_{k=0}^{n} A_0[\Phi(\tau_0 + kT_0 - t) - \Phi(kT_0 - t)] \cos 2\pi f t, \quad t \in T, \tag{2.7}$$

Figure 2.6 shows a graph of the implementation of a sequence of radio pulses in the form (2.7) with the corresponding parameters of this model.

Pulse width modulation (PWM) of a signal (2.5) or a signal with a PWM is described by the expression

$$u_{\mathrm{PWM}}(t) = \sum_{k=0}^{n} A_0[\Phi(\tau(t) + kT_0 - t) - \Phi(kT_0 - t)], \tag{2.8}$$

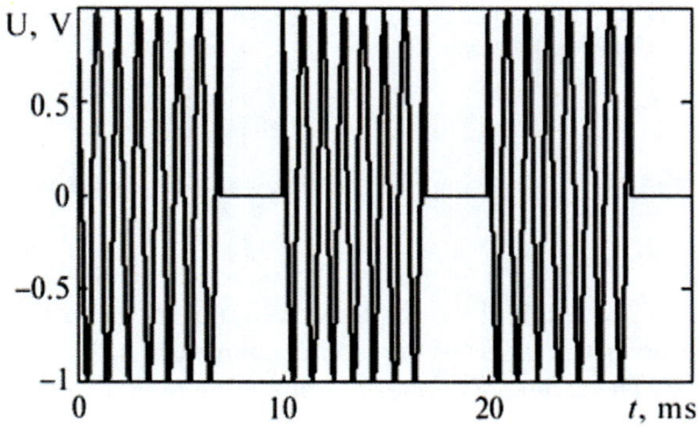

Fig. 2.6 Schedule for the implementation of the sequence of radio pulses

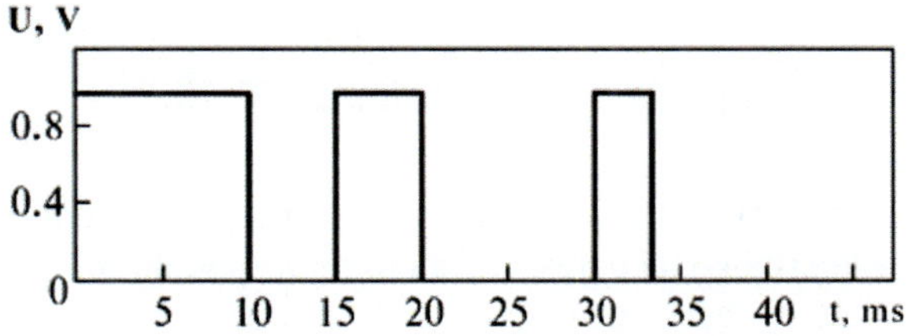

Fig. 2.7 PWM signal implementation graph

where the function $\tau(t)$ is the result of modulation by the information signal of the parameter τ_0 of the pulse sequence.

As an example of the function $\tau(t)$, it will be considering this:

$$\tau(t) = \begin{cases} \frac{\tau_0}{k+1}, & kT_0 \le t < (k+1)T_0, \quad k = \overline{0, n}, \\ 0, & \text{in other cases.} \end{cases}$$

The implementation graph of such a PWM signal for certain model parameters (2.8) is shown in Fig. 2.7.

Frequency-pulse modulation (FPM) of a signal of the form (2.5) or a FPM signal is described by the expression

$$u_{FPM}(t) = \sum_{k=0}^{n} A_0[\Phi(\tau_0 + kT(t) - t) - \Phi(kT(t) - t)], \ \tau_0 < \frac{T_0}{2}, \tag{2.9}$$

where function $T(t)$ is the result of modulation by an information signal of the parameter T_0 of model (2.5).

As an example of the function $T(t)$, it will be considering this:

$$T(t) = \begin{cases} \frac{(k+1)T_0}{2}, & kT_0 \le t < (k+1)T_0, \quad k = \overline{0, n}, \\ 0, & \text{in other cases.} \end{cases}$$

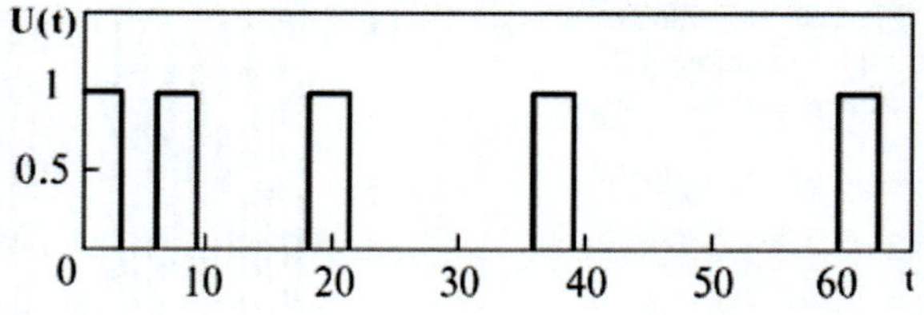

Fig. 2.8 FPM signal implementation graph

The graph of the implementation of the FPM signal for this example for certain numerical values of the model parameters is shown in Fig. 2.8.

The results of studies of pulsed and modulated pulsed signals are considered, for example, in [18].

2.2 Signals and Orthogonal Bases

Further it will be considering the known facts from the theory of orthogonal signals and orthogonal bases, and analyze the possibilities of their use for measurements.

It is known that two signals $u(t)$ and $v(t)$ are called orthogonal if their scalar multiplication, and therefore the mutual energy, are equal to zero:

$$(u, v) = \int_{-\infty}^{\infty} u(t)\, v(t)\, dt = 0.$$

An image of a Hilbert signal $u(t)$ of the form [20]

$$u(t) = \sum_{k=0}^{\infty} c_k \varphi_k(t), \quad t \in T \tag{2.10}$$

is called the distribution of the signal $u(t)$ along selected orthogonal basis $\{\varphi_k(t)\}$.

A series in the form (2.10) is also called the orthogonal series of the signal $\{u(t)\}$, where the sequence of real numbers $\{c_k\}$ is the expansion coefficients along the selected basis on the time interval of the orthogonality T.

One of the main ideas for using series (2.10) is the use of well-known systems of functions. Such functions are the classical orthogonal polynomials of Legendre, Chebyshev, Laguerre, Walsh with continuous time, a trigonometric system of functions, and others [21–24]. Orthogonal bases with a discrete argument are polynomials of Chebyshev, Kravchuk, Charles, Lagerra, Walsh functions with a discrete argument, which are widely used in digital processing of measurement data using computer technology [25–27].

The use of orthogonal signals and bases makes it possible to effectively solve a wide range of problems of measuring the characteristics and parameters of signals and has the following advantages.

- The volume and number of operations for determining the sequence of real numbers $\{c_k\}$ is fundamentally reduced, because there is no need to measure the characteristics of the signal as a function of time.
- Using the well-known orthogonal basis $\{\varphi_k(t)\}$ makes it possible to determine the convergence of series (2.10) with given accuracy indices.
- The use of images (2.10) is essentially an algorithm of simulation (mathematical and computer) modeling of the implementation of various types of signals, including test ones, with specified accuracy indicators.
- The rationale for the use of one or another orthogonal basis in the research problems of measuring signals is based on the results of measuring their characteristics.

It is known that series (2.10) is called the generalized Fourier series, and the coefficients of this series are determined by the expression

$$c_k = \int_{-\infty}^{\infty} s(t)\varphi_k(t)dt = (s, \varphi_k), \tag{2.11}$$

as the scalar multiplication of the signal $s(t)$ with one of the functions of the orthogonal basis $\{\varphi_k(t)\}$.

During choosing a specific orthogonal basis, series (2.10) are named as follows: for the Chebyshev orthogonal basis—the Fourier–Chebyshev series; for the Hermite orthogonal basis, the Fourier-Hermite series.

Let consider two characteristic examples of orthogonal bases, which are widely used in the theory and practice of measurements for studying the parameters and characteristics of information signals.

Example 1 The system of trigonometric functions with multiple frequencies, supplemented by a constant in time expression φ_0, has the following form

$$\begin{gathered}
\varphi_0(t) = 1/\sqrt{T}\ ; \\
\varphi_1(t) = \sqrt{2/T}\sin 2\pi t/T\ ; \\
\varphi_2(t) = \sqrt{2/T}\cos 2\pi t/T\ ; \\
\cdots\cdots\cdots\cdots\cdots\cdots \\
\varphi_{2k-1}(t) = \sqrt{2/T}\sin 2\pi kt/T\ ; \\
\varphi_{2k}(t) = \sqrt{2/T}\cos 2\pi kt/T,\ \ \ldots \\
\cdots\cdots\cdots\cdots\cdots\cdots
\end{gathered}$$

and forms an orthogonal basis on the time interval $\left[-T/2, T/2\right]$.

The graphs of the functions $\varphi_1(t)$ and $\varphi_2(t)$ are shown in Fig. 2.9. It can be shown that the scalar multiplication of any two functions of such a system is equal to zero, and the energy and, accordingly, the norm of each of them is equal to one.

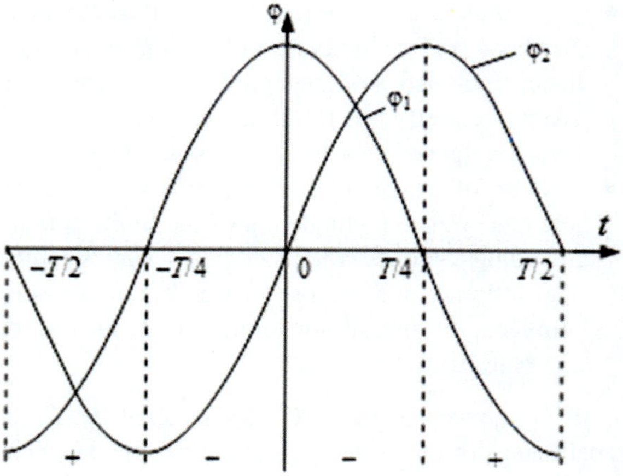

Fig. 2.9 Graphs of trigonometric functions $\varphi_1(t)$ and $\varphi_1(t)$

Example 2 In accordance with the sampling theorem, for Hilbert signal with a limited spectrum, an orthogonal basis is used, which is single and unique in the class of orthogonal bases.

Consider the features of such an orthogonal basis. Let single out a class of signals whose spectra are nonzero only within a finite frequency range. An example of such signals is a radio pulse with linear frequency modulation.

Let D be a finite frequency interval in which the spectrum $s(t)$ of some signal $u(t)$ is not equal to zero, that is, $s(f) \neq 0$, if $f \in D$; and $s(f) = 0$ if $f \notin D$. In general, the model of a signal with a limited spectrum is determined by the inverse Fourier transform formula: $u(t) = \frac{1}{2\pi} \int_D s(f)\, e^{i2\pi f t} df$.

The graph of such a signal $u(t)$ and its spectrum $s(t)$ is shown in Fig. 2.10a, b.

Depending on the choice of the interval D and function $s(t)$, various signals with a limited spectrum can be obtained. It will be considering oscillations which spectral density is constant and acquires real values within the frequency interval bounded by a certain upper frequency F, and accordingly $\Omega = 2\pi F$. Outside this interval, the spectral density is zero. Such signal $u(t)$ can be expressed as an orthogonal series [28]:

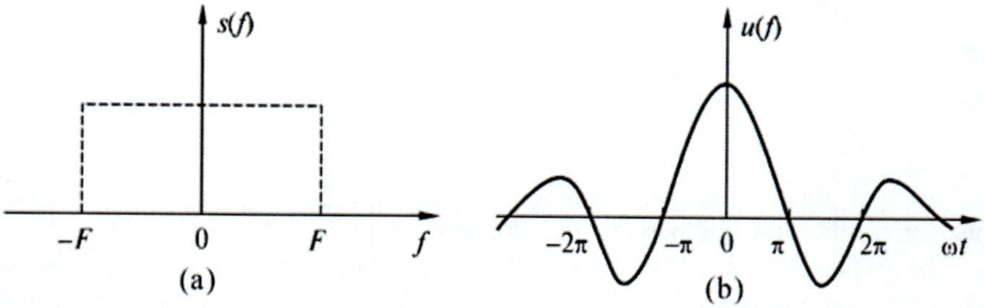

Fig. 2.10 Graphs of the spectrum $s(f)$ (**a**) of an ideal low-frequency signal $u(t)$ (**b**)

$$u(t) = \sum_{k=-\infty}^{\infty} u\left(\frac{k}{2F}\right) \frac{\sin 2\pi F(t - k/2F)}{2\pi F(t - k/2F)} = \sum_{k=-\infty}^{\infty} u(k\Delta t) \frac{\sin \Omega(t - k\Delta t)}{\Omega(t - k\Delta t)}, t \in T \tag{2.12}$$

This series is constructed using a system of orthogonal functions

$$\left\{ \frac{\sin kt}{kt}, k = \ldots, -2, -1, 0, 1, 2, \ldots \right\},$$

for which

$$\int_{-\infty}^{\infty} \frac{\sin \Omega(t - k\Delta t)}{\Omega(t - k\Delta t)} \cdot \frac{\sin \Omega(t - m\Delta t)}{\Omega(t - m\Delta t)} dt = \begin{cases} \Delta t, k = m, \\ 0, k \neq m. \end{cases}$$

The convergence of the series (2.12) is defined in the mean square value, i.e.

$$\lim_{n \to \infty} \left\{ \left(u(t) - \sum_{k=-n}^{n} u(k\Delta t) \frac{\sin \Omega(t - k\Delta t)}{\Omega(t - k\Delta t)} \right)^2 \right\} = 0. \tag{2.13}$$

Series (2.12) is the basis of the well-known sampling theorem, the essence of which is the transmission of the signal *u*(*t*) through communication systems not in the form of coefficients $\{c_k\}$ determined according to (2.11), but directly of the samples of the signal *u*(*t*) taken on a discrete set of time of the interval $-k\Delta t, -(k - 1)\Delta t, \ldots, -\Delta t, 0, \Delta t, \ldots k\Delta t$.

It is the property of using the samples of the signal *u*(*t*) on a uniform time lattice with an interval $\Delta t = 1/2F_B$ determines the basis for orthogonality (2.12).

The use of expression (2.13) plays an important role for measurement problems, in particular, during estimating the values of the accuracy indices of the approximation of the studied signal *u*(*t*) by the orthogonal series (2.12).

Interestingly, the series (2.12) in domestic and foreign publications is called differently. For the purpose of coordination, we propose the international name of the series (2.12) based on the history of its justification and use. It is proposed to name range (2.12) as Whittaker-Kotelnikov-Shannon. It is known that the famous mathematician Whittaker (USA, 1915) first proved the sampling theorem for a function with a limited spectrum, the Soviet scientist in the field of radio engineering Kotelnikov (USSR, 1933) suggested transmitting only the samples of the time signal *u*(*t*) via communication lines according to (2.12), and the well-known scientist in the field of information theory Shannon (USA, 1946) used series (2.12) to solve the problems of transmitting information through various channels. It should also be noted that in a number of publications, it is noted that the sampling theorem was first proved by the outstanding French mathematician O. Cauchy (France, 1789–1857).

2.3 Random Signal Models

It will be considered signal models that are described by random processes. In the study of a random process, which is considered as a sequence of random variables, the definitions and characteristics of a random variable are fundamental.

The random process $\{\xi(\omega, t), t \in T\}$ is a function of two variables, where t is interpreted as time. For each fixed $t = t_0$, it has a random variable $\xi(\omega, t_0)$. The random process $\xi(\omega, t)$ can be defined in 3 equivalent ways: as a set of random variables ordered with respect to variable t; as a set of numerical functions of time—realizations of the process, each of which is considered as one elementary event with the corresponding fixed $\omega \in \Omega$ and as a measured function on the multiplication of the spaces $\Omega \times T$.

Depending on the set of parameter $t \in T$, the random process has certain features and corresponding names:

- if $t \in T$ is given discretely and forms a numerical sequence $\{t_j, j = 0, 1, 2, \ldots)$ with a finite or countable number of its elements, then the random process $\{\xi(\omega, t_j), j = 0, 1, 2, \ldots\}$ is called a time series or a random sequence;
- if t is a continuous parameter, that is, T has a power of the continuum, then the process $\xi(\omega, t) = \xi(t)$ is random with continuous time.

In most cases, there are three types of influence of random factors on the formation and dynamics of a real-time signal in time:

- the dynamics of the signal obeys the exact laws that are uniquely determined by the initial conditions, which are random, that is, these conditions are not preserved during the repeated implementation of the signal. Such a case takes place for a harmonic signal with an initial random phase, the realization of which is formed only at the initial moment of observation time, i.e., for $t = 0$;
- the dynamics of the signal over a given time interval T is random, for example, the voltage of thermal noise at the output of the amplifier, the output voltage of the noise generator;
- the dynamics of the signal is due to the basic random pattern, which is distorted by a random disturbance that is valid during the entire time the signal is observed, for example, receiving a radar signal against a background of interference, finding a useful signal against a background of reverberation noise.

Random process is a complex mathematical object with various properties and characteristics. A conditional classification of random processes can be carried out, for example, based on such signs:

- types of probability distribution laws: Gaussian or Poisson processes;
- properties of process values: processes with independent increments, independent values, continuous or discrete, digital;
- properties of process characteristics: stationary, processes with fractional rational spectral densities, periodically correlated;

- types of images (integral, differential) processes: linear, harmonious and autoregression.

For solving the practical problems of random signal analysis, stationary and ergodic processes are most often used.

Stationary random processes are used as models of stochastic signals whose statistical characteristics have a certain stability over time. Among them are processes in the broad and narrow sense [29, 30].

The random process $\{\xi(\omega, t), t \in R\}$ for which

$$\mathbf{M}\{\xi(\omega, t)\} = a(t) \equiv a$$

and autocorrelation function

$$\mathbf{M}\{[\xi(\omega, t_1) - a][\xi(\omega, t_2) - a]\} = R(t_1, t_2) = R(t, t+\tau) \equiv R(t_2 - t_1) = R(\tau),$$

is called stationary in the broad sense.

Random process $\{\xi(\omega, t),\ t \in R\}$ for which a sequence of finite-dimensional distribution functions satisfies the condition

$$F(x_1, \ldots x_n; t_1, \ldots, t_n) = F(x_1, \ldots x_n; t_1 + \tau, \ldots, t_n + \tau), j = \overline{1, n}$$

is called stationary in the narrow sense.

Using the results of the study of random processes, which are presented in a significant number of works [31–33], a method for constructing an image of random processes is proposed, which allows to create temporal and spectral images of stationary and non-stationary, Gaussian and non-Gaussian random processes. The idea of depicting an arbitrary random function through a system of other random functions with well-studied and relatively simple in the probabilistic sense properties is widely used in the theory of random processes.

Based on the use of well-known methods of canonical decomposition, forming filters, updating and generating processes, stochastic integral images, a method is proposed that uses the image of a random process in the form of a stochastic integral:

$$\xi(\omega, t) = \int_{-\infty}^{\infty} \varphi(x, t) d\eta(\omega, x), t \in T, \tag{2.14}$$

where the nonrandom function $\varphi(x, t)$ and the random function $\eta(\omega, x)$ satisfy the specific conditions for specific processes.

Time images of the random process $\xi(\omega, t)$ can be constructed on the basis that the function $\varphi(x, t) = \varphi(\tau, t)$ has a physical interpretation as an impulse transition function of some linear system, in the general case with time-variable parameters. Other physical interpretations of the image are also known (2.14). For example, expression (2.14) has an interpretation as a superposition of pulses that occur at time τ and are described with variable time t in the pulse form $\varphi(\tau,t)$.

If in expression (2.14) instead of $\varphi(x, t)$ and $\eta(\omega, x)$ it will be used the function $e^{i2\pi ft}$ and $z(\omega, f)$, where the variable f has a physical interpretation as frequency, then it can be obtained a spectral image of a random process for which the core of integral image, unlike the time one, is the same.

Further it will be considering examples of constructive random processes used for creating modern information technologies for measuring the characteristics and parameters of signals of a stochastic physical nature.

Example 1 Linear random processes are models of measuring signals with the stochastic nature of the change in their values over time.

Linear Random Process (LRP). A random process is called linear, which is described by the stochastic integral (2.14) with the corresponding specification, namely [34]:

$$\xi(\omega, t) = \int_{0}^{\infty} \varphi(\tau, t) d\eta(\omega, \tau), t \in T \tag{2.15}$$

where $\varphi(\tau, t)$ is a nonrandom function of two variables integrated with a square in τ for each $t \in T$, and $\{\eta(\omega, t), \eta(0) = 0, t \in [0, \infty]\}$ is a random stochastic continuous process with independent or uncorrelated increments. In the plural of linear random processes, by analogy with stationary random processes, random processes are distinguished linear in a wide and narrow sense.

Linear random processes in the form (2.15) create:

- a class of linear processes in the narrow sense, when $\eta(\omega, \tau)$ is an inhomogeneous (homogeneous) process with independent increments and infinitely divisible distribution laws;
- a class of processes linear in the broad sense, when $\eta(\omega, \tau)$ is a process with uncorrelated (orthogonal) growths;
- a class of linear stationary in the narrow sense, when $\varphi(\tau,t) \equiv \varphi(t - \tau)$ and $\eta(\omega, \tau)$ is a homogeneous process with independent increments;
- a class of linear stationary processes in the broad sense, when $\varphi(\tau, t) \equiv \varphi(t - \tau)$ and $\eta(\omega, \tau)$ is a process with uncorrelated (orthogonal) growths.

Linear random process of the form (2.15) has the following physical interpretation. The input of the linear system (Fig. 2.11) is affected by a random process such as white noise $\zeta(\omega, t)$, as a generalized derivative of the generating random process

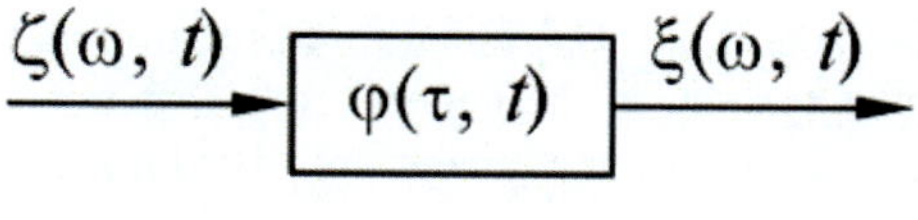

Fig. 2.11 Schematic representation of the formation of linear random process

$\eta(\omega, t)$. The response of this system is a linear random process ξ (ω, t) for which an n-dimensional sequence of characteristic functions is obtained [35].

This allows to study linear random processes in the full probabilistic volume, including the correlation or energy theory of the study as a special case.

The use of linear random processes in the study of measuring signals makes it possible to:

- create a wide range of models in the time domain of signals in the problems of measurement, control, diagnostics and forecasting;
- leave it in the class of linear random processes through linear transformations of linear random processes;
- analyze nonlinear transformations of linear random processes in the links and modules of systems and measuring instruments;
- substantiate the algorithmic software for processing measurement data, both analog and digital measurement systems;
- determine all the numerical characteristics of linear random processes during evaluating the characteristics of the measuring signals;
- summarize the well-known methods for studying random processes, namely, the methods of canonical decomposition of formative filters, updating and generating processes.

Thus, in the class of linear processes, it is possible to distinguish linear in the narrow and broad sense, stationary, Gaussian, periodic and periodically correlated processes.

The random process $\eta(\omega, t)$ is always real, and the linear process $\xi(\omega, t)$ will then be complex-valued when its kernel of the integral image (2.15) is such, that is, the function

$$\varphi(\tau, t) = \mathrm{Re}\varphi(\tau, t) + i\mathrm{Im}\varphi(\tau, t).$$

It's noted that the theory of random processes with independent increments is more studied in comparison with other classes of random processes based on the use of fundamental results that contributed to its development. This is a study of L. Bachelier (1900), N. Wiener (1923) on Brownian motion, B. de Finetti (1929) and A. N. Kolmogorov (1932) of homogeneous random processes with independent increments that have finite variance, and finally, the analysis of P. Levy (1934) of the so-called "additive random functions".

The results of studies of random processes with independent increments are closely related to the classical results of the analysis of sums of independent random variables and, thus, to the theory of infinitely divisible distribution laws, including the results of solving the central limit theorem of the sum of independent random variables.

To justify the spectral image of a random process [36, 37], a harmonized random process is the basis of the provisions. This made it possible to extend the methods of harmonic analysis of numerical functions to the class of random processes, including non-stationary ones.

Example 2 A Hilbert process $\{\xi(\omega, t), T \in t\}$, is called a harmonized random process, which admits images

$$\xi(\omega, t) = \int_{-\infty}^{\infty} e^{ift} dz(\omega, f), \ \omega \in \Omega, \ t \in T, \tag{2.16}$$

where f and t are real variables, and $z(\omega, f)$ is a complex-valued random function which correlation function

$$F(f_1, f_2) = \mathbf{M}\{z(\omega, f_1) z^*(\omega, f_2)\} \tag{2.17}$$

has bounded variation on the plane $f_1 \times f_2$,, and $z^*(\omega, f_2)$ is a complex conjugate random function.

The correlation function of a harmonized process is determined by the expression

$$R(t_1, t_2) = \int_{-\infty}^{\infty} \int_{-\infty}^{\infty} e^{i(f_1 t_1 - f_2 t_2)} d_{f_1} d_{f_2} F(f_1, f_2), t_1, t_2 \in T. \tag{2.18}$$

Harmonized random processes are statistically equivalent to linear processes and in some cases represent a different form of recording a linear random process, which reveals its harmonious structure. But not every linear process is harmonized and not every harmonized process is linear, that is, these two classes of processes intersect, but do not coincide.

In the class of harmonized random processes, it can be distinguished Gausses, stationary in the broad sense, and periodically correlated.

It was formulated a number of statements related to linear and harmonized random processes [38]:

1. For linear random process in the narrow sense of $\{\xi(\omega, t)\}$ with a generating homogeneous process with independent increments $\eta(\omega, t)$, the description is complete in the probability sense based on a certain sequence of finite-dimensional characteristic functions.
2. Linear random process is stationary in the narrow sense if the function $\varphi(\tau, t)$ depends only on the difference of arguments $t - \tau$, and the generating random process $\eta(\omega, t)$ is a homogeneous process with independent increments.
3. Linear random process is stationary in the broad sense if $\varphi(\tau, t) \equiv \varphi(t - \tau)$, and the generating process $\eta(\omega, t)$ is a process with uncorrelated (orthogonal) growths and satisfies the condition

$$\mathbf{M}\{\eta^2(\omega, \tau)\} = \sigma^2 d\tau, \tag{2.19}$$

where $\sigma \geq 0$ is a numerical value.

4. Harmonized random process is stationary in the broad sense if the generating random function $z(\omega, f)$ is a function with uncorrelated (orthogonal) increments for which

$$\mathbf{M}\{z(\omega, f)\} = 0, F(f_1, f_2) = F(\min(f_1, f_2); f_1 f_2 \in [0, \infty]), \quad (2.20)$$

5. Linear and harmonized random processes are Gaussian if the random process $\eta(\omega, \tau)$ and the random function $z(\omega, f)$, generating them, are Gaussian.
6. For Gaussian random processes, the following takes place: (a) the sets of linear non-stationary (stationary) in a narrow and wide sense completely coincide; (b) the set of non-stationary (stationary) in the broad sense of harmonized processes is the set of corresponding processes in the narrow sense [39].

Example 3 One of the non-stationary random processes that is used in measurement problems is a *periodically correlated random process* (PCRP). Further it will be considering its properties and characteristics.

Harmonized random process $\xi(\omega, t)$ is called periodically correlated with the period T_0 if its correlation function $R(t_1, t_2)$ is continuous and satisfies the condition

$$R(t_1, t_2) = \mathrm{R}(t_1 + T_0, t_2 + T_0), \quad (2.20)$$

where T_0 is some fixed numerical value.

It should be noted that knowledge of the period T_0 is fundamental in the study of PCRP. In essence, knowledge of $\mathrm{T0}$ is the "key" that allows to reveal the structure and properties of the characteristics of PCRP.

The structure and properties of PCRP are presented in [40–43]. Here it will be only given a number of statements about the use of PCRP, which are of great importance for measuring the characteristics of cyclic or rhythmic stochastic signals:

- PCRP is one of the most studied in the class of non-stationary random processes, therefore, the rationale for its use has both theoretical and practical basis;
- the practical use of PCRP in various subject areas of science and technology is due to the fact that, depending on the measurement tasks, it is possible to distinguish informationally important stationary components for processing the characteristics of which a powerful algorithmic-software resource for processing measurement data is used;
- one of the main measurement tasks arising during the study of PCRP is to evaluate the value of the period T_0 with correspondingly specified values of its uncertainty.

It will be turn to the consideration of multidimensional signals and fields.

2.4 Models of Multidimensional Signals and Fields

The models of measuring signals, considered above, are functions of time, that is, functions of one argument and are called *one-dimensional*. The functions of two arguments are called *two-dimensional*, and in the general case for the number of arguments $n \geq 2$ are called *multidimensional*. It will be briefly considering the main types of models of multidimensional signals and fields, which are objects of measurement and are widely used in various subject areas.

The class of models of multidimensional signals and fields should be divided into deterministic and random. This distribution allows to take into account the specificity and characteristic features of the measurement of the values and characteristics of the investigated signals:

- for deterministic multidimensional signals during measurements, it is sufficient to fix one of their realizations in space and time on the basis of the obtained measurement data and to determine the result and the measurement uncertainty function using the physical measure of the quantity being studied. In the future, the models of such signals will be called *multidimensional functions*, and for specific cases indicate the number of arguments. For example, a three-dimensional function is a deterministic function of three arguments;
- for random multidimensional signals, which are spatio-temporal fields, during measurements it is necessary to fix a number of their realizations (ensemble of realizations), and to obtain the corresponding results and characteristics of uncertainty, it is necessary to use the probabilistic measure of the research random function with the use of physical measures of the studied quantities. In the future, random-signal models will be referred to as *spatio-temporal random fields*.

Models of deterministic multidimensional signals. It was noted earlier that due to the increasing complexity of measurement tasks, the dimension of the models of the studied signals also increases, primarily due to the spatial coordinates $x \in R$, $(x, y) = \boldsymbol{\rho} \in R^2$ and $(x, y, z) = \boldsymbol{r} \in R^3$.

Examples of models of such signals are multidimensional functions of the form:

$$\{u(x, t) \in R, x \in R, t \in T\}\text{—two-dimensional;}$$

$$\left\{u(\boldsymbol{\rho}, t) \in R, \boldsymbol{\rho} \in R^2, t \in T\right\}\text{—three-dimensional;}$$

$$\left\{u(\boldsymbol{r}, t) \in R, \boldsymbol{r} \in R^3, t \in T\right\}\text{—four-dimensional.}$$

The study of such functions gives rise to a number of problems of measuring their characteristics. So, for example, the average value of function $u(\boldsymbol{r}, t)$ over time is a function of spatial coordinates and is determined by the formula

$$\overline{u}(\boldsymbol{r}) = \frac{1}{T}\int\limits_{0}^{T} u(\boldsymbol{r}, t)dt, t \in T.$$

The average value of the function $u(\boldsymbol{r}, t)$ over spatial coordinates is a function of time and is determined by the formula

$$\widetilde{u}(t) = \frac{1}{\mathrm{G}}\int\limits_{0}^{\mathrm{G}} u(\boldsymbol{r}, t)d\boldsymbol{r}, \ (x, y, z) = \boldsymbol{r} \in G \subseteq R^3.$$

Based on the using of quadratic transformation operators of such signals, their energy characteristics are investigated. For determining the characteristics and parameters of multidimensional deterministic signals, which by their physical nature are spatio-temporal signal fields, methods of functional analysis and mathematical physics are used. Models of multidimensional deterministic signals in the practice of measurements are idealized models that are formed by the fundamental laws of a particular subject area, for example, thermophysics, hydrodynamics, theoretical mechanics, and others. Examples of such signals are spatio-temporal electromagnetic, vibroacoustic, thermal, hydroacoustic and others.

Models of spatio-temporal random fields. Such signal models are studied by methods of the theory of random processes and fields. Methods of mathematical statistics based on the practical use of measuring tools and systems determine statistical estimates of their characteristics. Dimension, that is, the number of arguments of a random signal model in comparison with the determinate model increases by one. So, for example, a model of a deterministic four-dimensional signal is described by the function

$$\left\{u(\boldsymbol{r}, t), \ \boldsymbol{r} \in R^3, \ t \in T\right\},$$

and the model of a random same signal field is described by a spatio-temporal random field

$$\left\{\xi(\omega, \boldsymbol{r}, t), \omega \in \Omega, \boldsymbol{r} \in R^3, t \in T\right\}$$

In practice, in most cases of measuring the estimation of the spatio-temporal characteristics of the studied both deterministic and random signals, models are used which general form, depending on the set of arguments, is given in Table 2.2.

In Table 2.2 using the following notation: ω is an elementary random event from the event space Ω; T is the observation time interval, R is the set of real numbers.

In order to use these models in measurement problems under conditions of limited areas of the space G and finite time intervals T, the following multidimensional indicator function is used

Table 2.2 Space–time random field models

Types of functions	Arguments			
	$\omega \in \Omega$ $t \in T$	$\omega \in \Omega$ $x \in R$ $t \in T$	$\omega \in \Omega$ $(x, y) = \rho \in R^2 t \in T$	$\omega \in \Omega$ $(x, y, z) = r \in R^3 t \in T$
Deterministic	Number $\alpha \in R$, one-dimensional function $u(t) \in R$	Two-dimensional function $u(x,\ t) \in R$	Three-dimensional function $u(\boldsymbol{\rho},\ t) \in R$	Four-dimensional function $u(\boldsymbol{r},\ t) \in R$
Random	Value $\xi(\omega) \in R$, process $\xi(\omega) \in R$, as a two-dimensional random function	Three-dimensional spatio-temporal field $\xi(\omega,\ x, t) \in R$	Four-dimensional spatio-temporal field $\xi(\omega;\ \boldsymbol{\rho};\ t) \in R$	Five-dimensional spatio-temporal field $\xi(\omega;\ \boldsymbol{r};\ t) \in R$

$$I(\boldsymbol{r}, t) = \begin{cases} 1, & \boldsymbol{r} \in G \ and \ t \in T; \\ 0, & \boldsymbol{r} \notin G \text{ or } t \notin T. \end{cases}$$

Using a mathematical model of the form $\xi(\omega, \boldsymbol{r}, t) \cdot I(\boldsymbol{r},\ t)$ makes it possible to conduct studies for determining the spatio-temporal characteristics of signals, provided that measurements are made in various places of space at finite time intervals. It also provides a further comparative analysis of the measurement results in order to verify the adequacy of the proposed models, predicting the dynamics of changes in the basic characteristics of RO in space and time, etc.

In the case of the study of spatio-temporal signals based on the use of random models by means of measurement, as a rule, statistical estimates of the characteristics of signal fields are determined within the framework of the correlation (energy) theory. That is, for a random field $\xi(\omega, \boldsymbol{r}, t)$ are estimated:

- field expectation

$$a(\boldsymbol{r}, t) = \mathrm{M}\{\xi(\omega,\boldsymbol{r},t)\};$$

- field dispersion

$$\sigma^2(\boldsymbol{r}, t) = \boldsymbol{D}\{\xi(\omega, \boldsymbol{r}, t)\} = \mathrm{M}\left\{[\xi(\omega, \boldsymbol{r}, t) - a(\boldsymbol{r}, t)]^2\right\};$$

- field autocorrelation function

$$R(\boldsymbol{r}_1, \boldsymbol{r}_2, t_1, t_2) = \mathrm{M}\{[\xi(\omega, \boldsymbol{r}_1, t_1) - a(\boldsymbol{r}_1, t_1)] \times [\xi(\omega, \boldsymbol{r}_2, t_2) - a(\boldsymbol{r}_2, t_2)]\};$$

- structural function

$$B(\boldsymbol{r}_1,\boldsymbol{r}_2, t_1, t_2) = \mathrm{M}[(\xi(\omega, \boldsymbol{r}_1, t_1) - \xi(\omega, \boldsymbol{r}_2, t_2))^2],$$

Table 2.3 Characteristics of spatio-temporal fields

Types of random fields	Models and numerical characteristics				
	Model	Expected value	Dispersion	Autocorrelation function	Structural function
Inhomogeneous in spatial arguments and unsteady in time	$\xi(\omega; x; t)$ $\xi(\omega; \boldsymbol{\rho}; t)$ $\xi(\omega; \boldsymbol{r}; t)$	$a(x; t)$ $a(\boldsymbol{\rho}; t)$ $a(\boldsymbol{r}; t)$	$\sigma^2(x; t)$ $\sigma^2(\boldsymbol{\rho}; t)$ $\sigma^2(\boldsymbol{r}; t)$	$R(x_1, x_2; t_1, t_2)$ $R(\boldsymbol{\rho}_1, \boldsymbol{\rho}_2; t_1, t_2)$ $R(\boldsymbol{r}_1, \boldsymbol{r}_2; t_1, t_2)$	$B(x_1, x_2; t_1, t_2)$ $B(\boldsymbol{\rho}_1, \boldsymbol{\rho}_2; t_1, t_2)$ $B(\boldsymbol{\rho}_1, \boldsymbol{\rho}_2; t_1, t_2)$
Inhomogeneous stationary	$\xi(\boldsymbol{r}; t)$	$a(\boldsymbol{r})$	$\sigma^2(\boldsymbol{r})$	$R(\boldsymbol{r}_1, \boldsymbol{r}_2; t_2 - t_1)$	$B(\boldsymbol{r}_1, \boldsymbol{r}_2; t_2 - t_1)$
Homogeneous non-stationary	$\xi(\boldsymbol{r}; t)$	$a(\boldsymbol{r})$	$\sigma^2(\boldsymbol{r})$	$R(\boldsymbol{r}_1 - \boldsymbol{r}_2; t_2, t_1)$	$B(\boldsymbol{r}_1 - \boldsymbol{r}_2; t_2, t_1)$
Homogeneous and isotropic, non-stationary	$\xi(\boldsymbol{r}; t)$	$a(t)$	$\sigma^2(t)$	$R(\lvert\boldsymbol{r}_2 - \boldsymbol{r}_1\rvert; t_1, t_2))$	$B(\lvert\boldsymbol{r}_2 - \boldsymbol{r}_1\rvert; t_1, t_2)$
Homogeneous and stationary	$\xi(\boldsymbol{r}; t)$	a	σ^2	$R(\boldsymbol{r}_2 - \boldsymbol{r}_1; t_2 - t_1)$	$B(\boldsymbol{r}_2 - \boldsymbol{r}_1; t_2 - t_1)$
Homogeneous and isotropic, stationary	$\xi(\boldsymbol{r}; t)$	a	σ^2	$R(\lvert\boldsymbol{r}_2 - \boldsymbol{r}_1\rvert; t_2 - t_1)$	$B(\lvert\boldsymbol{r}_2 - \boldsymbol{r}_1\rvert; t_2 - t_1)$

where M{...} and D{...} respectively, are the operators for determining the mathematical expectation and variance of random functions.

The characteristics of spatio-temporal fields defined above are given in Table 2.3, taking into account their structure and species. Based on the analysis of publications [44–46], based on the results of measuring the characteristics of multidimensional deterministic and spatio-temporal random fields, the following conclusions can be drawn, which in a concise form reflect the specificity and characteristic features of obtaining results and characteristics of measurement uncertainty in various subject areas:

– The use of modern information technology measurement (Table 2.4) allows to get the measurement results not only one-dimensional, but also multidimensional signals and fields based on the creation and justification of models of such signals.

Table 2.4 Information technology measurement

Types of technology	Types of signal measurement models
Information technology for measuring 1D (ITB-1D)	$\alpha,\ u(x, t),\ \xi(\omega),$ $\xi(\omega, t);\xi(\omega\,; x\,; t)$
Information technology for measuring 2D (ITB-2D)	$u(\boldsymbol{\rho}, t),\ \xi(\omega;\ \boldsymbol{\rho}; t)$
Information technology for measuring 3D (ITB-3D)	$u(\mathbf{r}, t),\ \xi(\omega;\ \mathbf{r}; t)$

The need to obtain such measurement results is due to the pace of development of the information society, new tasks of solving scientific and technical problems in various subject areas.

- The creation of hardware and information support for the use of measurement tasks is associated with ensuring the unity of physical and probabilistic measures for assessing the characteristics of multidimensional signals and fields in accordance with the requirements of the concept of uncertainty.
- Each specific case of obtaining the results of measuring the values and characteristics of multidimensional signals and fields is an independent scientific and technical study using a significant amount of carrying out and performing theoretical, modeling and experimental studies.

The cited materials from the field of analysis of multidimensional signals and fields reflect only the methodological aspects of their research. In each case, the practical receipt of measurement results, their characteristics in space and time, requires the use of significant scientific, technical, financial and human resources.

References

1. Lauras, M., Marques, G., Gourc, D.: Towards a multi-dimensional project performance measurement system. Decis. Support Syst. **48**(2), 342–353 (2010). https://doi.org/10.1016/j.dss.2009.09.002
2. Bucci, G., Fiorucci, E., Landi, C.: Digital measurement station for power quality analysis in distributed environments. IEEE Trans. Instrum. Meas. **52**(1), 75–84 (2003). https://doi.org/10.1109/TIM.2003.809080
3. Wang, Y., Li, W., Lu, J.: Reliability analysis of wide-area measurement system. IEEE Trans. Power Delivery **25**(3), 1483–1491 (2010). https://doi.org/10.1109/TPWRD.2010.2041797
4. Sychev, I., Sycheva, I., Sobh, T., Elleithy, K., Mahmood, A. (eds.): Novel Algorithms and Techniques in Telecommunications and Networking. Springer, Berlin, pp. 201–206 (2010). https://doi.org/10.1007/978-90-481-3662-9_34
5. Da Silva, E.R.C., dos Santos, E.C., Jacobina, B.: Pulsewidth modulation strategies. IEEE Ind. Electron. Mag. **5**(2), 37–45 (2011). https://doi.org/10.1109/MIE.2011.941120
6. Hirt, W.: Ultra-wideband radio technology: overview and future research. Comput. Commun. **26**(1), 46–52 (2003). https://doi.org/10.1016/S1403-3664(02)00119-6
7. Vigneshvar, S., Sudhakumari, C.C., Senthilkumaran, B., Prakash, H.: Recent advances in biosensor technology for potential applications—an overview. Front. Bioeng. Biotechnol. **4**, 11 (2016). https://doi.org/10.3389/fbioe.2016.00011
8. Arefin, Md.T., Ali, M.H., Haque, A.K.M.F.: Wireless body area network: an overview and various applications. J. Comput. Commun. **5**(7), 53–64 (2017). https://doi.org/10.4236/jcc.2017.57006
9. Babak, V.P., Babak, S.V., Myslovych, M.V., Zaporozhets, A.O., Zvaritch, V.M.: Methods and models for information data analysis. In: Diagnostic Systems For Energy Equipments. Studies in Systems, Decision and Control, vol. 281, pp. 23–70. Springer, Cham, (2020). https://doi.org/10.1007/978-3-030-44443-3_2
10. Babak, V., Zaporozhets, A., Kuts, Y., Scherbak, L., Eremenko, V.: Application of material measure in measurements: theoretical aspects. In: Zaporozhets, A., Artemchuk, V. (eds.) Systems, Decision and Control in Energy II. Studies in Systems, Decision and Control. Springer, Cham, pp. 261–269 (2021). https://doi.org/10.1007/978-3-030-69189-9_15

11. Mordachev, V.: Mathematical models for radio signals dynamic range prediction in space-scattered mobile radiocommunication networks. In: Vehicular Technology Conference Fall 2000. IEEE VTS Fall VTC2000. 52nd Vehicular Technology Conference (Cat. No. 00CH37152), pp. 904–911, USA, (2000). https://doi.org/10.1109/VETECF.2000.887131
12. Marasanov, V., Sharko, A.: Mathematical models for interrelation of characteristics of the developing defects with the parameters of acoustic emission signals. Int. Front. Sci. Lett. **10**, 37–44 (2016). https://doi.org/10.18052/www.scipress.com/IFSL.10.37
13. Eremenko, V., Zaporozhets, A., Babak, V., Isaienko, V., Babikova, K.: Using hilbert transform in diagnostic of composite materials by impedance method. Period. Polytech. Electr. Eng. Comput. Sci. **64**(4), 334–342 (2020). https://doi.org/10.3311/PPee.15066
14. Babak, V.P., Babak, S.V., Myslovych, M.V., Zaporozhets, A.O., Zvaritch, V.M.: Simulation and software for diagnostic systems. In: Diagnostic Systems For Energy Equipments. Studies in Systems, Decision and Control, vol. 281, pp. 71–90. Springer, Cham (2020). https://doi.org/10.1007/978-3-030-44443-3_3
15. Bruckner, T., Holmes, D.G.: Optimal pulse-width modulation for three-level inverters. IEEE Trans. Power Electron. **20**(1), 82–89 (2005). https://doi.org/10.1109/TPEL.2004.839831
16. Zaporozhets, A.O.: Correlation analysis between the components of energy balance and pollutant emissions. Water Air Soil Pollut. **232**, 114 (2021). https://doi.org/10.1007/s11270-021-05048-9
17. Isermann, R., Munchhof, M.: Mathematical models of linear dynamic systems and stochastic signals. In: Identification of Dynamic Systems, pp. 33–74. Springer, Berlin (2010). https://doi.org/10.1007/978-3-540-78879-9_2
18. Simani, S., Fantuzzi, C., Patton, R.J.: Model-based fault diagnosis techniques. In: Model-based Fault Diagnosis in Dynamic Systems Using Identification Techniques. Advances in Industrial Control, pp. 19–60. Springer, London (2003). https://doi.org/10.1007/978-1-4471-3829-7_2
19. Narytnyk, T.: Telecommunication system construction principles for radio communication in the terahertz frequency range on the harmonic technology signal as a information carrier basis. In: 2018 International Scientific-Practical Conference Problems of Infocommunications. Science and Technology (PIC S&T), Ukraine, pp. 771–776 (2018). https://doi.org/10.1109/INFOCOMMST.2018.8632041
20. Babak, V., Eremenko, V., Zaporozhets, A.: Research of diagnostic parameters of composite materials using Johnson distribution. Int. J. Comput. **18**(4), 483–494 (2019)
21. Moradi, L., Mohammadi, F.: A comparative approach for time-delay fractional optimal control problems: discrete versus continuous Chebyshev polynomials. Asian J. Control **22**(1), 204–216 (2020). https://doi.org/10.1002/asjc.1858
22. Carrol, M.T., Rykken, E.: The involutory Laguerre transition matrix. Math. Maga. **92**(1), 19–23 (2019). https://doi.org/10.1080/0025570X.2019.1541688
23. Izadi, M.: A comparative study of two Legendre-collocation schemes applied to fractional logistic equation. Int. J. Appl. Comput. Math. **6**, 71 (2020). https://doi.org/10.1007/s40819-020-00823-4
24. Makarov, V., Gaponenko, V., Toporov, B., Kupriyanov, A.: Theoretical and applied aspects of orthogonal coding in computer networks technologies. In: Kravets, A. (eds.) Big Data-Driven World: Legislation Issues and Control Technologies. Studies in Systems, Decision and Control, vol. 181, pp. 47–58. Springer, Cham (2019). https://doi.org/10.1007/978-3-030-01358-5_5
25. Kim, T., Kim, D.S., Kwon, J., Dolgy, D.V.: Expressing sums of finite products of Chebyshev polynomials of the second kind and of Fibonacci polynomials by several orthogonal polynomials. Mathematics **6**(10), 210 (2018). https://doi.org/10.3390/math6100210
26. Yafaev, D.R.: A point interaction for the discrete Schrödinger operator and generalized Chebyshev polynomials. J. Math. Phys. **58**, 063511 (2017). https://doi.org/10.1063/1.4986414
27. Selezov, I.T.: Some analytical and numerical methods in the theory of wave propagation and diffraction. In: Wave Propagation and Diffraction. Foundations of Engineering Mechanics, pp. 1–24. Springer, Singapore (2018). https://doi.org/10.1007/978-981-10-4923-1_1
28. Wang, M.-L., Chang, R.-Y., Yang, S.-Y.: Generalization of generalized orthogonal polynomial operational matrices for fractional and operational calculus. Int. J. Syst. Sci. **18**(5), 931–943 (1987). https://doi.org/10.1080/00207728708964020

29. Sapehrian, B., Razzaghi, M.: Single-term Walsh series method for the Volterra integro-differential equations. Eng. Anal. Boundary Elem. **28**(11), 1315–1319 (2004). https://doi.org/10.1016/j.enganabound.2004.05.001
30. Shaffer, D., Lorson, P., Plunkett, Z., Ragai, I., Danesh-Yazdi, A., Ashour, O.: Development of experiment-based mathematical models of acoustic signals for machine condition monitoring. Proc. CIRP **72**, 1316–1320 (2018). https://doi.org/10.1016/j.procir.2018.03.269
31. Averina, T., Karachanskaya, E., Rybakov, K.: Statistical modeling of random processes with invariants. In: 2017 International Multi-Conference on Engineering, Computer and Information Sciences (SIBIRCON), Russia, pp. 34–37 (2017). https://doi.org/10.1109/SIBIRCON.2017.8109832
32. Deutsch, R.: Nonlinear Transformations of Random Processes. Courier Dover Publications (2017)
33. Atamanyuk, I.P., Kondratenko, V.Y., Kozlov, O.V., Kondratenko, Y.P.: The algorithm of optimal polynomial extrapolation of random processes. In: Engemann, K.J., Gil-Lafuente, A.M., Merigó, J.M. (eds.) Modeling and Simulation in Engineering, Economics and Management. MS 2012. Lecture Notes in Business Information Processing, vol. 115, pp. 78–87. Springer, Berlin, Heidelberg (2012). doi: https://doi.org/10.1007/978-3-642-30433-0_9
34. Chang, R.-Y., Yang, S.-Y., Wang, M.-L.: Solutions of linear dynamic systems by generalized orthogonal polynomials. Int. J. Syst. Sci. **17**(12), 1727–1740 (1986). https://doi.org/10.1080/00207728608926920
35. Wang, Z., Bovik, A.C.: Mean squared error: love it or leave it? A new look at signal fidelity measures. IEEE Signal Process. Mag. **26**(1), 98–117 (2009). https://doi.org/10.1109/MSP.2008.930649
36. Ioannidis, K., Andreadis, I.: A digital image stabilization method based on the Hilbert-Huang transform. IEEE Trans. Instrum. Meas. **61**(9), 2446–2457 (2012). https://doi.org/10.1109/TIM.2012.2188618
37. Fu, D., Shi, Y.Q., Su, W.: Detection of image splicing based on Hilbert-Huang transform and moments of characteristic functions with wavelet decomposition. In: Shi, Y.Q., Jeon, B. (eds.) Digital Watermarking. IWDW 2006. Lecture Notes in Computer Science, vol. 4283, pp. 177–187. Springer, Berlin (2006). https://doi.org/10.1007/11922841_15
38. Shih, D.-H., Kung, F.-C., Chao, C.-M.: Laguerre series approach to the analysis of a linear control system incorporating observers. Int. J. Control **43**(1), 123–128 (1986). https://doi.org/10.1080/00207178608933452
39. Sambariya, D.K., Prasad, R.: Optimal tuning of fuzzy logic power system stabilizer using harmony search algorithm. Int. J. Fuzzy Syst. **17**, 457–470 (2015). https://doi.org/10.1007/s40815-015-0041-4
40. Horvatic, D., Stanley, H.E., Podobnik, B.: Detrended cross-correlation analysis for non-stationary time series with periodic trends. EPL **91**(1), 18007. https://doi.org/10.1209/0295-5075/94/18007
41. Jacobina, C.B., Nogueira Lima, A.M., da Silva, E.R.C., Alves, R.N.C., Seixas, P.F.: Digital scalar pulse-width modulation: a simple approach to introduce nonsinusoidal modulating waveforms. IEEE Trans. Power Electron. **16**(3), 351–359 (2001). https://doi.org/10.1109/63.923767
42. Loh, P.C., Vilathgamuwa, D.M., Lai, Y.S., Chua, G.T., Li, Y.: Pulse-width modulation of Z-source inverters. In: Conference Record of the 2004 IEEE Industry Applications Conference, 2004. 39th IAS Annual Meeting, USA, p. 155 (2004). https://doi.org/10.1109/IAS.2004.1348401
43. Roberts, S., Osborne, M., Ebden, M., Reece, S., Gibson, N., Aigrain, S.: Gaussian processes for time-series modeling. Philos. Trans. R. Soc. A: Math. Phys. Eng. Sci. **371**, 20110550 (1984). https://doi.org/10.1098/rsta.2011.0550
44. Radosavljevic, V., Vucetic, S., Obradovic, Z.: Neural Gaussian conditional random fields. In: Calders, T., Esposito, F., Hüllermeier, E., Meo, R. (eds.) Machine Learning and Knowledge Discovery in Databases. ECML PKDD 2014. Lecture Notes in Computer Science, vol. 8725, pp. 614–629. Springer, Berlin (2014). https://doi.org/10.1007/978-3-662-44851-9_39

45. De Iaco, S., Posa, D.: Predicting spatio-temporal random fields: some computational aspects. In: Predicting Spatio-Temporal Random Fields: Some Computational Aspects. Computers & Geosciences, vol. 41, pp. 12–24 (2012). https://doi.org/10.1016/j.cageo.2011.11.014
46. Wikle, C.K., Hooten, M.B.: A general science-based framework for dynamical spatio-temporal models. TEST **19**, 417–451 (2010). https://doi.org/10.1007/s11749-010-0209-z

Chapter 3
Models and Measures for Measuring Random Angular Quantities

Abstract The models and measures for measuring random angular quantities (plane angles and phase shifts of cyclic signals) are considered, the basic concepts, terms, definitions and characteristics that are used in the statistical analysis of angular data are given. The definitions and characteristics of the integral probability distribution function (PDF) of random angles and the density of their distribution are given. A probabilistic model of a random angle on a unit circle is developed. Probabilistic measures of angular quantities are also determined. The most characteristic probability distributions of random angles are analyzed—von Mises, wrapped normal distribution, cardioid, triangular, wrapped Cauchy. The numerical characteristics of random angles are given—the characteristic function, the circular mean, circular dispersion, standard deviation and median. A comparative analysis of the numerical characteristics of random variables and random angles is performed. Definitions and analytical expressions are given for selective circular characteristics—trigonometric moments, mean, resulting vector length, dispersion, etc. The basic information about the models developed by the authors and measures of random phase shifts of cyclic signals for phase measurements are presented.

Keywords Random angle · Circular characteristics · Random angle distribution · Von Mises distribution · Wrapped normal distribution · Selective circular statistics · Probabilistic model of phase shifts of cyclic signals

Angular measurements have long been widely used in various fields of human activity. They have become an effective tool for understanding the world, comprehending the passage of time, the successes of astronomy, etc. Using the method of converting various physical quantities into angles, it was possible to study new phenomena and effects for science. In particular, this approach was used to study atmospheric electric charges (M. V. Lomonosov and G. W. Richmann), to discover the law of interaction of point electric charges (C. A. de Coulomb), to determine the value of the gravitational constant of the Earth (H. Cavendish), etc. [1–3].

V. P. Babak et al., *Models and Measures in Measurements and Monitoring*,
Studies in Systems, Decision and Control 360,
https://doi.org/10.1007/978-3-030-70783-5_3

The physical degree of a flat angle is a circle or part thereof, divided into equal intervals. Such measures are implemented in the form of limbs or goniometric scales (protractor, quadrant, sextant, etc.). For phase measurements, a measure of the phase shifts of the signals is implemented as a two-phase generators or more complex electron-optical means. Less common in the theory and practice of angular and phase measurements is probabilistic models and measures. This section is devoted to these issues.

3.1 Models of Deterministic and Random Angles in Measurements

Models of deterministic angles. The basic concepts in measurements of angles (goniometry) are formed in Euclidean geometry and are well known. First of all, this is the concept of a *flat angle*—a geometric figure formed by two different rays emanating from one point, which is called the vertex of the angle, as well as related concepts: central and adjacent angles, deployed and right angles, opposite and adjacent angles, and the like.

Angles can be compared. Two flat angles α and β are called equal (or congruent) if they can be combined in such a way that their respective sides and vertices coincide. In the system of Hilbert's axioms [4, 5], developed by him for Euclidean geometry, the congruence relation is denoted as follows:$\alpha \equiv \beta$.

On a plane from an arbitrary beam in a certain direction, you can set aside a single angle equal to this angle. So, the angle can be considered both as a model and as a measure of the rotation of a ray from its initial position to a predetermined one. Depending on the direction of rotation, the concept of positive and negative angles is introduced. It will be considered that the angles are positive, which formed by the movement of the beam in the counterclockwise direction.

To find out which of the angles α and β is large, it is necessary to combine in one plane the vertices and one pair of their sides. If the second side of one angle, for example, angle β, is located inside the angle α, it means that the angle α is larger than the angle β and denote this fact as $\alpha > \beta$ (Fig. 3.1).

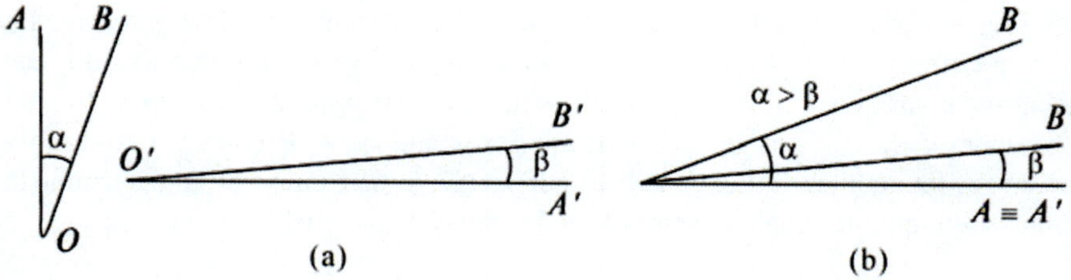

Fig. 3.1 Angle comparison: **a** before alignment, **b** after alignment

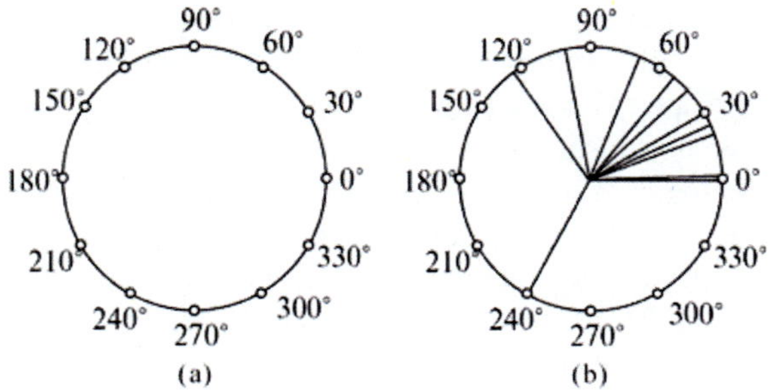

Fig. 3.2 Models of a flat angle on a unit circle in the case of reflection by points (**a**) and vectors (**b**)

In a geometric system, which is based on point-vector axiomatics, the angle is determined differently [6]. In this axiomatics, an angle is defined as a certain metric quantity, which is connected with two vectors through the operation of their scalar product. It is known that each pair of vectors $\overline{a}$ and $\overline{b}$ defines a certain angle φ—a number associated with the vectors of the following formula:

$$\cos\varphi = \left(\overline{a}, \overline{b}\right) / \, |\overline{a}|\left|\overline{b}\right|, \tag{3.1}$$

where $\left(\overline{a}, \overline{b}\right)$ is scalar multiplication of vectors.

The concept of an angle as a flat figure and as a specific metric is used in various geometric problems, where the angle is determined separately in a special way. For example, by the angle between the curves that intersect, it was understood the angle between the tangents to these curves at the points of intersection; the angle between the line and the plane is defined as the angle between this line and its rectangular projection on this plane; the angle between the crosses of the lines is the angle between the directions of these lines, that is, between the lines that are parallel to the original lines, and drawn through one point [7, 8].

More illustrative are models of angles on a circle of unit radius, or simply on a unit circle (Fig. 3.2).

In this case, the angles (or directions on the plane) are reproduced by points on the circle (Fig. 3.2a), or by vectors that end at these points, and begin in the center of the circle (Fig. 3.2b). Such a model reflects the main feature and differences of angular measurements compared to linear ones: for angles there is a mathematical operation "sum modulo 2π".

Random angle models. A new stage in the development of angularity is associated with the study of random angles. Statistical angle measurement refers to the section of mathematical statistics that studies random angles and other random variables that require the study of them on a circle. A historical review of the development of random angle analysis issues is given in [8–10]. The first studies in this direction concerned uniformly distributed random angles. In 1734, D. Bernoulli, considered the problem: is it possible to explain the proximity of the orbital planes of the 6 planets of the solar system by chance?

Different from the uniform distribution on the circle are studied only since the twentieth century. Von Mizes R. researched the distribution named after him (1918); wrapped Gaussian distribution was studied by Perrin F. (1928). In the twentieth century, Fisher R. A., Gumbel E. J., Durand D., Greenwood J. A., Watson G. S., Williams E. J., Rao C. R., Rao J. S., Pearson E. S., Mardia K. V. and others made a significant contribution to the development of statistical analysis of random angles.

In statistical angular measurements, as the basic mathematical model of a random angle $\Psi(\omega)$ defined on a probability space $(\Omega, \mathcal{A}, P)$—a dimensional probability space with a measure, is used, where Ω is the space of elementary events $\omega \in \Omega$, $\mathcal{A}$ is the σ-algebra of the subset Ω, P is the probability measure given on the subsets $\mathcal{A}$. Space $(\Omega, \mathcal{A}, P)$ generates the probability space of values of the random angle $\Psi(\omega)$ -$\left(X, \mathcal{B}, P_{\psi}\right)$, where $X \subseteq R$ is the set of the numerical axis R, $\mathcal{B}$ is the σ-algebra of the subset from X, and P_{ψ} is the probability measure (probability of random events $\Psi(\omega) \in \mathcal{B}$. Random angle $\Psi(\omega)$ is defined as a random variable function of argument $\omega \in \Omega$ in the space X.

In the general case, other more complex probabilistic models are used, for example, a two-dimensional vector of random angles, a process of random angles with discrete time, and others.

Definition 1 Real random variable.

$$\psi(\omega) = \Psi(\omega) - \left[\Psi(\omega)/2\pi\right]^{+} \cdot 2\pi, \omega \in \Omega, \tag{3.2}$$

where $\left[\Psi(\omega)/2\pi\right]^{+}$—integer part of a random angle $\Psi(\omega) \in R$ is called a random angle $\psi(\omega) \in [0, 2\pi)$, if the distribution function $\psi(\omega)$

$$G\left(x'\right) = P\left\{\omega \in \Omega : 0 < \psi(\omega) \le x'\right\}, x' \in [0, 2\pi), \tag{3.3}$$

has the next properties:

1. $G\left(x'\right)$ is monotonically non-decreasing on $x' \in [0, 2\pi)$ and is continuous on the right;
2. $G(2\pi) = 1$ and is continuous at a point $x' = 2\pi$;

$$G(0) = 0; \tag{3.4}$$

3. $G\left(x'_2\right) - G\left(x'_1\right) \ge 0$ if $x'_2 \ge x'_1$, $x'_1, x'_2 \in [0, 2\pi)$.

An analysis of expressions (3.4) shows the difference between the distribution function $G\left(x'\right)$ from the traditional distribution function on the number line R: the function $G\left(x'\right)$ is given on a finite interval of the argument $x' \in [0, 2\pi)$, but on different ranges of R. Therefore, the function $G\left(x'\right)$ is called the *PDF of the random angle* $\psi(\omega)$ on $[0, 2\pi)$.

Figure 3.3 shows graphs of continuous $G\left(x'\right)$ and discrete $G_D\left(x'\right)$ functions.

The probabilistic consideration of $\psi(\omega)$ is based on the use of the distribution function of $G\left(x'\right)$ on $[0, 2\pi)$.

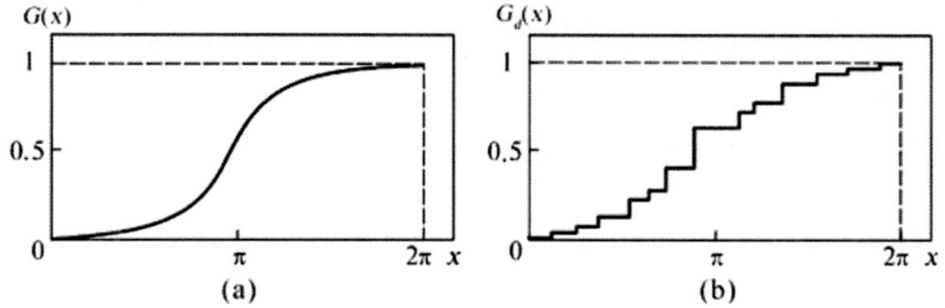

Fig. 3.3 Graphs of distribution functions: **a** $G(x')$ and **b** $G_D(x')$

In a significant number of practical problems, the probabilistic analysis of arbitrary random angles $\Psi(\omega) \in R$ reduces to the probabilistic analysis of its part $\psi(\omega) \in [0, 2\pi)$.

Definition 2 Real random variable $\Psi(\omega) > 2\pi$.

$$\Psi(\omega) = [\Psi(\omega)/2\pi] \cdot 2\pi + \psi(\omega), \quad \omega \in \Omega, \tag{3.5}$$

is called a random angle $\Psi(\omega)$ on R if $\psi(\omega)$ is a random angle on $[0, 2\pi)$ with a distribution function $G(x')$ on $[0, 2\pi)$ defined by Definition 1.

The generating function for PDF of random angles $\Psi(\omega)$ on R for $\forall k \in Z$ has the form [11]

$$F(x) = G(x') + [x/2\pi]^+ + C, \; x \in R, \; x' \in [0, 2\pi).$$

Function $F(x)$ has the following properties:

1. $F(x)$ is monotonously non-decreasing.
2. $F(x)$ is continuous on the right on $x \in R$.
3. $F(x') = F(x') - F(0_-), \; x' \in [0, 2\pi)$.
4. $F(x) - x/2\pi$ is a periodic function with period 2π.
5. $F(x + 2\pi) - F(x) \equiv 1, \; x \in R$.
6. For $0 \le x_2 - x_1 \le 2\pi$ is

$$P\{\omega \in \Omega : x_1 < \Psi(\omega) \le x_2\} = \begin{cases} 0, \; x_2 \le x_1; \\ F(x_2) - F(x_1)(x_1 < x_2 \le x_1 + 2\pi) \\ 1, \; x_2 > x_1 + 2\pi. \end{cases}$$

7. For the difference of angles within $2\pi - 0 < x_2 - x_1 \le 2\pi$, it can be right

$$\begin{cases} F(x_2) - F(x_1) = P\{x_1 < x' \le x_2\} \;\; \text{for} \;\; [x_2/2\pi]^+ = [x_1/2\pi]^+; \\ (F(x_2') - F(0)) + (F(x_1') - F(2\pi)) = \\ = P\{(0 \le x' \le x_2') + P(x_1' \le x' \le 2\pi)\}, \;\; \text{for} \;\; [x_2/2\pi]^+ = [x_1/2\pi]^+ + 1, \end{cases}$$

where $x_1' = x_1(\text{mod } 2\pi)$, $x_2' = x_2(\text{mod } 2\pi)$.

Constant C is usually equal to zero, but depending on the statement of the problem of angular measurements, it can acquire other numerical values.

From the above properties of the function $F(x)$, $x \in R$ it follows that the monotonically non-decreasing function $F(x)$ has the same increments

$$F(x + 2\pi(k+1)) - F(x + 2\pi k) = 1, \forall k \in Z,$$

which does not contradict the properties of the normalized probability measure.

Thus, on each finite interval $\{2\pi k\} \cup [0, 2\pi)$, $k \in Z$ of the number line R, the probability distribution of an arbitrary random angle $\Psi(\omega)$ of the form (3.5) for a particular k has the probability distribution $G(x')$ of a random angle on $[0, 2\pi)$. An example of the graph of the function $F(x)$ for the continuous case is shown in Fig. 3.4.

For a significant number of angular measurements, there is no fixed point of origin, therefore, an arbitrary number x_0 from the numerical axis R can be taken as the zero direction, and the results of angular measurements can be considered in the interval $[x_0, x_0 + 2\pi]$ or $[x_0 - \pi, x_0 + \pi]$.

Probability density $p(x)$ of a random angle $\psi(\omega) \in [0, 2\pi)$. The function $p(x)$ for absolutely continuous distribution laws on the circle has properties that coincide with the properties of the function $p(x)$ on the line, as well as distinctive properties.

For continuous functions $G(x')$ and $F(x)$ of random angles, it can be right

$$\begin{aligned} G(x_2) - G(x_1) &= F(x_2 + 2\pi k) - F(x_1 + 2\pi k) \\ &= \int_{x_1}^{x_2} p(y)dy, x_{1,2} \in [0, 2\pi), k \in Z. \end{aligned}$$

The function $p(x)$ is called the probability density of a random angle $\psi(\omega) \in [0, 2\pi)$ and has the following properties:

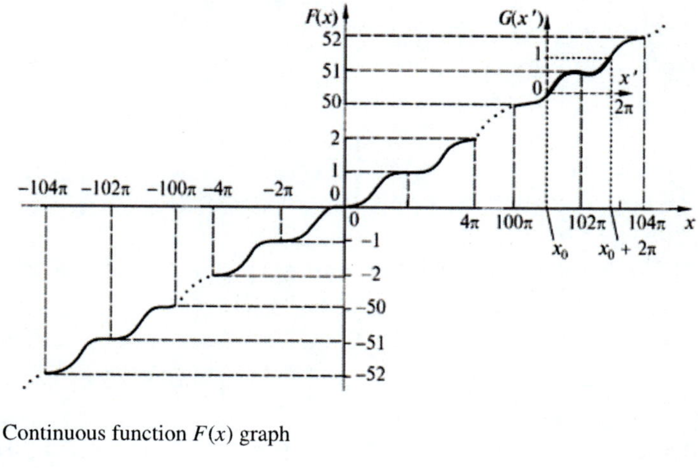

Fig. 3.4 Continuous function $F(x)$ graph

1. $p(x + 2\pi) = p(x)$, $p(x)$ is a periodic function with period 2π.
2. $p(x) \geq 0, x \in [0, 2\pi)$.
3. $\int_0^{2\pi} p(x)dx = \int_{2\pi k}^{2\pi(k+1)} p(x)dx = 1, k \in Z$.

Probabilistic model of a random angle on a unit circle. Consider a random angle $\Psi(\omega)$ with a domain of definition $\Omega = \{\omega \in [0, 2\pi)\}$ and a domain of values of $\Theta \subset R$. On the complex plane, it can be defined a vector

$$\eta(\omega) = \cos[\Psi(\omega)] + i \sin[\Psi(\omega)]. \tag{3.6}$$

Formula (3.6) establishes a unique correspondence between the directions in the space of the vector $\eta(\omega)$ and the part of the random angle $\Psi(\omega)$:

$$\psi(\omega) = [\Psi(\omega)]\text{mod } 2\pi = \Psi(\omega) - \left[\Psi(\omega)/2\pi\right]^{+}2\pi. \tag{3.7}$$

The realization of the random vector $\eta(\omega)$ is depicted on the plane in the Cartesian coordinate system xOy. The realization θ of the random angle $\psi(\omega)$ in this system coincides with the direction of the corresponding vector $\mathbf{r}_q$, therefore $\arg(\mathbf{r}_q) = \theta$. This vector starts at the point O and ends at the point with the coordinates ($x = \cos\theta$, $y = \sin\theta$). The angle θ is counted from the $_{\text{Ox}}$ axis in the counterclockwise direction (Fig. 3.5).

If θ changes to 2π, the end of the vector $\mathbf{r}$ describes on the xOy plane a complete circle of unit radius and returns to the same point in the circle. A circle with a unit radius is called a *normalized* or *unit circle.*

Divide the circle into a finite number l of equal arcs by the points of the circle $\pi(2q + 1)/l$, starting the q count from the Ox axis in the counterclockwise direction. These points form a finite partition D of the unit circle. For example, if $l = 4$, it can be obtained $D = \{\pi/4\pi, 3\pi/4, 5\pi/4, 7\pi/4\}$.

Let connect the origin with the indicated points on the circle by the vectors $\mathbf{r}$. The constructed radius-vectors form angles with the axis Ox

$$\theta_q = \pi(2q + 1)/l, q = \overline{0, (l - 1)}.$$

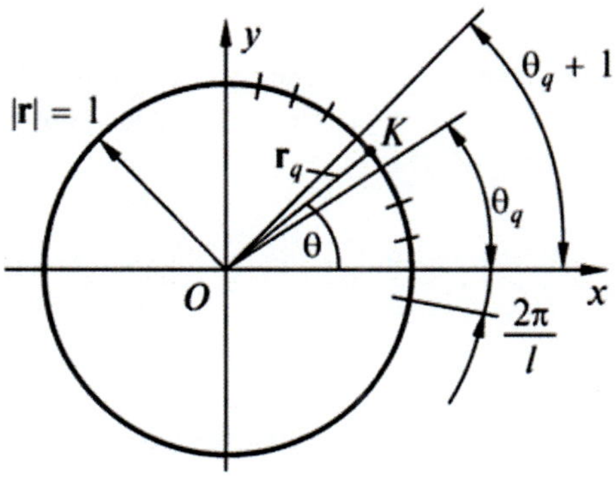

Fig. 3.5 Graphical representation of the results of angular observations on a circle

Let single out the elementary event ω_q that occurs as a result of the following conditions: for arbitrary real θ_q and such θ_{q+1} $\left(0 \le \theta_{q+1} - \theta_q < 2\pi\right)$, in the interval $\left[\theta_q, \theta_{q+1}\right)$ there is a number θ, which by modulo 2π is the value of $\psi(\omega)$. This statement can be written as follows

$$\omega_q = \left\{\psi(\omega) \equiv \theta(\text{mod}\, 2\pi),\ \theta \in \left[\theta_{q,}\theta_{q+1}\right)\right\},\ q = \overline{1, l-1}. \tag{3.8}$$

Thus, the continuum of values of the directions $\mathbf{r}$ in the interval $[0,\ 2\pi)$ is associated with a finite set of elementary events $\Omega = \left\{\omega_q,\ q = \overline{0, l-1}\right\}$ obtained by dividing the unit circle D into a finite number l of equal intervals. Operations with all subsets of Ω form the $\mathcal{A}$-algebra.

To complete the construction of the random angle model, it remains to set the probabilities P_q of events ω_q from the space Ω. Given the cyclical nature of the angles, we define the probabilities as follows:

$$P_q = \begin{cases} P\left\{\theta_q(\text{mod}\, 2\pi) \le \psi(\omega) < \theta_{q+1}(\text{mod}\, 2\pi)\right\}, \left[\theta_q/2\pi\right]^+ = \left[\theta_{q+1}/2\pi\right]^+; \\ P\left\{0 \le \psi(\omega) < \theta_{q+1}(\text{mod}\, 2\pi)\right\} + P\left\{\theta_q(\text{mod}\, 2\pi) \le \psi(\omega) < 2\pi\right\}; \\ \left[\theta_q/2\pi\right]^+ = \left[\theta_{q+1}/2\pi\right]^+ + 1. \end{cases} \tag{3.9}$$

The probability of an event $B \subset \mathcal{A}$ in general is defined as

$$P(B) = \sum_{\{\omega_q \in \theta\}} P\left(\omega_q\right). \tag{3.10}$$

To calculate the probabilities (3.9), it is necessary to introduce a function that generates a probability measure. With using $F(\theta)$, it can be defined (3.9) as

$$P\left(\theta_q \le \Psi(\omega) < \theta_{q+1}\right) = \begin{cases} 0, \quad \text{for} \quad \theta_{q+1} < \theta_q; \\ F\left(\theta_{q+1}\right) - F\left(\theta_q\right), \quad \text{for}\ \theta_q \le \theta_{q+1} < \theta_q + 2\pi; \\ 1, \quad \text{for} \quad \theta_{q+1} > \theta_q + 2\pi. \end{cases} \tag{3.11}$$

The distribution function $F(\theta)$ generates a measure of a random angle. For determining the probabilities, it can be used connection $F(x)$ with the corresponding density $p(\theta)$:

$$F\left(\theta_{q+1}\right) - F\left(\theta_q\right) = \int\limits_{\theta_q}^{\theta_{q+1}} p(x)dx, \tag{3.12}$$

where θ_q, θ_{q+1} are fixed angles corresponding to D.

The probability space $\{\Omega, \mathcal{A}, P\}$, where $P = P(B)$, and $B \in \mathcal{A}$ is the probability of event B, determines the area of definition and the measure of angular observations.

The set of events from Ω is displayed on the numerical axis

$$x(\omega) = \{\varphi(\omega_q),\ q = \overline{0, (l-1)}\},\ x(\omega) \in X \subseteq R.,$$

where $\varphi(\omega_q) = 2\pi q/l$—the length of the arcs between the semiaxis Ox and the vectors $\mathbf{r}_q$, $q = \overline{0, (l-1)}$.

The class of subsets X makes it possible to construct the algebra $\mathcal{B}$. Using expressions (3.11), (3.12), it is easy to find the probabilities $P(A \in \mathcal{B})$ for a certain probability distribution law. Thus, the obtained probability space $\{X, \mathcal{B}, \boldsymbol{P}_\varphi\}$, where $\boldsymbol{P}_\varphi = \{P(\text{A}), A \in \mathcal{B}\}$ determines the range of the probability model of a random angle $\psi(\omega)$.

3.2 Deterministic and Probabilistic Measures of Angular Quantities

Deterministic measures of an angle on a plane in the range $[0, 2\pi)$ **and operations with angles**. Let's consider the options for determining the degree of angle on the plane. The circle belongs to the class of closed planar curves and is generally described in a rectangular Cartesian coordinate system by the equation [12–14]

$$(x - x_0)^2 + (y - y_0)^2 = r^2, \tag{3.13}$$

where x_0, y_0 is the coordinates of the center, r is the radius of the circle. Figure 3.6 consists $x_0 = y_0 = 0$, and $r = 1$. For these conditions, the equation of the circle in the trigonometric form has a simple form –$(\sin\varphi)^2 + (\cos\varphi)^2 = 1$. Using φ-value, it can be determined the coordinates of the endpoint of the vector ρ in the Cartesian coordinate system by a pair of numbers $(\cos\varphi, \sin\varphi)$. For a circle the following basic relations hold: the circumference is equal to $C = 2\pi r$; the length of the arc corresponding to the central angle φ is $l = r\varphi$; the chord length, which corresponds to the central angle φ, is $a = 2\sqrt{2hr - h^2} = 2r\sin(\varphi/2)$, where h is the segment height; circle area—$S = \pi r^2$; sector area - $S_1 = 0.5lr = 0.5r^2\varphi$; the segment area—$S_2 = 0.5lr \pm S_\Delta$, where S_Δ is the area of the triangle with a vertex in the center of the circle and at the ends of the radius, bounding the corresponding sector, the " + " sign is taken if $\varphi > \pi$, the "–" sign if $\varphi < \pi$.

Each vector $\boldsymbol{\rho}$ (Fig. 3.6) can be assigned a one-to-one correspondence with a real number $\varphi \in [0, 2\pi)$, which can be formed as the value of the arc length $l = \varphi$ (Fig. 3.6a), which is cut out on a unit circle by the Ox axis and vector $\boldsymbol{\rho}$, or as the value of the sector area $S_1 = 0.5\varphi$ (Fig. 3.6b) bounded by a part of the Ox axis, vector $\boldsymbol{\rho}$, and the corresponding arc of the unit circle.

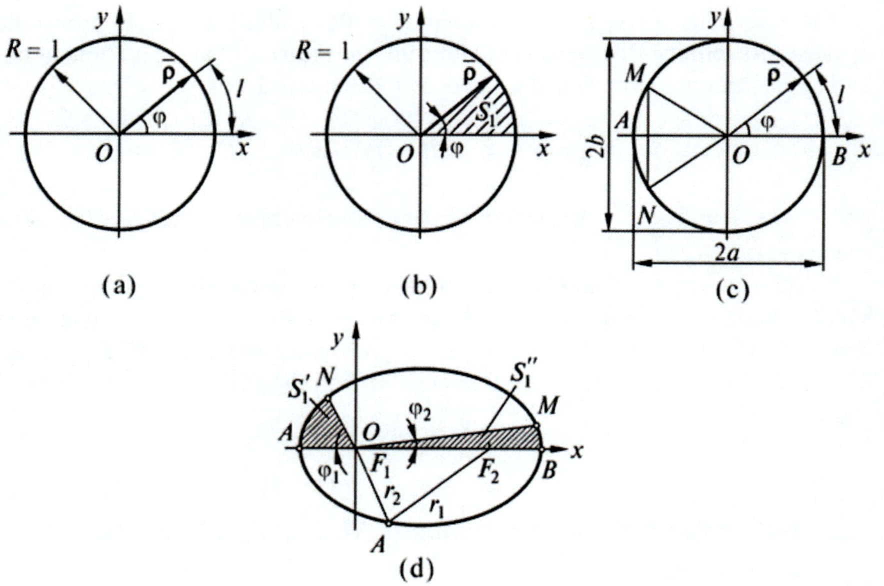

Fig. 3.6 Options for setting the angle: **a** length of the circular arc; **b** area of the circle sector; **c** length of the arc of the ellipse; **d** area of the ellipse sector

The task of setting a measure of angle on a circle with a uniform partition considered above is not the only one. As measures, one can consider other measures of random angles and other closed plane curves, for example, an ellipse (Fig. 3.6c). Such formation of a random angle can find application, for example, in geodesy during conducting angular observations on the earth's surface. It is known [15] that mathematically the shape of the Earth is close to the surface of an ellipsoid formed by the rotation of an ellipse around one of its axes. In geodesic and cartographic works, it uses the Krasovsky ellipsoid (by the name of the famous Soviet scientist F. M. Krasovsky (1878–1948), who in 1946 directed research on the exact determination of the size of the Earth's ellipsoid), for which the major axis is a = 6,378,245 m with the corresponding compression ratio 1: 298.3.

The ellipse, referred to the axes of symmetry, has the equation

$$x^2/a^2+y^2/b^2=1. \qquad (3.14)$$

The sum of the squares of the components x/a and y/b in (3.14) is equal to unity, which allows to take them for the sine and cosine of a certain angle φ. This replacement allows the use of another parametric representation of the ellipse

$$x=a\sin\varphi,\;\; y=b\cos\varphi,\; \varphi\in[0,2\pi). \qquad (3.15)$$

The length of the ellipse arc, corresponding to the angle φ, is calculated as

$$l = a \int_0^{\varphi} \sqrt{1 - \varepsilon^2 \sin^2 \varphi} d\varphi = aE(\varepsilon, \varphi). \tag{3.16}$$

where $\varepsilon = a^{-1}\sqrt{a^2 - b^2}$ is the numerical eccentricity of the ellipse, $E(\varepsilon, \varphi)$ is the designation of the elliptic integral of the 2nd kind [16].

A characteristic property of an ellipse is that the sum of the distances r_1 and r_2 from an arbitrary point of the ellipse to the points F_1, F_2—the foci of the ellipse, is equal to a constant value—2*a*, that is, its major semi-axis: $r_1 + r_2 = (a - ex) + (a + ex) = 2a$.

The area of the ellipse (Fig. 3.6c) is $S = \pi ab$, the area of the *AOM* sector is $S_1 = 0.5ab \arccos(x_M / a)$, and $S_2 = ab \arccos(x_M / a) - x_M y_M$ is the area of the segment *MAN*, x_M, y_M are coordinates of point *M*.

During the execution of orbital angular observations of space objects, it may be appropriate to locate the center of the coordinate system in one of the focal points of the ellipse (Fig. 3.6d) and to form the values of the random angle in proportion to the area of the corresponding segments. Such an assumption is based on Kepler's second law of planetary motion [17]: during unperturbed motion, the area described by the radius vector of a moving material point changes proportionally to time, that is, the radius vector of the planet at equal intervals describes equal areas. Figure 3.6d shows the elliptical orbit of the planet, in the focus of which F_1 the Sun is placed. The planet traverses the segments of the *BM* and *NA* trajectories at equal intervals of time; therefore, the areas of the *AON* and *BOM* sectors are equal. This leads to a measure of a random angle that is uneven over the segments of the arcs. In fact, as a result of the mutual influence of the planets of the solar system, the trajectories of the planets are complex spatial curves that can be approximated by a specific ellipse in just one or two revolutions.

It should be noted that the circle and ellipse do not exhaust all possible ways of constructing a random angle. For this purpose, other closed plane curves of the second order can be used. Such a case can occur in the case of measurements in systems based on unmanned aerial vehicles moving along closed paths [18–20].

Deterministic measure of angles greater than 2π. In the general case, the measured angles may extend beyond the half-interval $[0, 2\pi)$, that is, the range of values of the angle Φ can be the set of all real numbers. In this case, the value of such angles can be interpreted using the helix [21]. A graphical interpretation of the angles for this case is presented in Fig. 3.7. The angle value range is the set of *X* points of the axis of the $O\Phi$. The angle-number displaying occurs through a helix, which is given by the equations

$$x = \cos \Phi,\ y = \sin \Phi,\ \Phi = 2\pi n + \varphi,\ n \in Z.$$

The direction of the vector $\boldsymbol{\rho}$ in the xOy plane on Fig. 3.7 corresponds to an arc *AB* of length *l*. One of the numbers of $O\Phi$ axis of the object form of type $\varphi + 2\pi n$, $n \in Z$ is assigned to this arc through a helix line. The numerical value of *n* is specified by

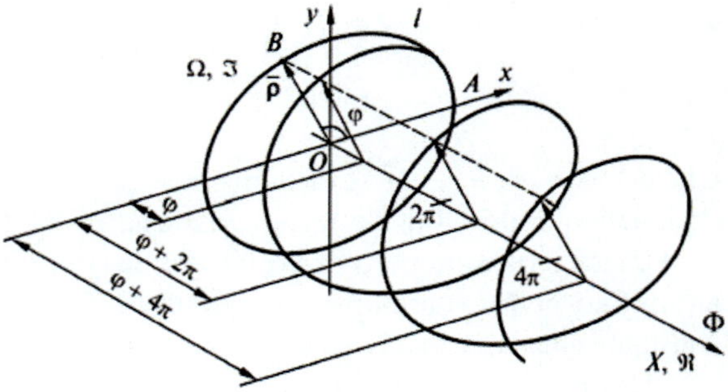

Fig. 3.7 Defining angles on a helix and on a circle

the initial conditions or by another method, which is determined by the conditions of the physical realization of the experiment of angular measurements.

The constructive form of the image of the angle $\Phi > 2\pi$ is presented in the form

$$\Phi = [\Phi/2\pi]^{+} \cdot 2\pi + \varphi, \tag{3.17}$$

where $[\Phi]^{+} \in Z$ is the integer (the number of full revolutions), and $\{\Phi\} = \varphi \in [0, 2\pi)$ is the fractional part of the angle Φ, is the basis of the formula for determining an arbitrary angle in angle measurements. In the practice of angular measurements, the main attention is paid to the angle $\varphi \in [0, 2\pi)$, but it is the structural form (3.17) that makes it possible to study arbitrary angles. From this expression it follows that

$$\varphi \equiv \Phi (\mathrm{mod}\ 2\pi), \tag{3.18}$$

that is, the fractional part of the angle $\Phi > 2\pi$ is determined by the operation of comparing the angle Φ modulo 2π.

A probabilistic measure of random angles is generated by their distribution functions. The probability measure of random variables and random angles is inherent in the properties of normalization, finiteness, and representativeness, which determines their practical value for measurements. Along with this, the probability distribution functions of random angles have certain features. One of the characteristic features of a circle as a space on which sets of angles are formed is the execution of an addition operation on a circle modulo 2π. This determines the periodicity property of the laws of the probability density distribution of a random angle, how they differ significantly from the probability distributions of random variables [22–26].

Laws of distributions of random angles. Figure 3.8 shows the probability density graph $p(\theta)$, $\theta \in (-\infty, \infty)$. The graph is constructed under conditions of a priori uncertainty of the location of the measurement interval $[\theta_i, \theta_i + 2\pi)$, where θ_i is the beginning of the interval, and for which the normalization condition is fulfilled: $\int_{\theta_i}^{\theta_i + 2\pi} p(\theta) d\theta = 1$.

Density $p(\theta)$ in Fig. 3.8 is periodic with a period of 2π. In statistical goniometry, such distributions are called single-frequency [27]. Multifrequency distributions have a period less than 2π by a certain integer number of times, that is, period $2\pi / j$, $j =$

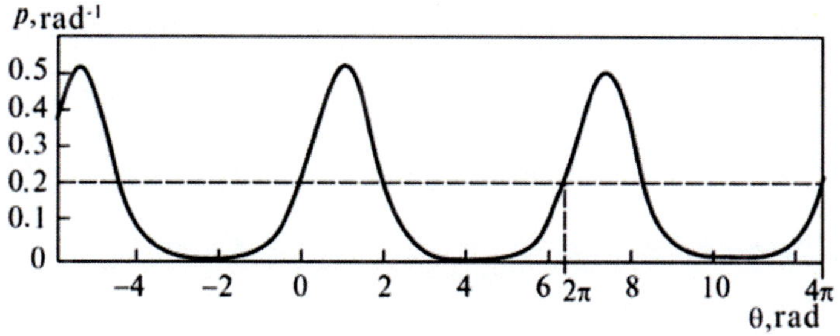

Fig. 3.8 The probability density graph of a random angle

2, 3, ... Next, it will be considering examples of characteristic single-frequency probability distributions of random angles.

Wrapped normal distribution. This distribution refers to the family of wrapped distributions [28–33] ("wrapped" on a unit circle) formed by the nonlinear transformation of a random variable $\xi(\omega)$ into a random angle $\psi(\omega)$ of the form

$$\psi(\omega) = [K\xi(\omega)]\text{mod}\,2\pi, \tag{3.19}$$

where K is the scale conversion coefficient. Transformation (3.19) leads to the transformation of the laws of distribution on a straight line into the laws of probability wrapped distribution of random angles. If the distribution $F(x)$ of a random variable $\xi(\omega)$ is given on a straight line, then the distribution $F(x)$ is wrapped around the unit circle to determine the corresponding distribution law $F_{2\pi}(\theta)$ of a random angle $\psi(\omega) \in [0,\ 2\pi)$. In the general case, it can be obtained

$$F_{2\pi}(\theta) = \sum_{j=-\infty}^{\infty} [F(\theta + 2\pi j) - F(2\pi j)],\quad \theta \in [0,\ 2\pi),\ j \in \mathrm{Z}. \tag{3.20}$$

If $\xi(\omega)$ has a probability distribution density $p(x)$, then a continuous random angle $\psi(\omega)$ also distributed continuously with a density

$$p_{2\pi}(\theta) = \sum_{j=-\infty}^{\infty} p(\theta + 2\pi j),\ j \in \mathrm{Z}. \tag{3.21}$$

In the general case, the function $p_{2\pi}(\theta)$ is asymmetric with respect to the middle of the interval $[0,\ 2\pi)$, and at its ends acquires the same values: $p_\nu(0) = \lim\limits_{\theta\to 2\pi} p_\nu(\theta)$. In many cases, the value of j in sum (3.21) can be limited to a range $\pm 1 . . . . \pm 5$.

The density of the wrapped normal distribution is given by

$$p_{2\pi}(\theta/\mu, \sigma) = \frac{1}{\sqrt{2\pi}\sigma} \sum_{j=-\infty}^{\infty} \exp\left(-\frac{[(\theta - \mu)(\text{mod}\,2\pi) + 2\pi j]^2}{2\sigma^2}\right), \tag{3.22}$$

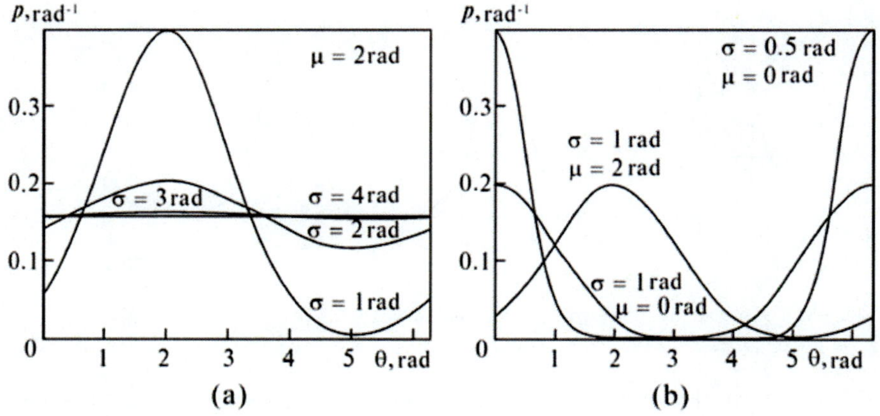

Fig. 3.9 Density of the wrapped normal probability distribution of random angles with various parameters

where μ is the mathematical expectation; σ is the average deviation of the random variable $\xi(\omega)$. This law has an important property: the sum of the independent angles $\sum_{i=1}^{n}\theta_i$, each of which has a distribution (3.22), also has the same distribution, but with different characteristics.

The characteristic distribution function (3.22) has the form

$$f_{2\pi}(n) = f_n = \alpha_n = \exp\left(-n^2\sigma^2/2\right), \beta_n = 0, n \in Z. \tag{3.23}$$

For the wrapped normal distribution, the central limit theorem holds on the circle: for independent random angles $\psi_1(\omega), \ldots, \psi_n(\omega)$, which have the same probability distribution function $F(\theta)$, the probability distribution of the normalized sum of angles $\psi_\Sigma(\omega) = \left[\frac{1}{\sqrt{n}}\sum_{j=1}^{n}\psi_j(\omega)\right] \bmod 2\pi$ in case $n \to \infty$ approaches to the wrapped normal distribution.

Examples of probability density graph of the wrapped normal distribution for various parameter values are shown in Fig. 3.9a, b. From these graphs it can be seen that the wrapped normal distribution is one-vertex and symmetrical with respect to $\theta = \mu (\text{mod } 2\pi)$.

If $\sigma \to \infty$ distribution turns into a uniform distribution with density $1/2\pi$, an increase of μ leads to a shift in the maximum of the function towards large angles. On the interval $[0,\ 2\pi)$, distribution (3.22) has two inflection points.

von Mises distribution. The von Mises probability distribution [34] density for a random angle $\psi(\omega)$ is determined by the formula

$$p_M(\theta|\mu, k) = \exp\{k\cos(\theta - \mu)\}/2\pi I_0(k), |\mu| < \infty, k > 0, \tag{3.24}$$

where $I_0(k)$ is the modified Bessel function of the first kind and zero order, μ is the circular average direction of a random angle; k is the concentration parameter

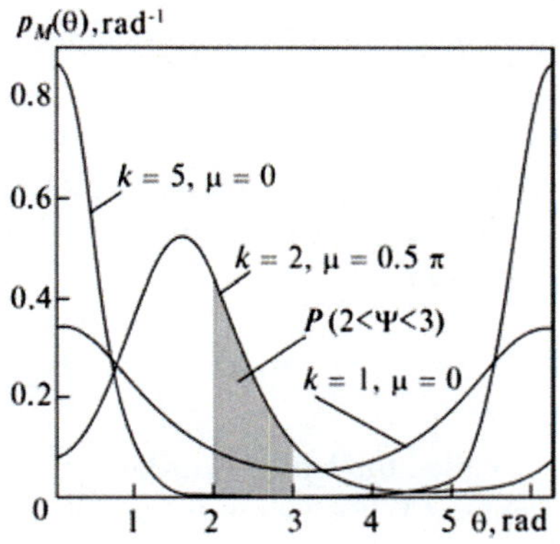

Fig. 3.10 Examples of von Mises probability density graphs

of a random angle in a vicinity of μ. The graphs of the functions $p_M(\theta)$ for various parameter values are shown in Fig. 3.10. With increasing parameter *k*, the von Mises distribution concentrates around μ (if $k = 2$ and $\mu = 0.5\pi$, the distribution is almost completely concentrated on an arc from 0 to 3 rad, and if $k \to 0$—turns into a uniform one).

The von Mises distribution is one-vertex and symmetric with respect to the value of μ, which is the mathematical expectation of this distribution.

The characteristic von Mises distribution function is defined as

$$f_n = \alpha_n = I_n(k)/I_0(k). \tag{3.25}$$

This distribution has the following important property: the most plausible estimate of the parameter μ is a circular average direction.

An appropriate choice of the parameters of the wrapped normal distribution allows a satisfactory approximation it by von Mises distribution.

Other typical circle distributions are given in Table 3.1.

Table 3.1 Typical distribution on the circle[a]

Title	Probability density[a] $p(\theta)$	Trigonometric moments
Uniform	$1/2\pi$	$f_n = \frac{e^{2\pi ni}-1}{2\pi ni} = \begin{cases} 1, & n = 0, \\ 0, & n \neq 0. \end{cases}$
Cardioid	$(2\pi)^{-1}[1 + 2\rho\cos(\theta - \mu)]$, $\lvert\mu\rvert < \infty$, $\lvert\rho\rvert < 0.5$	–
Triangular	$\frac{1}{8\pi}[4 - \pi^2\rho + 2\pi\,\rho\lvert\pi - \theta\rvert]$, $\rho \leq 4/\pi^2$	$\alpha_{2n-1} = \frac{\rho}{(2n-1)^2}$, $\alpha_{2n} = 0$, $\beta_n = 0$
Wrapped Cauchy[b]	$\frac{1}{2\pi} \cdot \frac{1-\rho^2}{1-2\rho\cos\theta+\rho^2}$, $\rho = e^{-a} \in [0, 1]$	$f_n = \rho^{\lvert n\rvert}$

[a]The concept of trigonometric moments will be discussed below
[b]The Cauchy distribution on the line has the next probability distribution density

$$p(x, a) = \frac{1}{\pi} \cdot \frac{a}{a^2 + x^2}, x \in R, a > 0.$$

The wrapped Cauchy probability distribution, like the wrapped normal distribution, has the property of an infinitely divisible distribution law.

The discrete probability distribution of random angles with probabilities, which has form

$$P[\psi \equiv (\nu + 2\pi q/l) \bmod 2\pi] = P_q, q = \overline{0, l-1}, \tag{3.26}$$

called the lattice distribution, and the value $2\pi/l$ is the lattice pitch. For probabilities (3.26), the normalization condition is satisfied: $\sum_q P_q = 1$. This distribution can be considered concentrated at the vertices of a regular l-gon inscribed in the unit circle. If $P_q = l^{-1}$, then the probability distribution turns into a uniform discrete one.

The characteristic function of the lattice distribution at $\nu = 0$ is equal to

$$f_n = \sum P_q \exp(2\pi q n i/l). \tag{3.27}$$

For a uniform discrete distribution, it can be written:

$$f_n = \begin{cases} 1, & n \equiv 0 \bmod 2\pi; \\ 0, & n \neq 0 \bmod 2\pi. \end{cases} \tag{3.28}$$

3.3 Numerical Characteristics of Random Angles

The characteristic function of a random angle $\Psi(\omega)$ probability distribution is defined as a sequence of values [11, 23, 28–33]

$$f_n = \mathbf{M}\{\exp(in\Psi(\omega))\} = \int_0^{2\pi} e^{inx} dG(x) = \int_{2\pi k}^{2\pi(k+1)} e^{inx} dF(x), \; k, n \in Z. \tag{3.29}$$

For continuous and discrete random angles with (3.29) it can be written, respectively, such expressions

$$f_n = \int_0^{2\pi} \exp(in\theta') p(\theta') d\theta', \quad f_n = \sum_{j=1}^{m} P_j \exp\left(in \frac{2\pi}{m}\right). \tag{3.30}$$

The characteristic function of random angles can be represented as a series of complex numbers

$$f_n = \alpha_n + i\beta_n = \rho_n \exp(i\mu_n), \tag{3.30a}$$

$$\alpha_n = \mathbf{M}\cos(n\psi) = \int_0^{2\pi} \cos(n\theta)dF(\theta), \beta_n = \mathbf{M}\sin(n\psi) = \int_0^{2\pi} \sin(n\theta)dF(\theta).$$

Since for a random angle $\Psi(\omega)$ the definition (3.29) makes sense only when the function $\exp(in\Psi(\omega))$ is periodic with period 2π, then n can only be an integer: $n = 0, \pm1, \pm2, \ldots$. Therefore, in the case of a random angle distributed on the circle, the characteristic function is a sequence of trigonometric moments calculated with respect to the zero direction.

The characteristic function has the following properties:

1. Characteristic function module is $|f_n| \le 1$.
2. For $n = 0$, it can be written $f_0 = 1$.
3. Characteristic function of a negative argument is equal to the complex conjugate characteristic function, i.e. $f_{-n} = f_n^*$, because $\alpha_{-n} = \alpha_n$, $\beta_{-n} = -\beta_n$.
4. Characteristic function of the sum of independent random angles $\psi_1(\omega), \ldots \psi_m(\omega)$ is equal to the product of their characteristic functions

$$f_n = \prod_{j=1}^{m} f_n^{(j)}, \tag{3.31}$$

   where $f_n^{(j)}$ is the characteristic function of the n-th order of the jth angle.
5. Characteristic function of the angle $[\psi(\omega) + \nu] \bmod 2\pi$, where $\nu = \text{const}$ is an arbitrary real angle, is

$$\mathrm{M}e^{in(\psi+\nu)} = e^{in\nu} f_n, \tag{3.32}$$

   that is, a change in the origin does not lead to a change in the trigonometric moment.
6. Probability distribution density of random angles is uniquely determined by its characteristic function

$$p(\theta) = \frac{1}{2\pi} \sum_{n=-\infty}^{\infty} f_n e^{-in\theta}. \tag{3.33}$$

Formula (3.33) is a Fourier series of $p(\theta)$. Its use in some cases allows to significantly simplify the expressions for the probability density distribution of random angles.

The characteristic function can also be calculated with respect to an arbitrary initial direction ν. In this case, it can be written

$$f_n(\nu) = \mathbf{M}e^{in(\psi-\nu)} = \alpha_n(\nu) + i\beta_n(\nu) = \rho_n(\nu)e^{i\mu_n(\nu)}. \tag{3.34}$$

Using the above properties of the characteristic function of random angles, we can prove the following relations:

$$\rho_n(\nu) = \rho_n,\, \mu_n(n) \equiv (\mu_n - n\nu)\bmod 2\pi. \tag{3.35}$$

Central trigonometric moments are determined for $\rho_1 > 0$ relative to the direction with a polar angle μ_1 according to the expressions

$$\begin{aligned} f_n(\mu_1) &= f_n(0)e^{-in\mu_1(0)};\\ a_n(\mu_1) &= \rho_n(0)\cos(\mu_n(0) - n\mu_1(0));\\ b_n(\mu_1) &= \rho_n(0)\sin(\mu_n(0) - n\mu_1(0)). \end{aligned} \tag{3.36}$$

For $n = 1$ it can written $a_1(\mu_1) = \rho_1(0)$, $b_1(\mu_1) = 0$.

Circular average deviation $\delta(\nu)$ of a random angle $\psi(\omega)$ relative to angle ν is defined as the mathematical expectation of a random angle

$$\min\{\psi'(\omega),\ 2\pi - \psi'(\omega)\}, \tag{3.37}$$

where $\psi'(\omega) \equiv (\psi(\omega) - \nu)(\bmod 2\pi)$, i.e.

$$\delta(\nu) = \int\limits_0^{\pi} \theta dF(\theta + \nu) + \int\limits_{\pi}^{2\pi} (2\pi - \theta) dF(\theta + \nu). \tag{3.38}$$

Circular dispersion of a random angle $\Psi(\omega)$ is defined as the value $\upsilon = 1 - \rho_1(0) = 1 - |f_1(0)|$. In general, it is considered as a characteristic of the deviation of a random angle $\psi(\omega) \equiv \Psi(\omega)(\bmod 2\pi)$ from a fixed angle $\varphi' \in [0, 2\pi)$:

$$\upsilon(\varphi') = \mathbf{M}\{1 - \cos(\psi(\omega) - \varphi')\} = 1 - \mathrm{Re}\mathbf{M}\{\exp i[\psi(\omega) - \varphi']\}. \tag{3.39}$$

In the case $\rho_1(0) > 0$, it can be written

$$\upsilon(\varphi') = 1 - \rho_1(0)\cos(\varphi' - \mu_1(0)) = 1 - \rho_1(0) + 2\rho_1(0)\left(\sin\frac{\varphi' - \mu_1(0)}{2}\right)^2. \tag{3.40}$$

This expression acquires a minimum value when $\varphi' = \mu_1(0)$. That is, the circular dispersion of a random angle $\psi(\omega) \equiv \Psi(\omega)(\bmod 2\pi)$ takes the minimum value that is $\upsilon = 1 - \rho_1(0)$ for $\varphi' = \mu_1(0)$, that is, when the circular average direction of a random angle $\psi(\omega)$ is taken as the direction φ'. The set of values of circular dispersion belongs to the interval $[0, 1)$.

Circular median. For single-vertex continuous distributions, the median is always uniquely determined. Such distributions on a circle differ in that in the interval $[0, 2\pi)$ there are two such angles φ_1 and φ_2; when a point moves in a unit circle from

φ_2 to φ_1 in both directions, the function $p(x') = dF(x')/dx'$ is monotonically non-decreasing.

Circle median of continuous division on a circle $G(\theta')$ of a random angle $\psi(\omega)$ is the value of the angle μ_m, which is one of the solutions of the equation $Q(\mu_m) = F(\mu_m + \pi) - F(\mu_m) - 0.5 = \int_{\mu_m}^{\mu_m+\pi} p(\theta)d\theta - 0.5 = 0$ and $Q(\mu_m - 0, 5\pi)$ is the maximum.

Another important property of the median of a random angle concerns its circular average deviation: in the case of a single-vertex distribution, the circular average deviation reaches a minimum at point μ_m.

Asymmetry. To estimate the asymmetry of the law of probability distribution of a random angle, an asymmetry coefficient is introduced:

$$\gamma_c = \beta_2(\mu_1)\Big/\upsilon^{3/2}. \tag{3.41}$$

The coefficient γ_c characterizes the asymmetry of the probability distribution law of a random angle relative to the direction $\theta = \mu_1$ and is equal to zero for symmetric probability distributions.

Excess. The dampness of the probability distribution curves of random angles in the vicinity of their mode characterizes the excess coefficient:

$$\gamma_e = \frac{\alpha_2(\mu_1) - (1 - \upsilon)^4}{\upsilon^2}. \tag{3.42}$$

The coefficient γ_e compares the curves of all the laws of probability distributions of random angles with the wrapped normal distribution for which $\gamma_e = 0$ (the excess coefficient is close to zero for the von Mises distribution as well).

Circular standard variance. The characteristic function allows to establish a relationship between the dispersion σ of a Gaussian random variable and the circular variance υ of a random angle with a wrapped normal distribution. For $n = 1$, it can be written $\exp(-0.5\sigma^2) = \rho_1 = 1 - \upsilon$, and then

$$\sigma = \sqrt{-2\ln(1 - \upsilon)}, \sigma \in [0, \infty). \tag{3.43}$$

The value of σ can be used as a measure of the dispersion of random angles (in a certain sense, it resembles the standard variance and, as a rule, is expressed in radians).

These numerical characteristics in comparison with the corresponding characteristics of random variables are given in Table 3.2.

Sample characteristics of random angular quantities. According to the results of observations and preliminary study of the measurement data, a sample of the angles $(\varphi_1, \ldots \varphi_j, \ldots \varphi_M)$, $\varphi_j \in [0, 2\pi)$ of volume M is obtained. This sample is considered as the realization of random angles having a certain continuous probability distribution density $p(x)$. Based on the values of $(\varphi_1, \ldots \varphi_j, \ldots \varphi_M)$, the sample

Table 3.2 Characteristic of random variables and angles

Characteristic	Random value	Random angle
Random value, angle	Real random variable is a function $\xi(\omega)$ with a domain of definition $\Omega = \{\omega\}$ and a domain of values $X \subset R$ such that for an arbitrary $x \in X$ set of those $\omega \in \Omega$, for which $\xi(\omega) < x$,is an event A from the set of random events $\Im$ given on a fixed probability space $(\Omega, \Im, P)$	Real random angle is a function $\Psi(\omega)$ with a domain of definition $\Omega = \{\omega\}$ and a domain of values $\Theta \subset R$ such that for any $\theta \in \Theta$ the fractional part of the angle $\psi(\omega) = \Psi(\omega) - [\Psi(\omega)/2\pi]^{+}2\pi = \Psi(\omega) \bmod 2\pi$ is a random variable
Characteristic function	Characteristic function of a random variable $\xi(\omega)$ is the mathematical expectation of a random function $\exp(iu\xi)$, i.e. $f(u) = \mathbf{M}e^{iu\xi} = \int_{-\infty}^{\infty} \exp(iux)dF(x),$ $u \in (-\infty, \infty).$	Characteristic function of a random angle $\Psi(\omega)$ (sequence of trigonometric moments with respect to the zero direction) is the sequence of mathematical expectations of function $\exp(jn\psi)$: $f_n = \mathbf{M}e^{in\psi} = \int_0^{2\pi} \exp(in\theta')dG(\theta') = \rho_n \exp(i\mu_n),$ where k—arbitrary integer; $n = 0, \pm 1, \pm 2, ..$
Quantiles of distribution laws	Solution of the equation $F(\alpha_\gamma) = \gamma$, where γ is the given probability $(0 < \gamma < 1)$ with respect to $\alpha_\gamma \in (-\infty, \infty)$ is called the quantile of the distribution $F(x)$ of level γ, where $\gamma = \int_{-\infty}^{\alpha_\gamma} p(x)dx$	Solution of the equation $G(\theta'_\gamma) = \gamma$, where γ is the given probability $(0 < \gamma < 1)$ with respect to $\theta'_\gamma \in [0, 2\pi)$ is called the quantile of the distribution $F(\theta)$ of level γ, where $\gamma = \int_0^{\theta'_\gamma} p(\theta)d\theta$
Trigonometric moment module of the first order	–	Resulting vector length, which is the mathematical expectation of a random vector (cos Ψ, sin Ψ), is $\rho_1 = \lvert f_1 \rvert$
First initial moment (average)	Mathematical expectation of a random variable $\xi(\omega)$ with a distribution function $F(x)$ is the number determined by the Stieltjes integral $\mathbf{M}\xi = \int_{-\infty}^{\infty} xdF(x)$	Circular mean value of a random angle $\Psi(\omega)$ with a distribution function $G(\theta')$, for which $f_1 = \mathbf{M}e^{i\Psi} \neq 0$, is called the angle $\mu_1 = \operatorname{Arg} f_1$

(continued)

Table 3.2 (continued)

Characteristic	Random value	Random angle
Spread of values of the random variable and random angle	Dispersion with the distribution function $F(x)$ is the mathematical expectation of the square of the deviation of the values of $\xi(\omega)$ from its mathematical expectation $\mathbf{M}\xi$: $\mathbf{D}\xi = \mathbf{M}(\xi - \mathbf{M}\xi)^2 = \int_{-\infty}^{\infty}(x - \mathbf{M}\xi)^2 dF(x)$.	Circular dispersion of a random angle $\Psi(\omega)$ with the distribution function $G(\theta')$ is called the quantity $\upsilon = 1 - \rho_1 = 1 - \lvert f_1 \rvert$, which characterizes the deviation of the value of the random angles from its average value
Median	Median of the continuous distribution $F(x)$ of a random variable $\xi(\omega)$ is the value of $x = Me$ for which it is equally likely that the random variable is greater or less than Me, i.e. $P(\xi < Me) = P(\xi > Me)$	Circle median of continuous division on a circle $G(\theta')$ of a random angle $\psi(\omega)$ is the value of the angle μ_m, which is one of the solutions of the equation $Q(\mu_m) = F(\mu_m + \pi) - F(\mu_m) - 0.5 = = \int_{\mu_m}^{\mu_m+\pi} p(\theta)d\theta - 0.5 = 0$, and $Q(\mu_m - 0.5\pi)$ is the maximum
Mode	Mode is the value of Mod of a random variable $\xi(\omega)$, for which the distribution density $p(x = \text{Mod})$ has a maximum value	Mode is the value of Mod of a random angle $\psi(\omega)$, for which the distribution density $p(\theta' = \text{Mod})$ has a maximum values

numerical characteristics of the angles (or difference in phase characteristics of the signal) are determined as follows.

The sample trigonometric moment of order n relative to a given direction $\alpha \in [0, 2\pi)$ is determined by the formula:

$$\hat{f}_n(\alpha) = M^{-1}\sum_{j=1}^{M} e^{in(\phi_j - \alpha)} = \hat{a}_n(\alpha) + i\hat{b}_n(\alpha) = \hat{r}_n(\alpha)e^{i\hat{m}_n(\alpha)}, \tag{3.44}$$

and the sample cosine and sine moments of order n—according to the formulas:

$$\hat{a}_n(\alpha) = M^{-1}\sum_{j=1}^{M}\cos\left[n\left(\varphi_j - \alpha\right)\right], \hat{b}_n(\alpha) = M^{-1}\sum_{j=1}^{M}\sin\left[n\left(\varphi_j - \alpha\right)\right]. \tag{3.45}$$

Sample characteristic function is a complex-valued sequence $(\hat{f}_n(0),\ n = 0, \pm 1, \pm 2, \ldots)$, all of whose selective trigonometric moments are defined with respect to the zero direction $\alpha = 0$.

The use of selective sample trigonometric moments in problems of approximating the distribution of angular data according to (3.44) was considered in [28], in [35] it was determined the signal-to-noise ratio in an additive mixture of a harmonic signal and Gaussian noise.

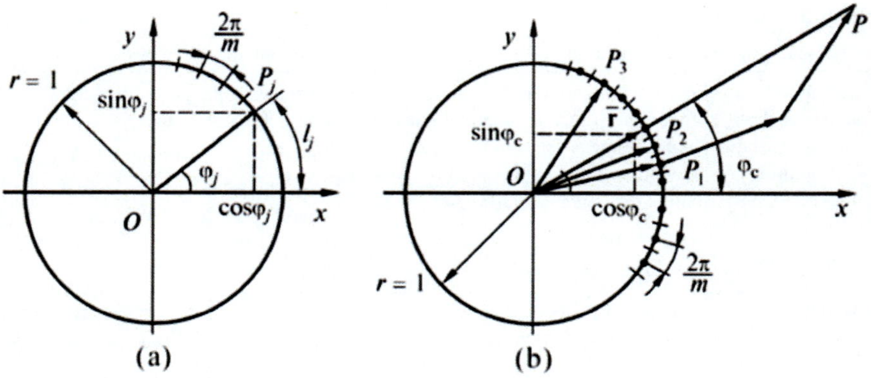

Fig. 3.11 Graphic image on the circle of the result of one measurement (**a**) and averaging of the sample of angles of the volume M = 3 (**b**)

Sample circular average. The result of a separate observation φ_j can be represented by the corresponding flat angle φ_j, which corresponds to an arc of length l_j between the positive abscis axis and the vector $\overrightarrow{OP_j}$ on the circle of unit radius $R = 1$ (Fig. 3.11a). The vector $\overrightarrow{OP_j}$ has Cartesian $\left[\cos\varphi_j, \sin\varphi_j\right]$ and polar $\left(1, \varphi_j\right)$ coordinates.

Measurements are performed with a certain step $2\pi/m$. The discrete nature of the results of angular measurements corresponds to the partition of the circle into m class intervals. Therefore, the points P_j are the midpoints of class intervals.

Any constructive characteristic L of a circular average, according to which the results of angular measurements are processed, must satisfy the additivity condition:

$$\{L(\varphi_1 - \nu, \ldots \varphi_M - \nu)\}(\text{mod } 2\pi) \equiv \{L(\varphi_1, \ldots \varphi_M) - \nu\}(\text{mod } 2\pi), \tag{3.46}$$

that is, for any angle $\nu \in [0, 2\pi)$, the fractional parts (modulo 2π) of the numbers $\{L(\varphi_1, \ldots \varphi_M) - \nu\}$ and $L(\varphi_1 - \nu, \ldots \varphi_M - \nu)$ must coincide. In other words, the angle specified by the characteristic $L(\varphi_1, \ldots \varphi_M)$ must additively depend on the initial angle ν. This requirement is satisfied by an estimate in the form of a sample circular average, which is defined as the direction of the sum of all unit vectors $\overrightarrow{OP}_1, \ldots \overrightarrow{OP}_M$. In Fig. 3.11, case M = 3 is considered. The total vector $\overrightarrow{OP} = \sum_{j=1}^{3} \overrightarrow{OP}_j$ has Cartesian coordinates $\left(\sum_{j=1}^{3} \cos\varphi_j, \sum_{j=1}^{3} \sin\varphi_j\right)$ and is characterized by a sample circular average angle φ_c. The physical meaning of φ_c from the point of view of mechanics is explained as follows. All unit vectors end with points P_j of the unit circle. If it will assign the same "mass" M^{-1} to all these points, then the coordinates of the "center of mass" of this system will be determined as follows

$$C = M^{-1}\sum_{j=1}^{M}\cos\varphi_j, \quad S = M^{-1}\sum_{j=1}^{M}\sin\varphi_j. \tag{3.47}$$

Recalculation of the coordinates of the vector **r** from the Cartesian system to the polar one is performed according to the formulas

$$r = |\mathbf{r}| = \sqrt{C^2 + S^2}. \tag{3.48}$$

$$C = r\cos\varphi_c, \quad S = r\sin\varphi_c. \tag{3.49}$$

The value of **r** is called the sample resultant length of the vector resultant (SRL) **r**.

The vectors **r** and $\overrightarrow{OP}$ are located in space at the same angle φ_c to the axis Ox. The value of φ_c is calculated by the formula

$$\varphi_c = \mathbf{L}[S, C] = arctgS/C + 0.5\pi\{2 - (signS)(1 + signC)\}. \tag{3.50}$$

If $r = 0$, value $\varphi_c \in [0, 2\pi)$ is not uniquely determined.

For $r > 0$, sample circular average (SCA) of the angle φ_c satisfies the requirement (3.46), and r does not depend on the origin of the angles. Indeed, if each of the vectors $\overrightarrow{OP_j}$ (Fig. 3.11b) is returned in space by an angle $\nu \in [0,\ 2\pi)$, this will only lead to a rotation of the vector **r** in space by the same angle ν, but the value of r will not change. After turning **r**, the new coordinates of the end of the vector are defined as

$$\overline{C} = r\cos(\varphi_c - \nu), \quad \overline{S} = r\sin(\varphi_c - \nu). \tag{3.51}$$

It is easy to verify that

$$\overline{C} = M^{-1}\sum_{j=1}^{M}\cos\left(\varphi_j - \nu\right) = r\cos(\varphi_c - \nu),$$

$$\overline{S} = M^{-1}\sum_{j=1}^{M}\sin(\varphi - \nu) = r\sin(\varphi_c - \nu), \tag{3.52}$$

i.e. $\varphi_c(\varphi_1 - \nu, \ldots, \varphi_M - \nu) \equiv \{\varphi_c(\varphi_1, \ldots, \varphi_M) - \nu\}(\text{mod}2\pi)$,

$$r(\varphi_1 - \nu, \ldots, \varphi_M - \nu) = r(\varphi_1, \ldots, \varphi_M),$$

which proves the additivity property of SCA.

If it assumes $\nu = \varphi_c$, then, taking into account (3.52), it can be written

$$\sum_{j=1}^{M} \sin(\varphi_j - \varphi_c) = 0, \tag{3.53}$$

The resulting Eq. (3.53) is used to verify the correct determination of φ_c.

Applied issues of SCA were considered in [36, 37].

Sample circular variance. It will define the deviations in the space of the direction of the vector $\overline{OP}_j$ from an arbitrary direction ν as

$$\Delta\varphi_j = \min\left\{(\varphi_j - \nu)', 2\pi - (\varphi_j - \nu)'\right\} = \pi - \left|\pi - (\varphi_j - \nu)'\right|, \ \Delta\varphi_j \geq 0, \tag{3.54}$$

where $(\varphi_j - \nu)'$ is the remainder of the angle $(\varphi_j - \nu)$ defined by modulus 2π,

$$(\varphi_j - \nu)' = (\varphi - \nu) - \left[(\varphi_j - \nu)/2\pi\right]^{+} 2\pi. \tag{3.55}$$

A convenient form of representing the scattering measure is a function of the form $f(\Delta\varphi) = 1 - \cos\Delta\varphi = 2\sin^2\Delta\varphi/2$ of the deviation angle $\Delta\varphi$, since it is positive and monotonic on the segment ν. Therefore, the value

$$V(\nu) = \frac{1}{M}\sum_{j=1}^{M}\left[1 - \cos(\varphi_j - \nu)\right] = \frac{2}{M}\sum_{j=1}^{M}\sin^2\frac{\varphi_j - \nu}{2} \tag{3.56}$$

is taken as the sample scattering characteristic of the sample of angles $\{\varphi_1, \varphi_2, \ldots \varphi_j, \ldots \varphi_M\}$ relative to the direction ν. From expression (3.56), taking into account (3.53), that the sample scattering characteristic with respect to the SCA ν has the form:

$$V(\varphi_c) = \frac{1}{M}\sum_{j=1}^{M}\left\{1 - \cos(\varphi_j - \varphi_c)\right\} = 1 - \frac{1}{M}\sum_{j=1}^{M}\cos(\varphi_j - \varphi_c) = 1 - r. \tag{3.57}$$

The value $V(\varphi_c) \in [0,\ 1]$ is called the sample circular variance (SCV) of the sample $\{\varphi_1, \varphi_2, \ldots \varphi_j, \ldots \varphi_M\}$. This characteristic is invariant with respect to the origin of the angles.

It follows from expression (3.56) that the choice of $\nu = \varphi_c$ minimizes the characteristic of circular scattering. Indeed, with (3.57) it can be written:

$$V(\nu) = M^{-1}\sum_{j=1}^{M}\left\{1 - \cos(\varphi_j - \nu)\right\} = V(\varphi) + 2r\sin^2\left((\varphi_c - \nu)/2\right),$$

which proves this statement.

The sample estimates of the characteristics of random angles are summarized in Table 3.3.

In the case of calculating the SRL of r based on the data intervals grouped into the class, it was obtained an estimate biased in the direction of lower values. To reduce the bias, the Sheppard correction type [38–40] is used: $r_g = rc_g$, where $c_g = \pi / m \cdot \sin(\pi/m)$. This correction can be neglected if $m > 40$, since $c_g(40) \approx 1.001$. The general questions of using angular statistics in measurements were considered in [41, 42], the modeling and processing of angular data in [42, 43], the use of a

Table 3.3 Sample random angle characteristics

Name of statistics	Content and definitions of statistics
Sample circular average (mean direction)	$\varphi_c = \mathrm{arctg}\frac{S}{C} + \frac{\pi}{2}\{2 - (\mathrm{sign}S) \times [1 + \mathrm{sign}C]$; $C = M^{-1}\sum_{j=1}^{M}\cos\varphi_j$; $S = M^{-1}\sum_{j=1}^{M}\sin\varphi_j$ M is a sampling angels volume
Resultant length of the vector resultant	$r = \sqrt{C^2 + S^2}$
Sample circular variance	$V = 1 - r$
Sample circular standard deviation	$\sigma = \sqrt{-2\ln(1 - V)} = \sqrt{-2\ln r}$
Sample median direction	It is the direction, which corresponds to the point of the circle P, the diameter PQ divides the statistic value in half, and in the vicinity of P we have the maximum concentration of values
Sample modal direction	It is the direction, which corresponds to the point of the circle, in the vicinity of which there is a maximum concentration of values of statistics
Sample circular sweep	The length of the smallest arc of statistics, which is determined from the variation series $T_j = \varphi_{j+1} - \varphi_j,\ j = 1, \ldots, M - 1$; $T_M = 2\pi - \varphi_M + \varphi$; $W = 2\pi - \max\{T_1, \ldots, T_M\}$
Sample trigonometric moments (n—integer)	$T_n(\alpha) = M^{-1}\sum_{j=1}^{M} e^{in(\theta_j - \alpha)} = a_n(\alpha) + ib_n(\alpha) = r_n(\alpha)e^{im_n(\alpha)}$; $a_n(\alpha) = M^{-1}\sum_{j=1}^{M}\cos\left[n\left(\varphi_j - \alpha\right)\right]$; $b_n(\alpha) = M^{-1}\sum_{j=1}^{M}\sin\left[n\left(\varphi_j - \alpha\right)\right]$; $r_n(\alpha) = \sqrt{a_n^2(\alpha) + b_n^2(\alpha)} = \sqrt{a_n^2(0) + b_n^2(0)} = r_n(0)$; $m(\alpha) = m_n(0) - n\alpha$
Sample asymmetry	$g_1 = \frac{b_2(m)}{V^{3/2}} = \frac{r_2\sin[m_2(0) - 2m]}{V^{3/2}}$
Sample kurtosis	$g_2 = \frac{a_2(m) - (1-V)^4}{V^2} = \frac{r_2\cos m_2(0) - 2n - (1-V)^4}{V^2}$

sample circular median in [9], and the use of SRL for detection signals of ultrasonic non-destructive testing—in [44–48].

3.4 Models and Measures of Random Angles in Phase Measurements

Models and measures of random angles in phase measurements are considered in the works [49, 50].

Probabilistic model of phase shifts of harmonic signals. It will be considering this problem in the formulation for the case of discrete measurements.

Let a lattice be given on observation interval $[0,\ T_c]$:

$$S = \{t_1, t_2, \ldots t_n\}, \tag{3.58}$$

where n—volume of the lattice, the set of elements of which is ordered and the non-equilibrium $0 \leq t_1 < t_2 < \cdots t_j < \cdots < t_n \leq T_c$ is performed for it.

The elements of the lattice S are evenly spaced and form an arithmetic progression $t_j = t_1 + (j-1)\Delta t,\ \ j = \overline{1, n}$, where Δt is the lattice spacing.

Two functions are defined on the lattice S:

$$\begin{gathered} u_1(t_j) = U_1 \cos(2\pi f t_j + \varphi),\ t_k \in S,\ \ \varphi \in [0,\ 2\pi); \\ u_2(t_j) = U_2 \cos(2\pi f t_j), \end{gathered} \tag{3.59}$$

which are the images of the corresponding analog signals with a discrete argument (3.58). The set S is the domain of functions (3.59), and the set of all values of the functions $\{u_1(t_j),\ u_2(t_j),\ j = \overline{1, n}\}$ is the set of their values. Functions (3.59) relate to a linear normed function space L_n, i.e., $u_1(t_j), u_2(t_j) \in L_n$. For elements of the space L_n, the operation of the scalar product of functions

$$(u_m, u_l) = n^{-1} \sum_{k=1}^{n} u_m(t_k) u_l(t_k) h(t_k), \tag{3.60}$$

where m, l—numbers of space L_n elements; $h(t_j)$—weight function, $h(t_j) \geq 0$. The norm of each element of the space L_n is also specified:

$$\|u_m\| = \sqrt{(u_m, u_m)}. \tag{3.61}$$

In the space L_n of functions of a real argument, there is a shift operator:

$$\mathbf{A}_{\Delta t} u_m(t_j) = u_m(t_j + \Delta t) = u_m(t_{j+1}). \tag{3.62}$$

Consistently applying the shift operator (3.62) by q times, it can be written

$$\underset{(q)}{\mathbf{A}_{\Delta t}\ldots\mathbf{A}_{\Delta t}}\,u_m(t_j) = \mathbf{A}_{q\Delta t}u_m(t_j) = u_m(t_j + q\Delta t) = u_m(t_j + \tau), \tag{3.63}$$

where $\{\tau = q\Delta t,\ q \in \overline{0, \infty}\}$—arbitrary shift of the signal in time on the set S.

The creation of quadrature signals requires the fulfillment of the condition: $T/\Delta t = 4s$, where $s \in N$. In this case, the sinusoidal sequences $u_1(t_j)$, $u_2(t_k)$ are periodic with period $4s\Delta t$ for all $t_j \in T$, for example, $u_{1,2}(t_j) = u_{1,2}(t_j + 4\Delta t)$ if $s = 1$.

Let consider the normalized scalar multiplication of signals (3.59) with a weight function:

$$h(t_j) = g(t_j - a)g(b - t_j) = \begin{cases} 1, & t_j \in [a, b),\ a < b,\ a \in [0, T_c - b + a); \\ 0, & t_j \notin [a, b),\ b \in (b - a, T_c], \end{cases} \tag{3.64}$$

where $g(t_j)$—Heaviside unit step function. It will be written:

$$C = \frac{(u_1(t_j), u_2(t_j))}{\|u_1(t_j)\|\,\|u_2(t_k)\|} = \frac{U_1U_2\sum_{j=1}^{n}\cos(2\pi f t_j + \varphi)\cos 2\pi f t_j g(t_j - a)g(b - t_j)}{\|u_1(t_j)\|\,\|u_2(t_j)\|} = \cos\varphi,$$

where the norms of the functions $u_1(t_j)$, $u_2(t_j)$ are defined as

$$\|u_1(t_j)\| = \sqrt{\frac{\Delta t}{b - a}(u_1(t_j), u_1(t_j))} = \frac{U_1}{\sqrt{2}};$$

$$\|u_2(t_j)\| = \sqrt{\frac{\Delta t}{b - a}(u_2(t_j), u_2(t_j))} = \frac{U_2}{\sqrt{2}}.$$

Next, apply the shift operator $\mathbf{A}_{\Delta t}$ to the function $u_2(t_j)$ and consider the normalized scalar multiplication of this function with $u_1(t_j)$:

$$S = \frac{(u_1(t_j), u_2(t_j + 3g\Delta t))}{\|u_1(t_j)\|\,\|u_2(t_j)\|} = \frac{U_1U_2\sum_{j=1}^{n}\cos(2\pi f t_j + \varphi)\sin 2\pi f t_j g(t_j - a)g(b - t_j)}{\|u_1(t_j)\|\,\|u_2(t_j)\|} = \sin\varphi.$$

A pair of numbers C and S obtained in this way can be considered as the coordinates of the unit vector **z**, which starting at point O and describing the unit circle in the Cartesian coordinate system xOy.

In the presence of noise and interference that inevitably accompanies the measurement process, the obtained values of S_n and C_n differ from the ideal values of S and C, and the points built on the xOy plane with coordinates (S_n, C_n) in the general case will not belong to the unit circle, since $|\mathbf{z}_n| = \sqrt{S_n^2 + C_n^2} \neq 1$. To reduce the experimental results to the unit circle, it was performed normalization by the formulas

$$\widehat{C} = C_n \Big/ \sqrt{C_n^2 + S_n^2}; \quad \hat{S} = S_n \Big/ \sqrt{C_n^2 + S_n^2}. \tag{3.65}$$

Thus, an unambiguous correspondence was established between the position in space of the vector $\mathbf{z}_n$ and the phase shift φ between the signals (3.59), which for this case allows the use of a probabilistic model of random angles in the plane.

The signal parameters $u_1(t_j)$ are defined as:

$$\widehat{\varphi} = \mathbf{L}[S_n, C_n] = \operatorname{arctg}\frac{S_n}{C_n} + \frac{\pi}{2}\{2 - [\operatorname{sign}(S_n)] \cdot [1 + \operatorname{sign}(C_n)]\}, \tag{3.66}$$

$$U_1 = \sqrt{2}\|u_1(t_j)\|. \tag{3.67}$$

During performing phase measurements with the conversion of a linear random variable $\xi(\omega)$ with a domain of definition $x \in (-\infty, \infty)$ to a random phase shift of signals $\psi(\omega) = K\xi(\omega)(\bmod 2\pi)$ with a domain of definition $\theta \in [0, 2\pi)$, where K is the conversion coefficient, it is advisable to apply the probability distribution laws from the family of wrapped that correspond to this for a probabilistic description of random phase shifts transformation.

In many cases, the distribution of random phase shifts is satisfactorily approximated by a wrapped normal probability distribution. This assumption can be substantiated given the fact that the random character of $\psi(\omega)$ is due to the action of a significant number of independent factors. According to the central limit theorem on the circle, the distribution law of the sum of many random independent angles tends to a wrapped normal, which is satisfactorily approximated by the von Mises distribution.

Probabilistic model of phase shifts of random narrow-band processes. The formation and transmission of information signals in phase systems is accompanied by exposure of noise and interference. Considering that the input circuits of the systems are linear links with a limited bandwidth $\Delta f << f_0$, it can be assumed that there is a noise in the form of a narrow-band random process at the input of the system along with the information signal. In phase metering [51, 52], the value of f_0 is considered known and equal to the frequency of the reference, deterministic signal. The narrow-band stationary real random process can be given in the form

$$\xi(\omega, t) = A_\xi(\omega, t) \cos \Phi_\xi(\omega, t), \ \omega \in \Omega, t \in (-\infty, \infty), \tag{3.68}$$

where $A_\xi(\omega, t)$, $\Phi_\xi(\omega, t)$ are respectively, the envelope and phase (amplitude and phase characteristics) of the process $\xi(\omega, t)$.

Using the model (3.68) corresponds to the real conditions of the phase-virulent experiment. Let consider the following problem.

On a finite time interval $[0, T_c]$, processes are studied

$$\begin{aligned} \eta(\omega, t) &= u(t) + \xi(\omega, t), \\ u_0(t) &= U_0 \cos(2\pi f_0 t), t \in [0, T_c]. \end{aligned} \tag{3.69}$$

The process $\eta(\omega, t)$ is a random process and is defined as an additive mixture of the information signal

$$u(t) = U \cos(2\pi f_0 t + \varphi) = U_c \cos(2\pi f_0 t) - U_s \sin(2\pi f_0 t), U > 0, t \in T_c, \tag{3.70}$$

and a stationary random Gaussian process $\xi(\omega, t)$ in the form of (3.68) with characteristics $\mathbf{M}\xi(\omega, t) = 0$, and $\mathbf{D}\xi(\omega, t) = \sigma^2$, which is the response of a narrow-band linear system with a resonant frequency f_0 on white noise. In formula (3.70), $U_c = U \cos \varphi$, $U_s = U \sin \varphi$ are the quadrature components of the signal, $U = \sqrt{U_c^2 + U_s^2}$, $\varphi \in [0, 2\pi)$ is the initial phase associated with U_c, U_s by the relation $\varphi = \mathbf{L}[U_s, U_c]$.

The random process $\xi(\omega, t)$ belongs to the class of second-order processes, that is, $\mathbf{M}\xi^2(\omega, t) < \infty, \forall t \in T_c$, and allows to apply the Hilbert transforms [4, 5, 43] to it.

The random process $\eta(\omega, t)$ has phase $\Phi_\eta(\omega, t)$.

It is necessary to construct a probabilistic model of the phase shift $\Delta\varphi(\omega, t) = [\Phi_\eta(\omega, t) - \Phi_0(t)] \bmod 2\pi$ between the phases of the processes (3.69), where $\Phi_0(t) = 2\pi f_0 t$ is the phase of the reference signal $u_0(t)$.

Let research the process $\eta(\omega, t)$, given by the expression (3.69). The random component $\eta(\omega, t)$ is the process $\xi(\omega, t)$, which is determined by expression (3.68), in which the functions $A_\xi(\omega, t)$ and $\Phi_\xi(\omega, t)$ are uniquely determined with using the Hilbert transform [4, 5, 43, 53].

At coinciding time instants, the random processes $\xi(\omega, t)$ and $\xi_h(\omega, t)$ are uncorrelated, which follows from the analysis of their scalar multiplication:

$$\mathbf{M}\{\xi(\omega, t) \cdot \xi_h(\omega, t)\} = R_{\xi\xi_h}(0) = 0. \tag{3.71}$$

Since the processes $\xi(\omega, t)$ and $\xi_h(\omega, t)$, under the condition of the problem, are Gaussian, their statistical independence follows from their non-correlation.

If the process $\xi(\omega, t)$ passes through a narrow-band linear system with a central frequency f_0, its phase can be represented as

$$\Psi_\xi(\omega, t) = 2\pi f_0 t + \psi(\omega, t), \tag{3.72}$$

and the process $\xi(\omega, t)$—in the form:

$$\xi(\omega, t) = \xi_c(\omega, t)\cos(2\pi f_0 t) - \xi_s(\omega, t)\sin(2\pi f_0 t), \tag{3.73}$$

where $\xi_c(\omega, t)$ and $\xi_s(\omega, t)$ are real independent Gaussian stationary random processes with zero mathematical expectation and dispersion σ^2:

$$\xi_c(\omega, t) = A(\omega, t)\cos\varphi(\omega, t);\ \xi_s(\omega, t) = A(\omega, t)\sin\varphi(\omega, t). \tag{3.74}$$

Taking into account (3.70) and (3.72), the measuring signal has the form

$$\eta(\omega, t) = [U_c + \xi_c(\omega, t)]\cos(2\pi f_0 t) - [U_s + \xi_s(\omega, t)]\sin(2\pi f_0 t). \tag{3.75}$$

Let construct the Hilbert image of the random process $\eta(\omega, t)$

$$\eta_{\mathrm{H}}(\omega, t) = [U_s + \xi_s(\omega, t)]\cos(2\pi f_0 t) + [U_c + \xi_c(\omega, t)]\sin(2\pi f_0 t). \tag{3.76}$$

To move from random processes to random variables, let consider the sets of values of processes (3.75), (3.76) at discrete time instants $t_j = jT, j = \overline{0, T_c/T}$. Given that the random process $\xi(\omega, t)$ is stationary under the conditions of the problem, that its components $\xi_c(\omega, t)$ and $\xi_s(\omega, t)$ are also stationary processes. Therefore, the sampled values of $\xi_c(\omega, t_j)$ and $\xi_s(\omega, t_j)$ obtained at fixed times t_j can be considered as random variables. For time moment t_j, it can be written a random variables

$$\eta_j(\omega) = U_c + \xi_c(\omega);\quad \eta_{\mathrm{H},j}(\omega) = U_s + \xi_s(\omega). \tag{3.77}$$

The pairs of instantaneous values of random processes $\eta(\omega, t)$ and $\eta_{\mathrm{H}}(\omega, t)$ obtained at coinciding time instants form random vectors $\overline{\eta}(\omega) = \left(\eta_j(\omega), \eta_{\mathrm{H},j}(\omega), j = \overline{0, T_c/T}\right)$, which are displayed on the plane in the Cartesian coordinate system xOy by straight line segments [54]. Vectors start at the center of the coordinate system and end at points $\left(\eta_j(\omega), \eta_{\mathrm{H},j}(\omega)\right)$. Figure 3.12 shows the realization $\boldsymbol{\rho}_j$ of a random vector $\boldsymbol{\eta}$ and denotes a vicinity of radius 3σ, which, with

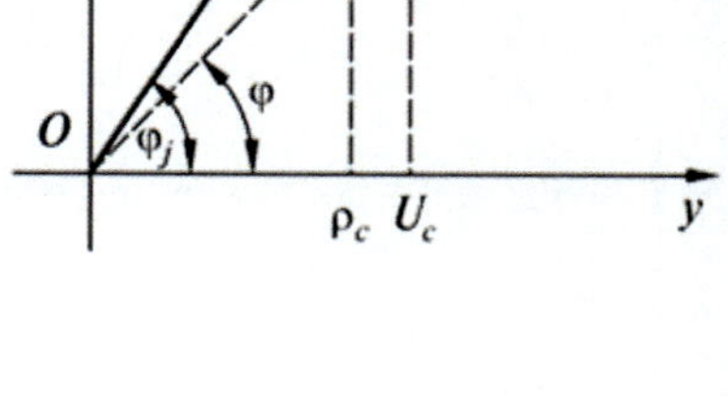

Fig. 3.12 Graphic image of a random vector implementation

a probability of 0.997, contains a set of points with coordinates $\left(\eta_j(\omega), \eta_{\mathrm{H},j}(\omega)\right)$ for the initial phase $\varphi_j = \mathbf{L}[U_s, U_c]$.

The angle $\varphi_j = \arg\left[\boldsymbol{\rho}_j\right]$ determines the direction of the vector $\boldsymbol{\rho}_j$ in the space xOy. It is counted from the Oy axis in the counterclockwise direction.

If we change the values of φ in the interval $[0, 2\pi)$, then the angle $\psi_j(\omega) = \arg\left[\boldsymbol{\eta}_j(\omega)\right]$ takes all the values in the intervals $[0, 2\pi)$, and the 3σ-vicinity forms a ring with an inner radius $U - 3\sigma$, an outer radius $U + 3\sigma$, and an average radius U, that is, a ring of thickness 6σ, as shown in Fig. 3.13.

The average radius U has the meaning of the mathematical expectation of the modulus of a random vector $\boldsymbol{\eta}(\omega)$. The area of the ring is $S = 12\pi U\sigma$. The ends of the random vectors $\boldsymbol{\eta}_j(\omega)$ can occupy an arbitrary position within the area of the ring. These vectors are characterized by two components, therefore, the probabilistic model must be considered as a set of models of angles and moduli of random vectors (or their quadrature components).

Probabilistic model of the angle of random vector. Let divide the ring (Fig. 3.13) into a finite number l of identical parts. To this end, first let divide the circle of the average radius U into a finite number l of equal arcs by the points of the circle $\pi(2q + 1)/l$, $q = \overline{0, (l - 1)}$, starting the counting of q from the Oy axis in a defined direction (in Fig. 3.13 by the letter A denotes one of the split points). Since the narrow-band random process (3.76) allows a range of envelope values to vary within $\pm\, 3\sigma$, it makes no sense to split the circle into arcs much less than 6σ. Therefore, it can be choose the number of partition intervals as the nearest integer to the value of $\pi R/\sigma$. The indicated points form a finite partition D_θ of a circle of radius U.

If we connect the origin with the points of division of the circle, then the central corners are formed by the value $\Delta\theta$. These corners correspond to the partition of the ring into l identical parts of area $s = 6\Delta\theta U\sigma$.

Let fix an elementary event ω_q, which corresponds to such a set of conditions: for arbitrary real θ_q and θ_{q+1}, such that $0 \le \theta_{q+1} - \theta_q < 2\pi$, in the interval $\left[\theta_q, \theta_{q+1}\right)$ there is a number θ comparable by modulo 2π with the value $\psi(\omega)$. This statement is formally given by such an expression

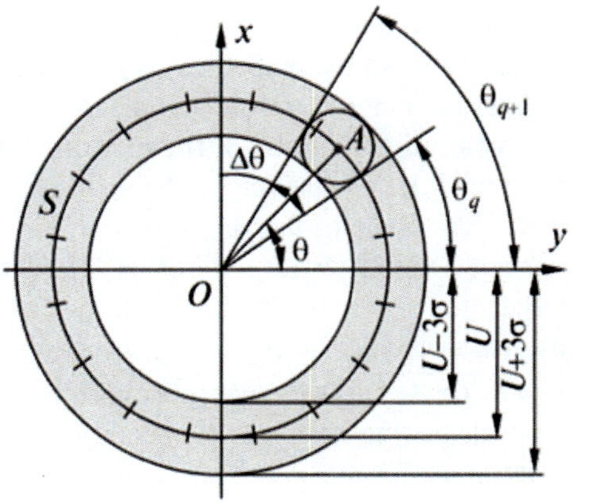

Fig. 3.13 Graphical construction of a probabilistic model of an argument of a random vector on the xOy plane

$$\omega_q = \left\{E_{\psi(\omega)} : \psi(\omega) \equiv \theta(\mathrm{mod}\ 2\pi), \theta \in \left[\theta_q, \theta_{q+1}\right)\right\}, q = \overline{1, l-1}, \tag{3.78}$$

where $E_{\psi(\omega)}$ set of values $\psi(\omega)$.

Thus, the continuum of values of the directions $\arg[\boldsymbol{\eta}(\omega)]$ in the interval $[0, 2\pi)$ is associated with a finite set of elementary events $\Omega = \left\{\omega_q, q = \overline{0, l-1}\right\}$, which is obtained by dividing D_θ ring into a finite number l of identical parts. Operations with all subsets of Ω form the algebra $\mathcal{B} = \beta(D_\theta)$.

To complete the construction of the model, let set the probabilities P_q of the events $\omega_q \in \Omega$. Given the cyclic nature of the angles, it can be written

$$P_q = \begin{cases} P\left\{\theta_q\left(\mathrm{mod}\, 2\pi\right) \le \psi < \theta_{q+1}\left(\mathrm{mod}\, 2\pi\right)\right\} \text{ для } \theta_{q+1} > \theta_q; \\ P\left\{0 \le \psi < \theta_{q+1}\left(\mathrm{mod}\, 2\pi\right)\right\} + P\left\{\theta_q\left(\mathrm{mod}\, 2\pi\right) \le \psi < 2\pi\right\},\ \theta_{q+1} < \theta_q. \end{cases} \tag{3.79}$$

For probabilities (3.79), the normalization condition is satisfied: $\sum_{1=0}^{l-1} P_q = 1$. The probability of an arbitrary event $B \in \mathcal{B}$ in general form can be written as

$$P(B) = \sum_{\{i:\omega_q \in B\}} P\left(\omega_q\right). \tag{3.80}$$

The probability space $\{\Omega, \mathcal{B}, P\}$, where $P = P(\mathrm{B})$, $B \in \mathcal{B}$ is the probability of the event B, defines the domain of definition of the model of angular observations of signals in a statistical measurement experiment.

The set of events from Ω is displayed on the numerical axis as $s\left(\omega_q\right)$, which are equal to the ratio of length of the average radius arc to the radius U—$s\left(\omega_q\right) = 2\pi q/l, q = \overline{1, (l-1)}$. So, the set of events Ω maps to the number set $x(\omega) = \left\{s\left(\omega_q\right), q = \overline{0, (l-1)}\right\}, x(\omega) \in X$. The subsets $x(X)$ form the algebra $\mathcal{B}_\psi$.

For calculating the probability $P\left(A \in \mathcal{B}_\psi\right)$ of arbitrary events A, it is necessary to justify the choice of the distribution density of the random angles. Since the random vector $\boldsymbol{\eta}(\omega)$ is considered as a vector with independent Gaussian quadrature components, the phase distribution of such a vector is satisfactorily approximated by the wrapped normal probability distribution law or the von Mises distribution [34].

The probability space $\left\{X, \mathcal{B}_\psi, P_\psi\right\}$ is obtained, where $P_\psi = \left\{P(\mathrm{A}), \mathrm{A} \in \mathcal{B}_\psi\right\}$, defines the range of values and the measure of the random angle $\psi(\omega)$.

Probabilistic model of the modulus of a random vectors. All random vectors have modules that belong to the interval $\left|\boldsymbol{\rho}_j(\omega)\right| \in (U + 3\sigma, U - 3\sigma)$, that is, they end in the plane of the ring shown in Fig. 3.13. Let construct a system $(m-1)$ of concentric circles with radii $\rho_g = U + 3\sigma\left(2g/m - 1\right)$, $g = \overline{1, (m-1)}$. Adjacent circles together with the inner and outer circles form a system m of circular rings of thickness $2\Delta\rho = 6\sigma/m$ and average radius $\rho_{g.\mathrm{av}} = U + 3\sigma\left((2g+1)/m - 1\right)$, $g = \overline{1, (m-1)}$. Two adjacent concentric circles are shown in Fig. 3.14.

It can be singled out the elementary event ω_g that occurs as a result of the following conditions: for arbitrary real ρ_g and ρ_{g+1} such that $\rho_{g+1} \in [U - 3\sigma, U + 3\sigma)$, $g =$

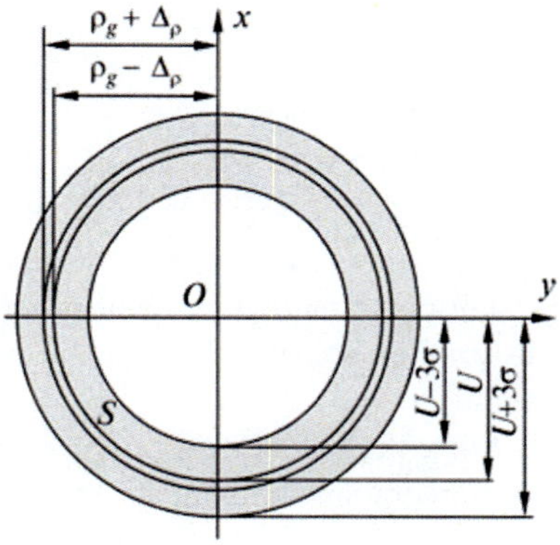

Fig. 3.14 Ring breaking with concentric circle system

$\overline{0, (m-1)}$, in the interval $[\rho_g, \rho_{g+1})$ there is a number ρ equal to $\eta(\omega) = |\boldsymbol{\eta}(\omega)|$, that is, equal to the length of the vector $\boldsymbol{\eta}(\omega)$. This statement can be written as

$$\omega_g = \{E\eta : \eta(\omega) = \rho, \rho \in [\rho_g, \rho_{g+1})\}, g = \overline{0, (m-1)}. \tag{3.81}$$

Thus, the continuum of values of the modules $|\boldsymbol{\eta}(\omega)| \subset [U - 3\sigma, U + 3\sigma)$ is associated with a finite set $\Omega_\rho = \{\omega_g, g = \overline{0, (m-1)}\}$ of elementary events obtained by breaking of ring D_ρ of thickness 6σ into a finite number m of rings of equal thickness $6\sigma/m$. All possible unions ω_g together with an empty set form an algebra $\Re_\rho = \beta(D_\rho)$.

At the end of the construction of the model, it remains to set the probabilities P_g of the events $\omega_g \in \Omega_\rho$. These probabilities are determined as follows:

$$P_g = P[\rho_g \le |\overline{\eta}(\omega)| < \rho_{g+1}], g = \overline{0, (m-1)}. \tag{3.82}$$

The probability of event $B_\rho \in \Re_\rho$ in general is defined as

$$P(B_\rho) = \sum_{\{j:\omega_g \in B_\rho\}} P(\omega_g). \tag{3.83}$$

For probabilities (3.82), the normalization condition $\sum_{1=0}^{l-1} P_g = 1$ is satisfied.

For calculating the probabilities (3.82), it is necessary to introduce a function that generates a probability measure. If the probability density $p_\rho(x)$ of the modulus of the random vector is known, the probability (3.82) is calculated as

$$P_\rho(\rho_g \le |\overline{\eta}(\omega)| < \rho_{g+1}) = \int_{\rho_g}^{\rho_{g+1}} p_\rho(x)dx. \tag{3.84}$$

The probability space $\{\Omega_\rho, \mathfrak{R}_\rho, P_\rho\}$, where $P_\rho = P(B_\rho)$, $B_\rho \in \mathfrak{R}_\rho$ is the probability of event B_ρ, defines the range of model values and the measure for the modulus of a random vector.

The subsets of Ω_ρ is displayed on the numerical set $x(\omega) = \{\rho(\omega_g), g = \overline{0, (l-1)}\}$, $x(\omega) \in X$, where $\rho_{g.\mathrm{av}} = U + 3\sigma(2g + 1/m - 1)$, $g = \overline{1, (m-1)}$ determined by the mean radius. The subsets X form the $\mathcal{B}_\rho$-algebra.

For obtaining the probability $P(A \in \mathcal{B}_\rho)$ of arbitrary events A, it is necessary to justify the choice of the probability density distribution of the moduli of a random vector.

Since the random vector $\boldsymbol{\eta}_j(\omega)$ is considered as a vector with independent Gaussian quadrature components, the distribution of the vector module has the probability distribution density of the generalized Rayleigh distribution [55, 56].

The obtained probabilistic spaces $\{X, \mathcal{B}_\varphi, P_\varphi\}$ and $\{X, \mathcal{B}_\rho, P_\rho\}$ determines the area of definition and the measure when measuring phase shifts of narrowband random processes.

References

1. D'Ambrosio, V., Spagnolo, N., Der Re, L., Slussarenko, S., Li, Y., Kwek, L.C., Marucci, L., Walborn, S.P., Aolita, L., Sciarrino, F.: Photonic polarization gears for ultra-sensitive angular measurements. Nat. Commun. **4**, 2432 (2013). https://doi.org/10.1038/ncomms3432
2. Dich, L.Z.: Accuracy problems and scale theory. Interpretation of measurement errors. Meas. Tech. **43**, 461–470 (2000). https://doi.org/10.1007/BF02503533
3. Alimov, Y.I., Kravtsov, Y.A.: Is probability a "normal" physical quantity? Soviet Phys. Uspekhi **35**(7), 606 (1992). https://doi.org/10.1070/PU1992v035n07ABEH002250
4. Von Plato, J.: Formalization of Hilbert's geometry of incidence and parallelism. Synthese **110**, 127–141 (1997). https://doi.org/10.1023/A:1004959405270
5. Feldman, M.: Hilbert transform in vibration analysis. Mech. Syst. Sign. Process. **25**(3), 735–802 (2011). https://doi.org/10.1016/j.ymssp.2010.07.018
6. Kuts, Y.V., Shengur, S.V., Scherbak, L.N.: Circular measurement data modeling and statistical processing in LabView. In: 2011 Microwaves, Radar and Remote Sensing Symposium, Ukraine, pp. 317–320 (2011). https://doi.org/10.1109/MRRS.2011.6053664
7. Eder, W.E.: A viewpoint on the quantity "plane angle" . Metrologia **18**, 171 (1982). https://doi.org/10.1088/0026-1394/18/1/002
8. Mills, I.: On the units radian and cycle for the quantity plane angle. Metrologia **53**(3), 991 (2016). https://doi.org/10.1088/0026-1394/53/3/991
9. Fisher, N.I.: Statistical Analysis of Circular Data, 277 p. Cambridge University Press, Cambridge (2000)
10. Bendat, J.S., Piersol, A.G.: Random Data: Analysis and Measurement Procedures, vol. 729. Wiley, Hoboken (2011)
11. Babak, V., Eremenko, V., Zaporozhets, A.: Research of diagnostic parameters of composite materials using Johnson distribution. Int. J. Comput. **18**(4), 483–494 (2019)
12. Ovaska, S.J., Valivita, S.: Angular acceleration measurement: a review. In: IMTC/98 Conference Proceedings. IEEE Instrumentation and Measurement Technology Conference. Where Instrumentation is Going (Cat. No.98CH36222), USA, 1998, pp. 875–880. https://doi.org/10.1109/IMTC.1998.676850
13. Masterenko, D.A., Teleshevskii, V.I.: Features of numerical processing of measurement information for high-precision linear and angular measurements. Meas. Tech. **59**, 1254–1259 (2017). https://doi.org/10.1007/s11018-017-1125-3

14. Taubner, A., von Martens, H.-J.: Measurement of angular accelerations, angular velocities and rotation angles by grating interferometry. Measurement **24**(1), 21–32 (1998). https://doi.org/10.1016/S0263-2241(98)00042-6
15. Tilma, T., Sudarshan, E.C.G.: Generalized Euler angle parameterization for U(N) with applications to SU(N) coset volume measures. J. Geom. Phys. **52**(3), 263–283 (2004). https://doi.org/10.1016/j.geomphys.2004.03.003
16. Nair, P.S., Saunders, A.T., Jr.: Hough transform based ellipse detection algorithm. Pattern Recogn. Lett. **17**(10), 777–784 (1996). https://doi.org/10.1016/0167-8655(96)00014-1
17. Katsikadelis, J.T.: Derivation of Newton's law of motion from Kepler's laws of planetary motion. Arch. Appl. Mech. **88**, 27–38 (2018). https://doi.org/10.1007/s00419-017-1245-x
18. Zaporozhets, A., Kovtun, S., Dekusha, O.: System for monitoring the technical state of heating networks based on UAVs. In: Shakhovska, N., Medykovskyy, M. (eds.) Advances in Intelligent Systems and Computing IV. CCSIT 2019. Advances in Intelligent Systems and Computing, vol. 1080, pp. 935–950. Springer, Cham (2020). https://doi.org/10.1007/978-3-030-33695-0_61
19. Zaporozhets, A.: Review of quadrocopters for energy and ecological monitoring. In: Babak, V., Isaienko, V., Zaporozhets, A. (eds.) Systems, decision and Control in Energy I. Studies in Systems, Decision and Control, vol. 298. Springer, Cham (2020). https://doi.org/10.1007/978-3-030-48583-2_2
20. De Filippis, L., Guglieri, G., Quagliotti, F.: Path Planning Strategies for UAVS in 3D Environments. J. Intell. Rob. Syst. **65**, 247–264 (2012). https://doi.org/10.1007/s10846-011-9568-2
21. Nojima, K., Ogata, K., Tanama, M., Nishi, R., Ono, Y., Koide, T.: Bending fatigue strength of case-carburized helical gears (In the case of large helix angles). J. Mech. Sci. Technol. **31**, 5657–5663 (2017). https://doi.org/10.1007/s12206-017-1106-0
22. Raich, A., Cinar, A.: Diagnosis of process disturbances by statistical distance and angle measures. Comput. Chem. Eng. **21**(6), 661–673 (1997). https://doi.org/10.1016/S0098-1354(96)00299-2
23. Kurz, G., Gilitschenski, I., Hanebeck, U.D.: Nonlinear measurement update for estimation of angular systems based on circular distributions. In: 2014 American Control Conference, USA, pp. 5694–5699 (2014). https://doi.org/10.1109/ACC.2014.6858982
24. Kurz, G., Gilitschenski, I., Hanebeck, U.D.: Recursive nonlinear filtering for angular data based on circular distributions. In: 2013 American Control Conference, USA, pp. 5439–5445 (2013). https://doi.org/10.1109/ACC.2013.6580688
25. Eremenko, V., Zaporozhets, A., Isaenko, V., Babikova, K.: Application of wavelet transform for determining diagnostic signs. In: CEUR Workshop Proceedings, vol. 2387, pp. 202–214. https://ceur-ws.org/Vol-2387/20190202.pdf
26. Zaporozhets, A., Eremenko, V., Isaenko, V., Babikova, K.: Approach for creating reference signals for detecting defects in diagnosing of composite materials. In: Shakhovska, N., Medykovskyy, M. (eds.) Advances in Intelligent Systems and Computing IV. CCSIT 2019. Advances in Intelligent Systems and Computing, vol. 1080, pp. 154–172. Springer, Cham (2020). https://doi.org/10.1007/978-3-030-33695-0_12
27. Guggenheimer, H.: The Jordan and Schoenflies theorems in axiomatic geometry. Am. Math. Mon. **85**(9), 753–756 (1978). https://doi.org/10.1080/00029890.1978.11994693
28. Pewsey, A.: The wrapped skew-normal distribution on the circle. Commun, Stat. Theory Methods **29**(11), 2459–2472 (2000). shttps://doi.org/10.1080/03610920008832616
29. Pewsey, A.: Modelling asymmetrically distributed circular data using the wrapped skew-normal distribution. Environ. Ecol. Stat. **13**, 257–269 (2006). https://doi.org/10.1007/s10651-005-0010-4
30. Kent, J.T., Tyler, D.E.: Maximum likelihood estimation for the wrapped Cauchy distribution. J. Appl. Stat. **15**(2), 247–254 (1988). https://doi.org/10.1080/02664768800000029
31. Pewsey, A., Lewis, T., Jones, M.C.: The wrapped *t*-family of circular distributions. Aust. New Zealand J. Stat. **49**(1), 79–91 (2007). https://doi.org/10.1111/j.1467-842X.2006.00465.x
32. Kurz, G., Gilitschenski, I., Hanebeck, U.D.: Efficient evaluation of the probability density function of a wrapped normal distribution. In: 2014 Sensor Data Fusion: Trends, Solutions, Applications (SDF), Germany, pp. 1–5 (2014). https://doi.org/10.1109/SDF.2014.6954713

33. Ravindran, P., Ghosh, S.K.: Bayesian analysis of circular data using wrapped distributions. J. Stat. Theory Pract. **5**(4), 547–561 (2011). https://doi.org/10.1080/15598608.2011.10483731
34. Best, D.J., Fisher, N.I.: Efficient simulation of the von Mises distribution. Appl. Stat. Ser. C **28**(2), 152–157 (1979). https://doi.org/10.2307/2346732
35. Guo, Y., Shen, Y., Tan, J.: Stochastic resonance in a piecewise nonlinear model driven by multiplicative non-Gaussian noise and additive white noise. Commun. Nonlinear Sci. Numer. Simul. **38**, 257–266 (2016). https://doi.org/10.1016/j.cnsns.2016.02.032
36. McCool, J.I.: Systematic and random errors in least squares estimation for circular contours. Precis. Eng. **1**(4), 215–220 (1979). https://doi.org/10.1016/0141-6359(79)90102-8
37. Dergunov, A.V., Kuts, Y.V., Scherbak, L.N.: Comparative analysis of modern time-series analysis methods. In: 2011 Microwaves, Radar and Remote Sensing Symposium, Ukraine, pp. 378–381 (2011). https://doi.org/10.1109/MRRS.2011.6053679
38. Kushwaha, R., Singh, B: An EV battery charger based on PFC Sheppard Taylor converter. In: 2016 National Power Systems Conference (NPSC), India, pp. 1–6 (2016). https://doi.org/10.1109/NPSC.2016.7858944
39. Karaarslan, A.: The analysis of average sliding control method applied on Sheppard-Taylor power factor correction converter. Electr. Eng. **95**, 255–265 (2013). https://doi.org/10.1007/s00202-012-0257-3
40. Zhao, L., Zhang, B., Ma, H., Liu, X.: Research and experiment of Sheppard-Taylor topology. In: 31st Annual Conference of IEEE Industrial Electronics Society, 2005, p. 6. IECON 2005, USA (2005). https://doi.org/10.1109/IECON.2005.1569081
41. Braitenberger, E.: Uncertainty measures and uncertainty relations for angle observables. Found. Phys. **15**, 353–364 (1985). https://doi.org/10.1007/BF00737323
42. Nikolaidis, N., Pitas, I.: Nonlinear processing and analysis of angular signals. IEEE Trans. Sign. Process. **46**(12), 3181–3194 (1998). https://doi.org/10.1109/78.735295
43. Zayed, A.I.: Hilbert transform associated with the fractional Fourier transform. IEEE Signal Process. Lett. **5**(8), 206–208 (1998). https://doi.org/10.1109/97.704973
44. Babak, V.P., Babak, S.V., Myslovych, M.V., Zaporozhets, A.O., Zvaritch, V.M.: Methods and models for information data analysis. In: Diagnostic Systems For Energy Equipments. Studies in Systems, Decision and Control, vol. 281, pp. 23–70. Springer, Cham (2020). https://doi.org/10.1007/978-3-030-44443-3_2
45. Moore, K.C.: Making sense by measuring arcs: a teaching experiment in angle measure. Educ. Stud. Math. **83**, 225–245 (2013). https://doi.org/10.1007/s10649-012-9450-6
46. Matsakis, P., Wendling, L., Ni, J.: A general approach to the fuzzy modeling of spatial relationships. In: Jeansoulin, R., Papini, O., Prade, H., Schockaert, S. (eds.) Methods for Handling Imperfect Spatial Information. Studies in Fuzziness and Soft Computing, vol. 256, pp. 49–74. Springer, Berlin, Heidelberg (2010). https://doi.org/10.1007/978-3-642-14755-5_3
47. Korosec, M.: Technological information extraction of free form surfaces using neural networks. Neural Comput. Appl. **16**, 453–463 (2007). https://doi.org/10.1007/s00521-006-0071-9
48. Hasna, M.O., Alouini, M.S.: Harmonic mean and end-to-end performance of transmission systems with relays. IEEE Trans. Commun. **52**(1), 130–135 (2004). https://doi.org/10.1109/TCOMM.2003.822185
49. Kuts, Y.V., Yeremenko, V.S., Monchenko, E.V., Protasov, A.G.: Ultrasound method of multilayer material thickness measurement. In: AIP Conference Proceedings, 2009, 1096, 1115. https://doi.org/10.1063/1.3114079s
50. Von Martens, H.-J.: Generalization and analysis of the fringe-counting method for interferometric measurement of motion quantities. Measurement **25**(1), 71–87 (1999). https://doi.org/10.1016/S0263-2241(98)00071-2
51. Franke-Arnold, S., Barnett, S.M., Yao, E., Leach, J., Courtial, J., Padgett, M.: Uncertainty principle for angular position and angular momentum. New J. Phys. **6**, 103 (2004). https://doi.org/10.1088/1367-2630/6/1/103
52. Kuts, Y., Scherbak, L., Sokolovska, G.: Methods of processing broadband and narrowband radar signals. In: 2011 Microwaves, Radar and Remote Sensing Symposium, Ukraine, pp. 374–377 (2011). https://doi.org/10.1109/MRRS.2011.6053678

53. Eremenko, V., Zaporozhets, A., Babak, V., Isaienko, V., Babikova, K.: Using hilbert transform in diagnostic of composite materials by impedance method. Period. Polytech. Electr. Eng. Comput. Sci. **64**(4), 334–342 (2020). https://doi.org/10.3311/PPee.15066
54. Kendall, P.C., Bourne, D.E.: Vector Analysis and Cartesian Tensors. CRC Press, Boca Raton (1992)
55. Dey, S., Salehi, M., Ahmadi, J.: Rayleigh distribution revisited via ranked set sampling. METRON **75**, 69–85 (2017). https://doi.org/10.1007/s40300-016-0099-2
56. Dey, T., Dey, S., Kundu, D.: On progressively type-II censored two-parameter rayleigh distribution. Commun. Stat. Simul. Comput. **45**(2), 438–455 (2016). https://doi.org/10.1080/03610918.2013.856921

Chapter 4
Models and Measures for the Diagnosis of Electric Power Equipment

Abstract Examples of building models and measures in monitoring and diagnostics of electric power objects are given. It is known that accuracy and reliability of results of diagnostics of technical objects depends on many factors. And, not in the last place, from qualitatively formed training sets which on different parameters and characteristics correspond to certain technical conditions of investigated objects. The questions connected with the appearance of some physical processes and their mathematical models accompanying the work of electric power equipment units are considered. The results of mathematical models formation of training sets (measures) which correspond to different technical conditions and modes of robots of the investigated electric power equipment are given. The choice of diagnostic spaces, the coordinates of which are the estimates of parameters or functional characteristics of diagnostic signals, is justified. Known in statistics scattering ellipses are used as learning sets, the boundaries of which with a certain probability cover the data of the results of experiments obtained on real power equipment. A scheme and an algorithm implementing it are proposed, which allow to form learning sets that take into account both possible types of defects in individual nodes of electrical power equipment and their modes of operation (rotor speed of the electric machine, temperature of the diagnosed node, various degrees of electrodynamic and mechanical loading, etc.). This approach allows to use monitoring and diagnostics systems within the Smart Grid technology, which provides the possibility to diagnose power equipment in real time. For building training sets that correspond to both certain types of defects and modes of operation of power equipment units, the results of experimental studies obtained at the laboratory stands of the Institute of Electrodynamics of the National Academy of Sciences of Ukraine were used. As an example of practical application of the proposed models, the problem of constructing solving rules at vibrodiagnostics of rolling bearings of electric machines has been considered.

Keywords Electric power equipment · Diagnostic signals · Decision rules · Vibrodiagnostics

V. P. Babak et al., *Models and Measures in Measurements and Monitoring*,
Studies in Systems, Decision and Control 360,
https://doi.org/10.1007/978-3-030-70783-5_4

One of the most important performance characteristics of power equipment is its reliability. The main source of information about its technical condition is the monitoring results based on the measurement of the physical processes that accompany its work. Therefore, measurements of diagnostic signals provide the primary information for further determining the technical condition and, as a result, the reliability of the equipment, that is being diagnosed.

It is known [1–5] that the accuracy and reliability of the results of diagnosing technical objects depends on many factors and not least, on well-formed training sets, according to various parameters and characteristics, correspond to certain technical conditions of the research objects. The process of constructing such training sets (measures) is based on the accuracy of the results of measuring diagnostic signals, and also depends on other factors (the correct selection of the set of possible defects, taking into account the operating modes of the research equipment, the level and nature of interference, etc.).

This chapter discusses the formation of diagnostic signals in operating electric power equipment. The results of constructing mathematical models of the formation of training sets (measures) are presented, corresponding to various technical conditions and operating modes of the investigated electric power equipment (EE). Also considered are examples of constructing decisive rules for the diagnosis and classification of various types of defects in individual nodes of EE.

4.1 Physical Processes for Generating Diagnostic Signals

Solving problems of functional or test diagnostics involves specifying the research object. This is due to the specifics of the functioning and operation of these facilities. Let consider the formation of information signals and their use for the diagnosis of EE.

Basic energy equipment [6–13] is defined as equipment designed to generate (electricity, steam, hot water), convert (chemical energy of fuel burned into thermal energy of steam or hot water), transport or transfer mechanical energy of energy carrier (water, gas, steam compressed air, oxygen, nitrogen, etc.). Some materials and their properties are considered in this works [14–18].

The main power equipment is conventionally divided into:

- thermomechanical: boilers of steam and hot water; waste heat boilers (cooler boilers); steam and gas turbines; auxiliary equipment of boiler plants; air separation units; refrigeration units; gas distribution equipment; centrifugal and piston compressors; blowers (blowers and exhausters), coke blowers; smoke exhausters; pumps; tanks working under pressure (energy); pipelines of water (drinking, hot, technical, circulation, sludge, water reduction), gas (natural, blast furnace, coke, etc.), steam, heating, air, oxygen, nitrogen, hydrogen and other media; channels of rain, technological, sewage water; masts and supports, power lines; fittings (shutoff, regulating), service areas for piping fittings located at a height;

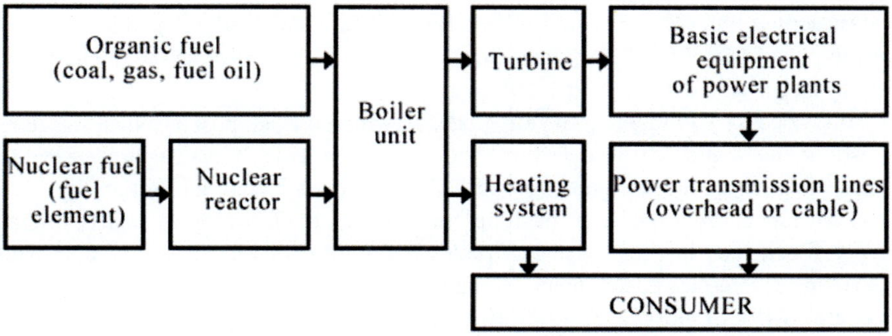

Fig. 4.1 Generalized scheme of the technological process of production, distribution and consumption of electric and thermal energy

- electrical: generators; engines transformers; synchronous compensators; switching equipment; power transmission lines and other network equipment; protection and automation controls equipment; computer facilities.

The generalized scheme of the technological process of production, distribution and consumption of electric and thermal energy is shown in Fig. 4.1.

It will be focused on the issues of measuring physical processes and diagnosing certain types of electrical equipment. At the same time, it will be noted that the physical processes that occur during operation of turbines attached to electric generator are largely similar, since rotating parts are connected.

Various methods for diagnosing EE [1, 3, 6, 7, 9, 19–22] are known and practically used. These methods are determined by the physical process, which generates a diagnostic signal, and is measured to obtain information about the technical condition of the research node.

Technical diagnostics of electric power equipment, as a rule, is carried out by non-destructive testing methods. Non-destructive testing, depending on the physical phenomena, which it based, is divided into types (DSTU 2865-94): magnetic, electric, eddy current, radio wave, thermal, optical, radiation, vibration, acoustic, penetrating substances, etc.

In this chapter, the main sources of diagnostic information are vibrational diagnostic signals that occur at the nodes of the investigated EE. These signals are measured both directly on operating equipment (functional diagnostics) and on equipment that is inoperative. In the latter case, the diagnosis is carried out using special, as a rule, shock effects (test diagnostics). Let dwell on the study of diagnostic signals that occur in the nodes of working electrical machines (EMs).

According to operating experience, the main cause of EM accidents and repairs is insulation failure both between the turns and between the winding and the casing. The second after the winding by the number of failures is the bearing assembly [23–27]. In the vast majority of cases, failures of generators and electric motors occur due to damage to the windings (85–95%). From 2 to 5% EMs fail occur because of bearing damage. The remaining nodes (brush-collector node, ventilation system, etc.) account for 1–2% of failures. Therefore, it will be focused on the diagnostics of rolling bearings and a clad magnetic core, since it is in the grooves of the magnetic core that the EM winding is located.

It is necessary to identify specific nodes in which it is supposed to measure diagnostic signals, and also to determine the most informative physical processes that allow diagnosing the technical state of EM with a certain degree of reliability.

EM, especially low-power machines, are characterized by a wide variety of their types. Each type has its own weak nodes and requires a special approach during developing a model of diagnostic signals and a diagnostic information-measuring system (IMS). In most cases, EM designs are far from complying with the principle of equal reliability of individual nodes. For example, in induction motors, the main source of failure is the stator winding [2, 4, 7]. In DC machines, the weak point is the brush-collector assembly. In low-power EMs, especially high-speed, a significant number of failures are associated with rolling bearings.

A significant impact on the results of diagnostics of EM nodes is exerted by their operating conditions and operating modes. So, for aviation applications [28, 29], the essential requirement is failure-free operation, that is, a low probability of accidental malfunctions; for power machines [30]—a significant resource, durability.

The main components of a typical EM that are usually to be diagnosed include [1, 2, 9, 23, 31–34]: rotor and stator windings, lined rotor and stator magnetic circuits, bearing units with rolling bearings, frame and places of its attachment to the foundation, brush-collector unit, elements of the cooling system (fan cooling, air ducts or channels for circulating coolant, etc.).

For obtaining information about the technical condition of the nodes of electrical equipment, various processes that occur in it during its operation are used, also as their numerical characteristics (parameters). To diagnose specific EM nodes, such physical processes are usually used, as well as their parameters and characteristics:

- *windings*: electrical resistance of inter-turn insulation; winding surface temperature; magnitude of magnetic induction;
- *charged magnetic core*: vibration (movement, speed, acceleration) of the frontal parts; magnitude of magnetic induction; the surface temperature of the magnetic circuit;
- *bearing assembly*: vibration (movement, speed, acceleration) of the bearing shield; bearing shield temperature;
- *frame and places of its attachment to the foundation*: acoustic emission; frame vibration;
- *brush-collector assembly*: transition resistance (conductivity) of the sliding contact; vibration (movement, speed, acceleration) of the brush holder; brush holder temperature;
- *EM cooling system elements*: aerodynamic noise of fan blades; vibration of the surfaces of the circulation channels for moving the EM refrigerant.

So, for almost all EM nodes, their vibration is used as an informational diagnostic signal. The surface temperatures of these units provide additional information regarding their technical condition. Acoustic emission processes are used to diagnose massive EM nodes (frames) that are influenced by significant electrodynamic loads.

Let dwell on the consideration of some issues related to primary research (measurement) of the above processes, which act as diagnostic signals in solving EM diagnostic problems.

Vibration of electrical machines. Among the disturbing effects that cause EM vibrations are:

- electromagnetic forces;
- forces due to the operation of rolling bearings;
- aerodynamic forces;
- forces due to mechanical imbalance of the rotors;
- forces due to the operation of the brush-collector assembly.

The intensity of vibration due to electromagnetic forces is mainly due to the fundamental frequency of the rotating magnetic field [35]. The frequencies of their vibrations are concentrated in the doubled frequency range of the network. The perturbing electromagnetic forces, acting in the air gap of the EM, can be represented in the form of sinusoidal waves of force, each of which has its own frequency and amplitude, that is, the system of forces excites vibrations, has a polyharmonic character and is mainly deterministic.

A particular influence on the nature of the vibrations of the entire EM is caused by the **vibrations of the rolling bearings**. In the case of installing accelerometers on the studied EM units (case, bearing shield, brush-collector assembly, etc.), the realizations of a random process, which is an additive mixture of vibrations generated by various disturbing forces, are measured. Therefore, an essential point in diagnosing a particular EM unit is the choice of accelerometer locations. So, for example, in the case of diagnosis of rolling bearings, accelerometers are usually located on bearing shields, and if the state of pressing of a charged magnetic core is diagnosed—on the stator pressure plate [27–31].

Aerodynamic noise is directly proportional to the increase in the amount of cooling air, which is supplied to the EM to remove generated heat. Emergence of aerodynamic noise related to:

- the operation of the fan, in which the air stream is dissected by the edges of the blades and the fan disk;
- rotation of the rotor, resulting of which is breakdown of the vortices from its surface from the opening of the air stream by the heads of the rotor windings;
- air flow, caused by the breakdown of vortices from stationary obstacles in the ventilation ducts;
- periodic pressure fluctuations in individual sections of the aerodynamic chain, as well as pulsations of the air flow, which leaving the radial ventilation ducts of the rotor and entering the radial ventilation ducts of the stator.

EM vibrations caused by rotor imbalance occur at frequencies that are multiples of the rotational speed, and have mainly deterministic nature. The main causes of rotor imbalance: poor balancing of the rotor assembly, thermal overheating, the presence of internal residual stresses in the rotor barrel from thermomechanical processing of forgings, disconnection of the mounting of the rotor iron on the shaft [23].

Noise and vibrations of the brush-collector apparatus are mainly due to the following technological and structural reasons: the technical condition of the collector associated with the protrusion of individual plates and gaskets, collector beating; physical processes occurring in a sliding contact; poor condition of the brushes and the brush holder (misalignment of the brush, unacceptable clearances between the brush and the brush holder, insufficient pressure on the brush, etc.), the operating mode of the EM [36, 37].

The research results presented in [38] show that the frequency range of vibroacoustic processes caused by the operation of the brush-collector apparatus varies from 1 to 8 kHz, and the spectral composition of these processes does not significantly depend on the rotation frequency.

4.2 Models for the Formation of Training Sets (Measures) for Diagnostics of Electric Power Equipment

The current stage in the development of technologies used in energy production is characterized by a significant increase in information exchange between elements of the energy system at all its hierarchical levels. Electric power systems (ES) of the developed countries of the world are switching to the use of smart grids built on the basis of the Smart Grid concept, which puts forward new requirements for the means of ensuring the reliability of both the power systems as a whole and its components. There is a need for the formation of an integrated multi-level control system that provides a high level of automation and reliability of the entire ES, covers electricity producers, transmission and distribution networks, consumers. An important place is obtaining relevant information on the actual condition of each element of the electric network and the exchange of this information between many participants, that in whole provides an increase in the reliability of the ES.

One of the key tasks in the electric power industry is the development of monitoring methods and technical tools for diagnosing the state of individual ES devices in real time, ensuring the synthesis of such diagnostic information, extracting from the large data array information that is critical for the system as a whole, and transferring it to a high hierarchy level [36, 37].

The goal can be achieved by creating an intelligent distributed multi-level system for monitoring the status and diagnostics of electric power facilities (EPFs).

In the course of the practical use of such systems, which are based on previous training and focused on the use of Smart Grid technologies, questions arise, related to the principles of constructing training sets and the subsequent organization of their use for determining the technical condition of a particular energy object or its unit. It should be noted that the monitoring and diagnostic systems built on Smart Grid technology must work in real time, that is, such a system must quickly find the appropriate set in the bank of training sets, which contains information as about bth the type of defect of this object and its work modes.

Let consider the issues of developing models for the formation of spaces of diagnostic features (measures), which correspond to various technical conditions of EE nodes that operate in different modes (EM rotor speed, temperature of the diagnosed nodes, various degrees of electrodynamic and mechanical load, etc.).

In the case of functional diagnostics in the formation of training sets, the problem of choosing diagnostic spaces arises. In modern mathematics, "… space is a logically mental form (or structure) that serves as an environment where other forms and certain constructions are realized …" [39]. In our case, by space it will be meant the set of any objects that are called points; they can be geometric shapes, functions, the state of the physical system, etc.

As the coordinates of the diagnostic spaces usually choose the parameters or functional characteristics of the diagnostic signals, it turned out to be the most sensitive to changes in the technical condition of the research objects [40, 41]. The dimensionality of the diagnostic space is directly related to the number of coordinates by which the diagnostic signals are measured with using sensors.

Let denote the space of diagnostic features by Ω. In the case of selecting statistical diagnostic models as a combination of diagnostic features, Ω usually includes certain statistical parameters and characteristics that are most informative in identifying the presence and classification of various types of defects in EO nodes.

Parameters and characteristics of diagnostic signals $\xi(\omega)$ can be obtained by considering them as realizations of random processes or fields

$$\xi(\omega) \begin{array}{l} \nearrow \quad \xi(\omega, t), \quad \omega \in \Omega, \quad t \in T \\ \searrow \quad \xi(\omega, r, t), \quad \omega \in \Omega, \quad r(x, y, z), \quad t \in T \end{array} \tag{4.1}$$

Given this view of the measurement of diagnostic signals, the formation of the diagnostic space is shown schematically in Fig. 4.2. At the top of Fig. 4.2 shows Ω—the space of the set of diagnostic features determined by the corresponding statistical parameters and characteristics. As shown by theoretical and experimental

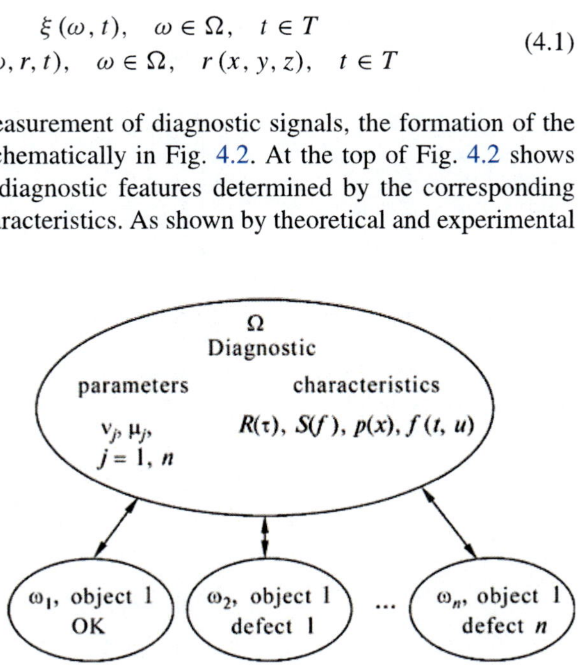

Fig. 4.2 Schematic representation of the formation of the diagnostic space

studies, among these parameters, the most informative are the initial ν_j and central μ_j moments (cumulants), $j = \overline{1, n}$ up to the n-th order, inclusive, and among the characteristics are correlation function $R(\tau)$, spectral power density $S(f)$, probability distribution density $p(x)$, and characteristic function $f(t,\ u)$ [20–22, 42–44].

The defining moment of diagnosing the technical condition of EE nodes is the classification of certain types of defects that can occur in the studied EE nodes. Note that an important point in the formation of training sets of such defects is the choice of a specific diagnostic object. Information content of certain diagnostic signs depends from the choice of an object (or a node included in its composition Fig. 4.2 at the bottom schematically shows the principle of constructing subspaces of sets of diagnostic signs Ω corresponding to the working condition of an object or the presence of certain types of defects (defect 1, defect 2, …, defect n) and are part of the space Ω

$$\omega_1, \omega_2, \ldots, \omega_n \in \Omega. \tag{4.2}$$

The given set of subspaces is constructed separately for each of the objects of diagnosis. Conditionally in Fig. 4.2, these subspaces are constructed for object 1, which can be selected as any object of EE, for example, powerful rotary EMs, transformers, auxiliary engines and the like.

The next step in creating training sets is a conditional breakdown of the selected diagnostic object into separate components (nodes). For example, the main elements for rotary EMs are a winding, a brush-collector assembly, rolling bearings (or bearings for powerful electric machines), a cooling assembly, and the like. Figure 4.3 shows an example of the formation of such sets $\omega_1^1, \omega_1^2, \omega_1^3$ for individual EM units, namely, windings, rolling bearings and brush-collector units, which account for about 95% of failures [45–47].

For a more detailed account of the type of defect for each of these nodes, it is possible to further divide of subspaces $\omega_1^1, \omega_1^2, \omega_1^3$ into subspaces that take into account the most typical defects for the research node. Figure 4.4 shows an example of the formation of training sets (measures) for typical defects of EM rolling bearings. Such defects, as a rule, include skew of the outer ring, lack of lubrication, defect (pitting) of the track of the inner or outer ring, breakdown of the separator [45–47]. The specified defects in Fig. 4.4 correspond to the subspaces $\omega_{11}^2, \omega_{12}^2, \omega_{13}^2, \omega_{14}^2$.

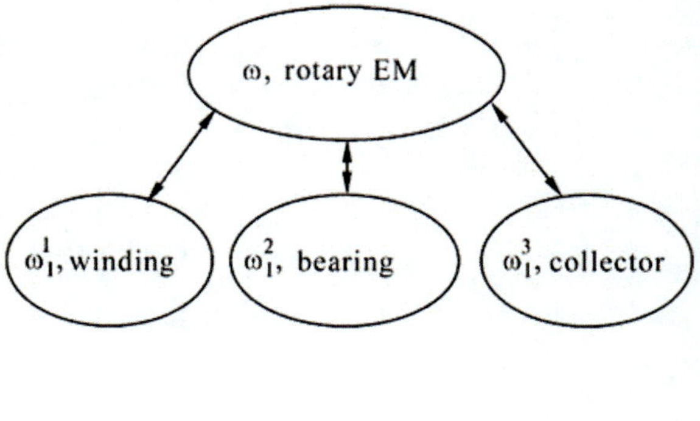

Fig. 4.3 Example of the formation of measures for individual rotary EM units

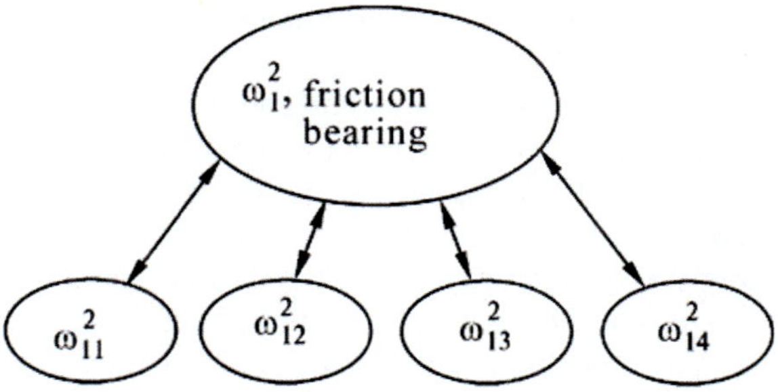

Fig. 4.4 Example of the formation of measures (training sets) for typical defects of EM rolling bearings

Similarly, the formation of training sets for other nodes of rotary EM is carried out. The number of aggregates that is formed for each of the EM nodes is determined by the types of defects for which it is necessary to diagnose a specific EM node.

The presented approach to the formation of training sets (measures) provides for their practical application in monitoring and diagnostic systems of modules built using Smart Grid technology. Such systems, in the presence of pre-formed training sets, provide diagnostic information in real time. And this, in turn, depending on the technical condition of the EE or its components, the presence and degree of catastrophic of the identified defects, makes it possible to reasonably make a decision on the further operation of the research objects.

An important point in the formation of a databank of training sets (measures) for diagnosing EE is taking into account its operating modes (rotation speed of the EM rotor, temperature of the nodes that are being diagnosed, various degrees of electrodynamic and mechanical load, etc.). This is due to the fact that in the case of the operation of the EE monitoring and diagnostics system within the framework of the Smart Grid concept, it is expected to obtain real-time information on the state of this equipment in real time, which, in turn, requires taking into account the operation mode of the EE for a suitable moment (or period) of time. This problem can be solved by creating a dynamic bank of training sets, which are diagnostic spaces built for certain technical conditions of the studied EE nodes and for certain modes of their operation.

If in the existing training sets for a certain object, both the set of possible defects and their operating modes are simultaneously taken into account, then the established set of diagnostic spaces can be represented in matrix form:

$$\Omega = \begin{Vmatrix} \omega_{11} & \omega_{12} & \dots & \omega_{1n} \\ \omega_{21} & \omega_{22} & \dots & \omega_{2n} \\ \dots & \dots & \dots & \dots \\ \omega_{k1} & \omega_{k2} & \dots & \omega_{kn} \end{Vmatrix}. \tag{4.3}$$

In the above form (4.3), the subspaces located in rows correspond to the same modes in which the investigated EE is operated, and each column corresponds to a certain technical condition of this node. In the aggregate of subsets of ω_{nk}, index

$j = \overline{1,k}$ denotes a certain operating mode of the EE, and index $p = \overline{1,n}$ denotes a certain type of defect.

Since the monitoring and diagnostics systems, that operating within the framework of the Smart Grid concept, carry out two-way information exchange between the diagnostic object and the operator, information about the technical condition of the equipment and its operation mode are immediately received to the last. Accordingly, from the bank of training units of the system, the appropriate standard is selected (diagnostic subspace ω_{kn}) for subsequent comparison and decision-making on the technical condition of the research unit. An illustrative scheme for constructing diagnostic spaces, in accordance with the matrix (4.3), has a rather complex form and is not considered in this book.

As already noted, as part of the application of the Smart Grid concept, it is supposed to carry out maintenance and repair of EE according to its actual state. It is possible to realize this by having pre-formed training sets with both possible types of defects in EE nodes and their operating modes. That is, for the prompt selection of a specific standard, the monitoring system needs to have almost instantly information on the technical condition and operation modes of the investigated object of EE. This can be realized on the basis of two-way exchange of information between the node that is being diagnosed and the central diagnostic system [36, 37, 48, 49], which is provided by Smart Grid technology.

4.3 Formation of Diagnostic Spaces Based on the Measurement of Information Signals

Probabilistic models of diagnostic signals, diagnostic signs based on the results of their analysis, real electric power facilities and experimental test benches, as well as laboratory samples of various components of electrical equipment are necessary components, which allows to directly move to experiments on the diagnosis of such objects [50, 51]. An experimental study of the proposed diagnostic features is carried out by the example of vibration diagnostics of EM bearing units, the state of pressing an iron package by a clad magnetic core, and the vibration characteristics of the main shaft of a wind electric unit (WEU) operating in various load conditions.

Briefly describe the test setups and objects on which experimental studies were carried out.

In experiments with vibration diagnostics of rolling bearings on experimental setup and in studying vibrations of the main shaft of WEU, a laboratory sample [52] of IMS of functional vibration diagnostics was used, and for diagnosing the state of pressing by a clad magnetic core—IMS of test vibration-shock diagnostics [52].

Installation for vibration testing of rolling bearings of EM. For further research, it is advisable to analyze the vibrations of the rolling bearing, "deciding" its vibration from the vibrations of the electric drive. To implement this idea, an

Fig. 4.5 Installation for vibration testing of rolling bearings

experimental installation for vibration testing of single rolling bearings was developed and manufactured at the IED NAS of Ukraine. A general view of the installation is shown in Fig. 4.5. Its main purpose was to experimentally check diagnostic signs in the presence of typical defects such as skew, lack of lubrication; investigation of damage to the outer or inner ring of the bearing due to metal chipping (so-called pitting).

The installation structurally consists of three main units: an electric drive, a massive shaft, a unit of mounting and measuring the vibrations of the test bearing. The rotation of the test bearing installed in the unit of mounting and measuring the vibrations (shown on the left side of Fig. 4.5) is provided by a P-51 type electric DC motor through a massive shaft. This 11 kW motor provides rotation of the bearing under test at any speed in the range from 10 to 1500 rpm. The use of a special coupling with rubber fingers allows to minimize vibration caused by the operation of the electric drive. In addition, reducing the vibration of the shaft of the experimental setup contributes to fixing it in bearings with sliding bearings made of fluoroplastic (teflon), as well as placing the shaft and the unit of mounting and measuring the vibrations of the test bearing on a massive plate.

The main purpose of the unit of mounting and measuring the bearing vibrations is the ability to artificially reproduce the main defects of the bearing and the placement of primary vibration-converting equipment (accelerometers). For measuring the vibration accelerations of the test bearing, ABC-017 type accelerometer was used, which made it possible to measure bearing vibrations in the frequency range of 20 Hz–30 kHz. It is mounted in the radial direction with respect to the tested bearing. A detailed description of the bearing test facility is given in [52].

Installation for checking the degree of pressing of the plates of a charged magnetic core. As a research object, a part of the clad magnetic core of a low-power transformer installed on a special test bench was used (Fig. 4.6).

The magnetic core plates were pressed using studs, the tightening of which was carried out with a torque wrench with a moment of 30 N·m. The vibration acceleration signal was measured using an accelerometer mounted on the research magnetic core. The used ABC 17 type accelerometer made it possible to measure a vibrational wave

Fig. 4.6 Model of a clad magnetic core

in the frequency band of 20 Hz–30 kHz, which is violated by an impact hammer (Fig. 4.7) in the body of a charged magnetic core.

The magnetic core plates were pressed using studs, the tightening of which was carried out with a torque wrench with a moment of 30 N·m. The vibration acceleration signal was measured using an accelerometer mounted on the research magnetic core. The used ABC 17 type accelerometer made it possible to measure a vibrational wave

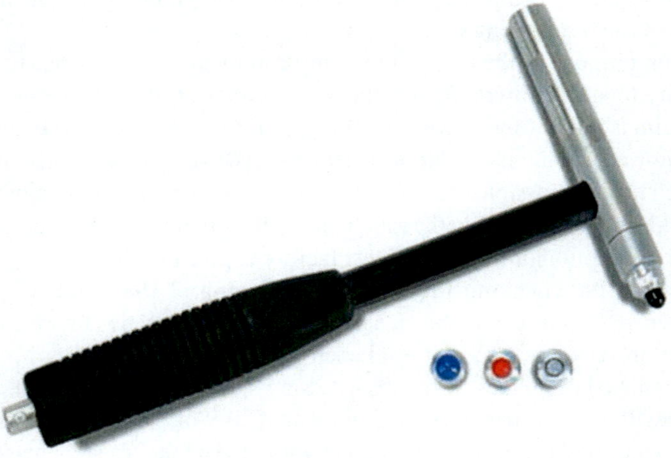

Fig. 4.7 Striking hammer

in the frequency band of 20 Hz–30 kHz, which is violated by an impact hammer (Fig. 4.7) in the body of a charged magnetic core.

The experiment for diagnosing the state of pressing the plates of the magnetic core was carried out as follows. Using the clamping nuts mounted on the studs, the package of the magnetic core plates was compressed with a force of 30 N·m. Then, the laboratory sample of the shock diagnostics IMS was turned on and tuned [52]. After tuning the IMS in the perpendicular direction to the plane of the plates, a blow was carried out, which was recorded by an accelerometer mounted on a hammer. The signal from the accelerometer was fed to the trigger circuit of the shock diagnostics IMS and the vibration signal was measured in the body of the magnetic core by the accelerometer, which was installed on the plates of the magnetic core.

After that, the nuts of the studs with which the plates were pulled together by the magnetic core plates were completely released, and the experiment was repeated in the indicated sequence.

WEU-20. Experimental studies on the measurement of the vibration signal were carried out on the main shaft of the wind turbine WEU-20 manufactured by Karbon LLC [53]. The main technical characteristics of the WEU-20: rated power—20 kW; rated wind speed—9 m/s; rated voltage—380/220 V; output voltage frequency—50 Hz; generator rotation speed—100 rpm; maximum working wind speed—25 m/s; maximum permissible wind speed—50 m/s; pulling speed—3 m/s; wind generator weight—$\leq$800 kg.

The sensor block was located on a directly visible place of the main wind turbine shaft. The measurements were carried out after the establishment of a stable mode of rotation of the wind turbine rotor. The information receiving and processing unit (IRPU) [53] was located at ground level at a distance of 40 m from the initial position of the sensor unit. Using the IRPU software module, the following is performed: control of the sensor unit; statistical processing of measured signals; training the system to determine a specific type of defect; constructing decision rules and making decisions on determining the technical condition of a diagnosed node. The experiments consisted in measuring vibration accelerations on the main shaft of the WEU at different levels of consumed electricity (power).

Let proceed directly to the question of the formation of training sets in accordance with (4.2) and (4.3). Recall that, according to expression (4.2), models are formed according to the type of defects, and it is with this method that training sets were formed, which were constructed according to the processing of vibrations of rolling bearings that were tested on the installation (Fig. 4.5). Since the details of the preparation and conduct of these experiments are described in detail in [53], it will be considering only the selected diagnostic space and measures formed according to different conditions of the rolling bearings.

It has been experimentally established that the most informative diagnostic features are the asymmetry coefficients k and kurtosis γ of the studied vibrations associated with the third and fourth moments of the distribution of random variables by known relations [54–56].

Table 4.1 Asymmetry and kurtosis coefficients

Bearing test conditions	Number of implementations selected for analysis	Interval estimates of the average value of the coefficients	
		k	γ
OK	60	0.12 ± 0.04	0.17 ± 0.06
No lubrication	89	0.114 ± 0.011	0.66 ± 0.12
Skew	75	0.105 ± 0.007	1.07 ± 0.09
Inner ring defect	85	0.087 ± 0.005	1.22 ± 0.11

Table 4.1 shows the observed values of the estimates of the asymmetry coefficients k and kurtosis γ of vibration of the bearing 309 ES2, individual copies of which had different technical conditions and were subsequently installed in the test setup.

Based on these data, the Pearson diagram [57] with the coordinates of (β_1, β_2) was chosen as a diagnostic space. Figure 4.8 shows, as an example, typical training sets in the form of scattering ellipses formed in a two-dimensional diagnostic space (β_1, β_2).

It is more convenient to diagnose defects with training sets not in a rectangular coordinate system (β_1, β_2), but in the polar system (ρ, φ), the transition to which is carried out by the relations

$$\rho = \sqrt{\beta_1^2 + (\beta_2 - 1)^2}, \tag{4.4}$$

$$\phi = arctg\frac{\beta_2 + 3}{\beta_1 + 4}. \tag{4.5}$$

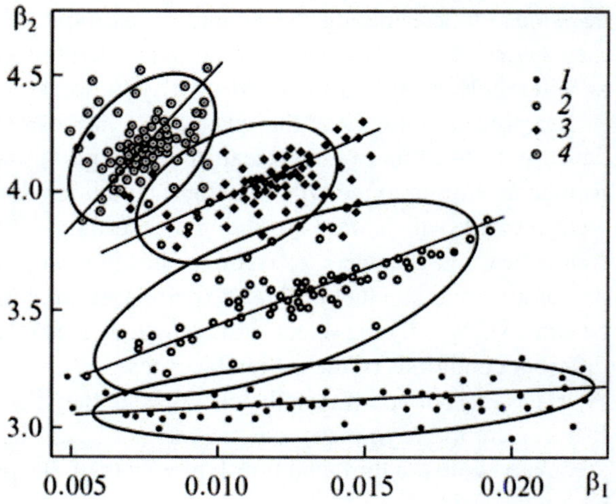

Fig. 4.8 Training sets in a two-dimensional diagnostic space (β_1, β_2): 1—OK bearing; 2—no lubrication; 3—skew $14' \pm 2.5'$; 4—defect of the inner ring

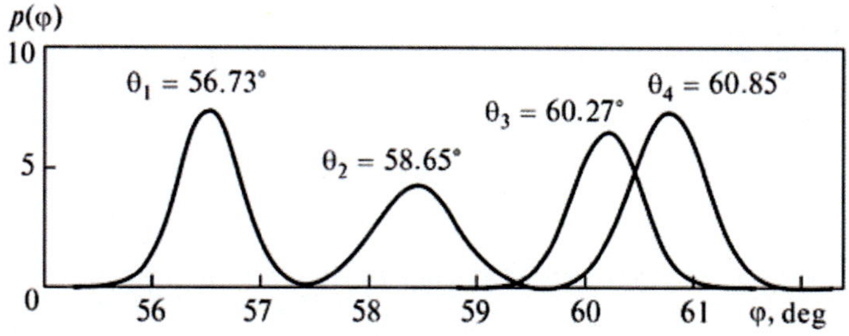

Fig. 4.9 Curves that smooth histograms plotted on points of scattering ellipses

From a comparison of Fig. 4.8 and Pearson diagrams [57], it can be found that the distribution parameter determination domains have the form of sectors whose vertices begin at the points with coordinates $\beta_1 = 0$, $\beta_2 = 1$ and $\beta_1 = 0$, $\beta_2 = 3$. This determines the transition, according to (4.4) and (4.5), to the new coordinate system. In this system, training sets corresponding to various technical conditions of bearings turn out to be elongated along the radius ρ and compressed along the angle φ. Therefore, the distribution density cross section in the plane of the angle φ for a fixed ρ is more sensitive to changes in the type of curve and its parameters than in any other plane.

The training sets presented in Fig. 4.8 can be transformed into a one-dimensional diagnostic space and they are histograms constructed from points that form a certain scattering ellipse, or curves smoothing these histograms (Fig. 4.9).

The training sets for monitoring the state of pressing the plates with a charged magnetic core were obtained using the model of a charged magnetic core (Fig. 4.6) and a hammer (Fig. 4.7).

The obtained experimental data are the results of vibration acceleration measuring on the surface of the investigated charge magnetic core. These measurements and further spectral processing of the obtained diagnostic signal were carried out using the IMS of vibration-shock diagnostics described in detail in [58–61].

The formation of training sets (standards) corresponding to certain degrees of compaction of the plates of the magnetic core was carried out by 2 cases:

- dependence only on the degree of pressing of the plates (model (2)), i.e., dependence on the defect;
- dependence on the degree of pressing and on the mode of operation (voltage supplied to the winding of the magnetic core, model (3)).

In the *first* case, the experiments consisted in measuring vibration signals excited by a shock hammer (Fig. 4.7) in an array of a charged magnetic core (Fig. 4.6), which was compressed with a force of 30 N·m and completely removed compression. After measurement, these implementations of vibrations were processed on the IMS of shock diagnostics using spectral analysis. According to the results of this processing, spectrograms $S(f)$ of the vibrations of the magnetic core were obtained, the study of which convincingly proved that the most effective diagnostic sign of the degree of

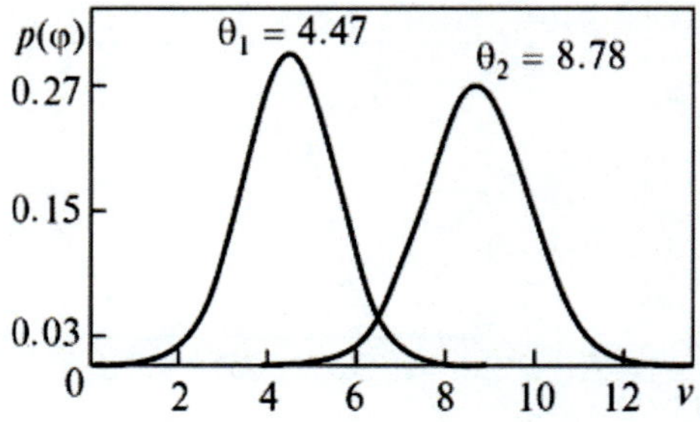

Fig. 4.10 Curves that smooth the histograms of the averaged number of υ maxima of the spectrogram *S*(*f*)

its pressing is the number of resonance maxima υ recorded at a certain amplitude level of the spectrogram *S*(*f*). These experiments are described in detail in [52, 53]. It is only necessary to recall that it was from the results of processing more than 200 spectrograms of vibration *S*(*f*) that the diagnostic spaces were selected, in which training sets were constructed, that made it possible to diagnose the degree of compaction of the magnetic core. Each such training set was a histogram of the distribution of resonance maxima for different states of the magnetic core. But the construction of decision rules for identifying the state of compaction of the magnetic circuit is more convenient to perform not by histograms, but by smoothing distribution curves. These curves are shown in Fig. 4.10.

In the *second* case, the charged magnetic core was tested in a dynamic mode, that is, voltage of various sizes was applied to its winding (Fig. 4.6), and the magnetic core plates were compressed with varying degrees of force. Subsequently, in cases of a combination of different test conditions, a shock wave was excited with a hammer in the body of the magnetic core, and measurements and statistical processing of the diagnostic signal were carried out with the help of the IMS of shock diagnosis.

As in the case considered above, the measurements were carried out for two extreme states of magnetic core pressing: "fluffiness" and compression using a torque wrench with a force of 30 N·m. Changing the operating mode was carried out by regulating the voltage on the magnetic coil winding. The voltage *U* and current *I* through the winding were measured. The vibration acceleration signals of a charged magnetic core were measured using an ABC-017 accelerometer, after which quantitative estimates of the parameters σ, k and γ (Table 4.2) were determined using the appropriate IMS diagnostic software, and spectrograms *S*(*f*) were constructed, some of which are shown in Fig. 4.11.

From the experimental values given in Table 4.2, it can be seen that with the fluffy magnetic core, the values of all the studied parameters change, and the standard deviation σ increases systematically, but there are no certain regular changes in the quantitative estimates of the asymmetry coefficients k and kurtosis γ. So, the parameters k and γ are not informative in the case of diagnosing this type of defect, and the parameter σ is informative.

Along with the dependence of the standard deviation σ on the state of compaction of the magnetic core, there was a clear pattern of growth in its value with increasing current through the winding. In this case, there may be cases when, for different

Table 4.2 Quantitative estimates of σ, k, γ parameters

Load level			State of pressing the magnetic core					
No.	Voltage U, B	Current I, A	Compression			Fluffiness		
			σ	k	γ	σ	k	γ
1	50	3.3	0.161×10^{-3}	0.257	−0.119	0.322×10^{-3}	0.023	−1.000
2	100	6.0	0.240×10^{-3}	0.108	−0.036	1.045×10^{-3}	−0.340	−1.100
3	150	9.1	0.493×10^{-3}	0.131	−0.131	4.322×10^{-3}	−0.028	−0.773

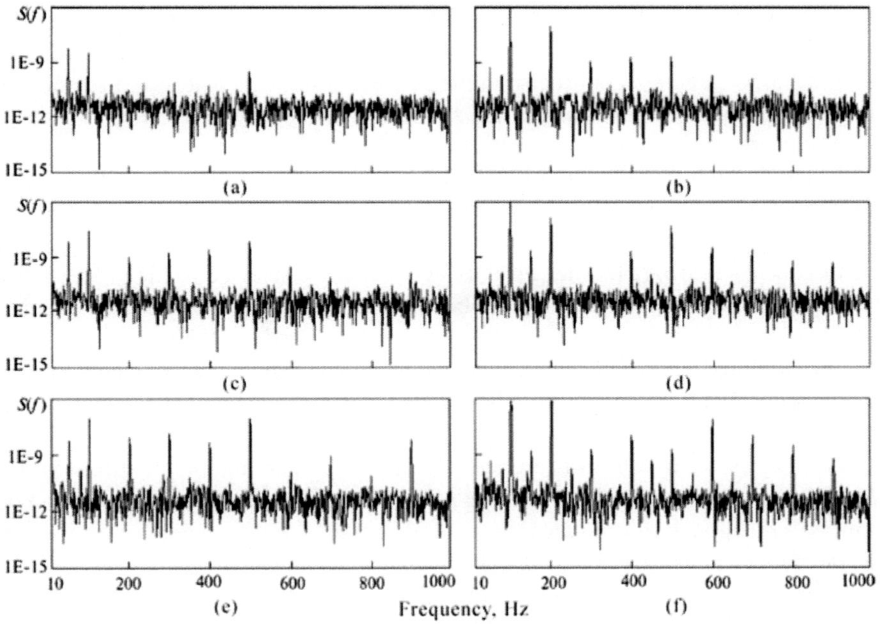

Fig. 4.11 Spectrograms of vibrations $S(f)$ of the magnetic circuit, depending on the degree of load of the winding: **a** voltage 50 V, current 3.5 A, compressed magnetic core; **b** voltage 50 V, current 3.5 A, fluffy magnetic core, **c** voltage 100 V, current 6 A, compressed magnetic core; **d** voltage 100 V, current 6 A, fluffy magnetic core, **e** voltage 150 V, 9.1 A current, compressed magnetic core, **f** 150 V voltage, 9.1 A current, fluffy magnetic core

pressing states, close values of σ are observed, but at different transformer loads. Therefore, it is impossible to unambiguously determine the state of compaction of the transformer magnetic core, using only information about the standard deviation σ of the vibration acceleration signal. It is necessary to additionally take into account the transformer loads by measuring electrical parameters—current or voltage.

The results of experimental studies on the degree of compaction of the magnetic core and the current mode of the winding of the magnetic core proved the informativeness (informativity) of the spectral analysis of the studied vibrations. An analysis of the obtained spectrograms (Fig. 4.11) showed that in the case of fluffy magnetic core, additional spectral components with a high amplitude appear.

For example, at low load (experiment No. 1—the current through the winding is about 3.5 A; graphs 1 and 2 in Fig. 4.11), there are two clearly defined frequency components in the vibrations of the compressed magnetic core in the frequency range 50 and 100 Hz, and also less intense component with a frequency of about 500 Hz. During fluffing of magnetic core, the intensity of these components changes and quite intense peaks at frequencies of 200, 300, 400 Hz and a number of less intense appear in the spectrogram. At a higher load (experiment No. 3—the current through the winding is about 9.1 A; graphs 5 and 6 in Fig. 4.11) for a compressed magnetic core, the number of significant frequency components of the spectrogram is 7, and with a fluffy magnetic core their number increases to 9. In other experiments the number of spectral peaks changed in a similar way.

Therefore, the number of significant maxima of the spectrograms $S(f)$ can be considered an informative sign for diagnosing the degree of compaction of the transformer magnetic core, but only with taking into account the degree of its load. At the stage of training the diagnostic system on the basis of studies of a number of obtained spectrograms, the level is automatically selected, which determines which frequency components should be considered significant. Next, the number of frequency maxima, that exceed this level, is calculated, and this number is used to form training sets within the same operating mode of the research object. System training is carried out sequentially for various modes of equipment loading, which is diagnosed.

According to the results of further research of a certain number of spectrograms $S(f)$ for each load stage of the winding of the magnetic circuit, training sets can be formed to diagnose and monitor the state of pressing of the magnetic core, which works with different degrees of load.

During the experimental studies, the information content of the vibration RMS σ of acceleration signal, as well as the number of significant frequency maxima in the spectrogram of this signal, as diagnostic signs for determining the degree of compaction of the transformer magnetic core, were confirmed, as well as the need to take into account the mode of functional diagnostics of operating parameters, such as voltage U or current I.

Experimental research of WEU-20. In these experiments, a conditionally operable WEU-20 wind turbine was used. The main purpose of the experiments was to prove the need to take into account the WEU operating modes during forming training sets, and later during diagnosing certain defects in its nodes.

Realizations of the information signal were obtained by measuring them on the reachable part of the WEU shaft with using a sensor. A wireless channel was used to transmit the measured data to the receiving blocks of the vibrodiagnostics IMS, which made it possible to measure diagnostic signals directly on the moving parts of the WEU-20.

Table 4.3 Experimental results

No.	Load, kW	Estimates of statistical parameters		
		$\sigma \times 10^{-2}$	$k \times 10^{-2}$	γ
1	0.7	0.29 ± 0.04	0.10 ± 0.04	1.238 ± 0.023
2	5.0	0.57 ± 0.07	0.099 ± 0.021	1.74 ± 0.07
3	9.0	0.64 ± 0.05	0.12 ± 0.03	1.50 ± 0.06

Table 4.4 Bearing vibration parameters

Parameter	Test bearing number			
	1	2	3	4
OK bearing				
φ_1	55.28°	56.92°	57.10°	56.03°
φ_2	56.53°	57.89°	55.95°	58.91°
$\frac{1}{2}\sum_{j=1}^{2}\phi_j$	55.91°	5741°	56.53°	57.02°
No lubrication (no grease)				
φ_1	55.99°	60.58°	59.70°	60.04°
φ_2	59.26°	56.10°	57.35°	60.36°
$\frac{1}{2}\sum_{j=1}^{2}\phi_j$	57.63°	58.34°	58.53°	60.20°

The WEU tests were carried out in 3 load modes (power consumption): 700, 5 and 9 kW. The recorded samples in the vibration diagnostics IMS were processed using a histogram analysis program to obtain estimates of the standard deviation σ, the asymmetry coefficients k and kurtosis γ. The results of these experiments are given in Table 4.3.

The data given in Table 4.4 were obtained by processing 150 implementations measured on the main shaft of WEU-20 for each load stage of the wind turbine.

According to the results of experimental studies, the information content of the σ RMS of the vibration acceleration signal is confirmed. So, in accordance with the data in Table 4.3, there is a steady increase in the σ values together with an increase in the degree of load at the output of the WEU. That is, the parameter σ can be used as a diagnostic feature and should be taken into account during forming training sets for further vibration diagnostics of WEU nodes.

4.4 Rules for Determining the Technical Condition of Electric Power Facilities

The next and final step in the diagnostics of EE is the construction of decisive rules for the diagnosis and classification of certain types of defects in the nodes of EE. Let briefly dwell on these issues.

For constructing the decisive rules for diagnosing specific defects in bearings, it will be using training sets formed according to the results of a histogram analysis of the studied vibrations in the diagnostic space φ in the form of smoothing histogram curves (Fig. 4.9).

Distributions of φ ratings corresponding to various technical conditions of the test bearing are shifted relative to each other and have different mathematical expectations: serviceable bearing—$\varphi = 56.73°$; lack of lubrication—$\varphi = 58.65°$; skew 14′ ± 2.5′—$\varphi = 60.27°$; defect of the inner ring—$\varphi = 60.83°$. The obtained curves refer to the XIII, IV or VII types of curves from the Pearson system, that is, they can be approximated by the normal distribution law [34, 62, 63].

For building decision rules for diagnosing these defects in the bearing, it can be used the classical dual-alternative procedure for testing statistical hypotheses according to Neumann-Pearson, which is described by relatively simple mathematical relations for a normal distribution.

For diagnosing the technical condition of bearings and classify defects in them, two alternatives can be built by solving the rules. They are reduced to testing the following hypotheses:

- H_0: $\theta_1 = 56.73°$ (OK) ↔ H_1: $\theta_2 = 58.65°$ (no lubrication);
- H_0: $\theta_1 = 56.73°$ (OK) ↔ H_1: $\theta_3 = 60.27°$ (skew $14' \pm 2.5'$);
- H_0: $\theta_1 = 56.73°$ (OK) ↔ H_1: $\theta_4 = 60.83°$ (ring defect);
- H_0: $\theta_2 = 58.65°$ (no lubrication) ↔ H_1: $\theta_3 = 60.27°$ (skew $14' \pm 2.5'$);
- H_0: $\theta_3 = 60.27°$ (skew) ↔ H_1: $\theta_4 = 60.83°$ (ring defect).

Based on the hypotheses formulated, it was considered a specific example of diagnosing the technical condition of bearings 309 ES2 installed on a test installation. It will be creating a decisive rule for diagnosing the lack of lubrication in the bearing. Let begin the solution of the problem through the planning of the experiment.

Diagnosing the lack of lubrication in the bearing 309 ES2 reduces to testing the main hypothesis $H_0 = 56.73°$ (the bearing is OK) against a simple alternative: $H_1 = 58.65°$ (no lubrication). Having accepted the probability of errors of the first and second kind, $\alpha = 0.05$ and $\beta = 0.01$ and finding for them the values (according to the tables [52]) of the standard normal distribution of the corresponding quantiles $u_{1-\alpha} = -1.645$ and $u_{1-\beta} = -2.326$, it will be determining the required number of observations and the threshold value, given the dispersion is $\sigma^2 = 0.3933$:

$$n = \frac{\left(u_\alpha + u_\beta\right)^2 \sigma^2}{(\theta_2 - \theta_1)^2} = \frac{6.202}{3.6864} \approx 2,$$

$$C = \frac{(\theta_2 - \theta_1)(u_\alpha - u_\beta)}{2(u_\alpha + u_\beta)} + \frac{(\theta_1 + \theta_2)}{2} = 57.52^\circ$$

Having obtained all the necessary data for constructing the decisive rule as a result of the experimental design, let us diagnose the absence of lubrication in specific samples of bearings of the 309 ES2 type installed on the test setup.

For checking the technical condition of the bearings, 4 rolling bearings were randomly selected, which were subsequently tested at the installation. In the process of testing, 2 samples containing the vibrations of the testing bearings in each of the experiments were measured and processed with a histogram analysis program with using vibration diagnostics IMS. Estimates of the values of the parameter φ of the vibration of the installed bearings are given in Table 4.4.

According to [52], the ratio of threshold inequality for this case has the form

$$\frac{1}{2}\sum_{j=1}^{2} \phi_j \leq C.$$

The left side of the given threshold inequality, after substituting the data of Table 4.4 for the 1-4th samples of the tested bearings takes values equal to 55.91°; 57.41°; 56.53°; 57.02°. The right-hand side of the indicated inequality after substituting the data obtained during the planning of the experiment into it is 57.52°. Thus, based on the obtained results, hypothesis $H0$ is accepted—all 4 copies of bearings 309 ES2, which were tested at the installation, are OK.

After that, lubrication was removed from all tested bearings. Then, similar experiments were carried out to measure and process the vibrations of bearings operating without lubrication for all 4 of the examined specimens of bearings. Estimates of the vibration parameter of the tested bearings are given in the lower part of the Table 4.4.

After the substitution of the data from Table 4.4 in the given threshold inequality, its left part for the 1-4th tested bearings takes the values of 57.63°; 58.34°; 58.53°; 60.20°, that is, for all 4 samples of bearings 309 ES2 the left side of the threshold inequality is larger than the right. So, based on the results obtained, a hypothesis is accepted—there is no lubrication in the bearings. Similarly, according to the hypotheses formulated above, decoupling rules are constructed for diagnosing other types of defects.

Let dwell on the issues of choosing diagnostic spaces and constructing training sets, which correspond to varying degrees of compaction of a charged magnetic core (Fig. 4.6). In paragraph 3, diagnostic signs were substantiated, statistical estimates of which can be obtained on the basis of statistical spectral analysis. The energy spectrum, which is determined as a result of such an analysis, characterizes the frequency distribution of the shock vibrational wave, which is violated in the research magnetic core by a special shock hammer (Fig. 4.7).

Based on the consideration of spectrograms of shock vibrational waves propagating through the body of a charged magnetic core [64, 65], it can be noted that the most informative diagnostic features that allow diagnosing the degree of compaction of a charged magnetic core are: amplitudes A_f of the main maxima of the spectrogram $S(f)$; frequency f_n of the main maxima of the spectrogram $S(f)$; the number v_j of the main maxima of the spectrogram $S(f)$ in a fixed frequency band.

For diagnosing the degree of compaction of a lined magnetic core, let construct a two-alternative decision rule, which reduces to testing hypotheses: H_0: $\theta_1 = 4.47$ (a magnetic core pressed with a force of 30 N·m) versus H_1: $\theta_2 = 8.78$ (there is no compressive force in the magnetic core).

Let us verify the main hypothesis. Having taken the value of errors of the first $\alpha = 0.05$ and second kind $\beta = 0.01$ and finding for them from the tables [53] of the standard normal distribution the corresponding quantiles $u_{1-\alpha} = -1.645$ and $u_{1-\beta} = -2.326$, it was determined the required number of observations N and the threshold C level using the expressions given in [66] during describing the rules for testing statistical hypotheses using the Neumann-Pearson method.

The required number of observations, given the fact that $u_{1-\alpha} = -u_\alpha, u_{1-\beta} = -u_\beta$ is

$$N = \frac{\left(u_\alpha + u_\beta\right)^2 \sigma^2}{(\theta_2 - \theta_1)^2} = \frac{15.769 \cdot 4.36}{(4.31)^2} = 3.70 \approx 4,$$

$$C = \frac{(\theta_2 - \theta_1)\left(u_\alpha - u_\beta\right)}{2\left(u_\alpha + u_\beta\right)} + \frac{(\theta_1 + \theta_2)}{2} = -\frac{2.935}{7.942} + 6.625 = 6.255,$$

where σ is the largest dispersion value among the curves shown in Fig. 4.10. For this task, $\sigma = 2.09$.

Having obtained, as a result of the experiment planning, all the necessary data for constructing the decision rule, it will be diagnosed the degree of compaction by laminating magnetic cores similar to those shown in Fig. 4.6. For checking the degree of pressing, 5 such magnetic cores were randomly selected, sequentially mounted and tested at the stand. In the process of testing with the help of a laboratory sample of shock diagnostics IMS, measured in 4 samples containing vibrations of a lined magnetic core. The number υ of the maxima of the spectrogram $S(f)$ of these vibrations at the level of 0.5 amplitude is given in Table 4.5.

The left side of the above threshold inequality after substituting the data from Table 4.5 for the 1-5th samples of the tested magnetic cores takes the values of 1.75; 4,5; 3.75; 3, 4.75. The right side of the threshold inequality after substituting the data obtained during the planning of the experiment into it is 6.255. Thus, based on the obtained results, the hypothesis $H0$ is accepted—the research magnetic cores are compressed with a force of 30 N·m.

After that, the compressive force was removed from all the tested magnetic cores and similar experiments were conducted to measure and process vibrations caused by shock, across all 5 research magnetic cores. Estimates of the parameter υ of the vibrations of the tested magnetic cores are given in Table 4.5. After substituting the data from Table 4.5 into the expression for the threshold roughness, its left side for the 1–5th research magnetic cores takes the value 6.75; 9; 7, 10, 8.25, that is, for

Table 4.5 Maximums of the spectrogram $S(f)$ of vibrations

Parameter	Number of the research magnetic core				
	1	2	3	4	5
Magnetic core, compressed with a force of 30 N·m					
υ_1	3	5	4	3	5
υ_2	1	3	4	4	6
υ_3	1	5	3	3	4
υ_4	2	5	4	2	4
$\frac{1}{4}\sum_{j=1}^{4}\upsilon_j$	1.75	4.5	3.75	3	4.75
Magnetic core without pressure					
υ_1	8	11	6	9	7
υ_2	7	8	7	10	9
υ_3	7	9	8	10	8
υ_4	5	8	7	11	9
$\frac{1}{4}\sum_{j=1}^{4}\upsilon_j$	6.75	9	7	10	8.25

all 5 samples, the left side of the threshold inequality is larger than the right, so the hypothesis $H1$ is accepted—there is no pulling force in the magnetic cores.

So, from the considered example it follows that the number v_j of the maxima of the vibrations spectrogram of the charged magnetic core can be used as a diagnostic feature to determine the degree of its pressing.

Similarly, decoupling rules for diagnosing the state of compaction of the magnetic core during its operation in dynamic mode, that is, at the current load of its winding (Fig. 4.6), can be created. However, in this case, during constructing decision rules, it is necessary to use training sets corresponding to a certain degree of load, which, for example, are given in Table 4.3.

The relative simplicity of the application of the decision rules discussed above in combination with the laboratory sample of the shock diagnostics IMS makes the considered method of shock diagnostics of clad magnetic cores suitable for practical use.

Summing up the content of the material presented in this chapter, it can be stated that the proposed approach to the formation of training sets (standards) allows to systematize the use of these training sets during working as part of diagnostic IMS using Smart Grid technologies. Depending on the operating conditions of the research equipment, thanks to formalized rules for working with vectors (model (4.2)) or matrices (model (4.3)), the IMS can quickly select the necessary standard, without waiting until the working equipment reaches certain conditions for which the bank data is an appropriate learning set.

References

1. Gertsyk, S.M., Gyzhko, Y.I., Zvarich, V.M., Myslovych, M.V., Ostapchuk, L.B., Sysak, R.M.: Use of autonomous measuring systems for diagnosing of electrical equipment with regard to its operating modes. Tekhnichna Elektrodynamika **5**, 116–120 (2018). https://doi.org/10.15407/techned2018.05.116
2. Babak, V.P., Babak, S.V., Myslovych, M.V., Zaporozhets, A.O., Zvaritch, V.M.: Principles of construction of systems for diagnosing the energy equipment. In: Diagnostic Systems For Energy Equipments. Studies in Systems, Decision and Control, vol. 281, pp. 1–22. Springer, Cham (2020). https://doi.org/10.1007/978-3-030-44443-3_1
3. Zaporozhets, A.A., Eremenko, V.S., Serhiienko, R.V., Ivanov, S.A.: Development of an intelligent system for diagnosing the technical condition of the heat power equipment. In: 2018 IEEE 13th International Scientific and Technical Conference on Computer Sciences and Information Technologies (CSIT), Ukraine, pp. 48–51 (2018). https://doi.org/10.1109/STC-CSIT.2018.8526742
4. Hertsyk, S., Gyzhko, Y., Zvaritch, V., Myslovych, M., Ostapchuk, L.: Features of the construction of autonomous diagnostic systems of electrical equipment, taking into account the modes of its operation. In: 2019 IEEE 20th International Conference on Computational Problems of Electrical Engineering (CPEE), Ukraine, pp. 1–3 (2019). https://doi.org/10.1109/CPEE47179.2019.8949148
5. Zaporozhets, A.: Development of software for fuel combustion control system based on frequency regulator. In: CEUR Workshop Proceedings, vol. 2387, pp. 223–230 (2019). https://ceur-ws.org/Vol-2387/20190223.pdf
6. Zaporozhets, A.: Methods and means for the control of the fuel combustion process. In: Control of Fuel Combustion in Boilers. Studies in Systems, Decision and Control, vol. 287, pp. 1–33. Springer, Cham (2020). https://doi.org/10.1007/978-3-030-46299-4_1
7. Babak, V.P., Babak, S.V., Myslovych, M.V., Zaporozhets, A.O., Zvaritch, V.M.: Technical provision of diagnostic systems. In: Diagnostic Systems For Energy Equipments. Studies in Systems, Decision and Control, vol. 281, pp. 91–133. Springer, Cham (2020). https://doi.org/10.1007/978-3-030-44443-3_4
8. Zaporozhets, A.: Analysis of control system of fuel combustion in boilers with oxygen sensor. Period. Polytech. Mech. Eng. **63**(4), 241–248 (2019). https://doi.org/10.3311/PPme.12572
9. Zaporozhets, A., Eremenko, V., Serhiienko, R., Ivanov, S.: Methods and hardware for diagnosing thermal power equipment based on smart grid technology. In: Shakhovska, N., Medykovskyy, M. (eds.) Advances in Intelligent Systems and Computing III. CSIT 2018. Advances in Intelligent Systems and Computing, vol. 871, pp. 476–489. Springer, Cham (2019). https://doi.org/10.1007/978-3-030-01069-0_34
10. Zaporozhets, A.: Hardware and software implementation of modules of the system of the fuel combustion control process. In: Control of Fuel Combustion in Boilers. Studies in Systems, Decision and Control, vol. 287, pp. 61–87. Springer, Cham (2020). https://doi.org/10.1007/978-3-030-46299-4_3
11. Zaporozhets, A., Kovtun, S., Dekusha, O.: System for monitoring the technical state of heating networks based on UAVs. In: Shakhovska N., Medykovskyy M.O. (eds.) Advances in Intelligent Systems and Computing IV. CSIT 2019. Advances in Intelligent Systems and Computing. Springer, Cham, pp. 935–950 (2020). https://doi.org/10.1007/978-3-030-33695-0_61
12. Babak, V.P., Mokiychuk, V.M., Zaporozhets, A.A., Redko, A.A.: Improving the efficiency of fuel combustion with regard to the uncertainty of measuring oxygen concentration. Eastern-European J. Enterp. Technol. 6, **8**(84), 54–59 (2016). https://doi.org/10.15587/1729-4061.2016.85408
13. Zaporozhets A., Khaidurov V., Tsiupii T.: Optimization models of industrial furnaces and methods for obtaining their numerical solution. In: Zaporozhets A., Artemchuk V. (eds.) Systems, Decision and Control in Energy II. Studies in Systems, Decision and Control. Springer, Cham, pp. 121–139 (2021). https://doi.org/10.1007/978-3-030-69189-9_7

14. Romaka, V.A., Rogl, P.-F., Frushart, D., Kaczorowski, D.: Mechanism of the generation of donor–acceptor pairs in heavily doped n-ZrNiSn with the Ga acceptor impurity. Semiconductors, pp. 294–304 (2018). https://doi.org/10.1134/S1063782618030193
15. Romaka, V.A., Rogl, P., Romaka, V.V., Kaczrowski, D., Stadnyk, Yu.V., Korzh, R.O., Krayovskyy, V.Ya, Kovbasyuk, T.M.: Features of the band structure and conduction mechanisms of n-HfNiSn semiconductor heavily Lu-doped. Semiconductors **49**, 290–297 (2015). https://doi.org/10.1134/S1063782615030185
16. Romaka, V.A., Rogl, P., Romaka, V.V., Kaczorowski, D., Stadnyk, Yu.V., Krayovskyy, V.Ya., Horyn, A.M.: Features of conductivity mechanisms in heavily doped compensated V_{1-x} Ti_x FeSb semiconductor. Semiconductors **50**, 860–868 (2016). https://doi.org/10.1134/S1063782616070204
17. Romaka, V.A., Rogl, P.-F., Romaka, L.P., Krayovskyy, V.Ya., Stadnyk, Yu.V., Kaczorowski, D., Horyn, A.M.: Features of structural, energy and kinetic characteristics of $TiNiSn_{1-x}Ga_x$ thermoelectric material. J. Thermoelectricity **3**, 24–33 (2016)
18. Romaka, V.A., Romaka, L.P., Stadnyk, Yu.V., Krayovskyy, V.Ya, Romaka, V.V., Horyn, A.M.: Research on electrical conductivity mechanisms of thermoelectric material based on n-ZrNiSn doped with Ga. J. Thermoelectricity **4**, 44–58 (2016)
19. Babak, V.P., Babak, S.V., Myslovych, M.V., Zaporozhets, A.O., Zvaritch, V.M.: Methods and models for information data analysis. In: Diagnostic Systems For Energy Equipments. Studies in Systems, Decision and Control, vol. 281, pp. 23–70. Springer, Cham (2020). https://doi.org/10.1007/978-3-030-44443-3_2
20. Eremenko, V., Zaporozhets, A., Babak, V., Isaienko, V., Babikova, K.: Using hilbert transform in diagnostic of composite materials by impedance method. Period. Polytech. Electr. Eng. Comput. Sci. **64**(4), 334–342 (2020). https://doi.org/10.3311/PPee.15066
21. Babak, S., Babak, V., Zaporozhets, A., Sverdlova, A.: Method of statistical spline functions for solving problems of data approximation and prediction of objects state. In: CEUR Workshop Proceedings, vol. 2353, pp. 810–821 (2019). http://ceur-ws.org/Vol-2353/paper64.pdf
22. Zaporozhets, A., Eremenko, V., Isaenko, V., Babikova, K.: Approach for creating reference signals for detecting defects in diagnosing of composite materials. In: Shakhovska N., Medykovskyy M.O. (eds.) Advances in Intelligent Systems and Computing IV. CSIT 2019. Advances in Intelligent Systems and Computing. Springer, Cham, pp. 154–172 (2020). https://doi.org/10.1007/978-3-030-33695-0_12
23. Kankar, P.K., Harsha, S.P., Kumar, P., Sharma, S.C.: Fault diagnosis of a rotor bearing system using response surface method. Eur. J. Mech. A. Solids **28**(4), 841–857 (2009). https://doi.org/10.1016/j.euromechsol.2009.03.004
24. Wu, S.-D., Wu, P.-H., Wu, C.-W., Ding, J.-J., Wang, C.-C.: Bearing fault diagnosis based on multiscale permutation entropy and support vector machine. Entropy **14**(8), 1343–1356 (2012). https://doi.org/10.3390/e14081343
25. Sohaib, M., Kim, C.-H., Kim, J.-M.: A hybrid feature model and deep-learning-based bearing fault diagnosis. Sensors **17**(2), 2876 (2017). https://doi.org/10.3390/s17122876
26. Dybala, J., Zimroz, R.: Rolling bearing diagnosing method based on empirical mode decomposition of machine vibration signal. Appl. Acoust. **77**, 195–203 (2014). https://doi.org/10.1016/j.apacoust.2013.09.001
27. Boudiaf, A., Moussaoui, A., Dahane, A., Atoui, I.: A comparative study of various methods of bearing faults diagnosis using the case western reserve university data. J. Fail. Anal. Prev. **16**, 271–284 (2016). https://doi.org/10.1007/s11668-016-0080-7
28. Eremenko, V., Zaporozhets, A., Isaenko, V., Babikova, K.: Application of wavelet transform for determining diagnostic signs. In: CEUR Workshop Proceedings, vol. 2387, pp. 202–214. https://ceur-ws.org/Vol-2387/20190202.pdf
29. Babak, V., Eremenko, V., Zaporozhets, A.: Research of diagnostic parameters of composite materials using Johnson distribution. Int. J. Comput. **18**(4), 483–494 (2019)
30. Jedrzejewski, J., Kowal, Z., Kwasny, W., Modrzycki, W.: High-speed precise machine tools spindle units improving. J. Mater. Process. Technol. **162–163**, 615–621 (2005). https://doi.org/10.1016/j.jmatprotec.2005.02.149

31. Lim, W.Q., Zhang, D.H., Zhou, J.H., Belgi, P.H., Chan, H.L.: Vibration-based fault diagnostic platform for rotary machines. In: IECON 2010—36th Annual Conference on IEEE Industrial Electronics Society, USA, pp. 1404–1409 (2010). https://doi.org/10.1109/IECON.2010.5675477
32. Gyzhko, Y.I., Myslovych, M.V., Sysak, R.M.: Issues of improving of the accuracy of diagnostic parameters estimations in the spectral processing of vibration signals. Tekhnichna Elektrodynamika **2**, 127–128 (2012)
33. Yan, R., Gao, R.X.: Rotary machine health diagnosis based on empirical mode decomposition. J. Vib. Acoust. **130**(2), 021007 (2008). https://doi.org/10.1115/1.2827360
34. Tandon, N., Parey, A.: Condition monitoring of rotary machines. In: Wang, L., Gao, R.X. (eds.) Condition Monitoring and Control for Intelligent Manufacturing. Springer Series in Advanced Manufacturing, pp. 109–136. Springer, London (2006). https://doi.org/10.1007/1-84628-269-1_5
35. Baranski, M.: Vibration diagnostic method of permanent magnets generators—detecting of vibrations caused by unbalance. In: 2014 Ninth International Conference on Ecological Vehicles and Renewable Energies (EVER), Monaco, pp. 1–6 (2014). https://doi.org/10.1109/EVER.2014.6844134
36. Zaporozhets, A.A., Sverdlova, A.D.: Peculiarities of application of smart grid technology in systems for monitoring and diagnostics of heat-and-power engineering objects. Techn. Diagn. Non-Destr. Test. **2**, 33–41 (2017). https://doi.org/10.15407/tdnk2017.02.05
37. Zaporozhets, A., Babak, V., Sverdlova, A., Isaienko, V., Babikova, K.: Development of a system for diagnosing heat power equipment based on IEEE 802.11s. In: Zaporozhets, A., Artemchuk, V. (eds.) Systems, Decision and Control in Energy II. Studies in Systems, Decision and Control. Springer, Cham, pp. 141–151 (2021). https://doi.org/10.1007/978-3-030-69189-9_8
38. Hertsyk, S.M., Gorodzha, A.D., Myslovych, M.V., Podoltsev, O.D., Sysak, R.M., Troshchynskyi, B.O.: Models of wave processes in objects of limited form and their use for diagnostics of electrotechnical equipment. Tekhnichna Elektrodynamika **2**, 86–94 (2018). https://doi.org/10.15407/techned2018.02.086
39. Dergunov, A.V., Kuts, Y.V., Scherbak, L.N.: Comparative analysis of modern time-series analysis methods. In: 2011 Microwaves, Radar and Remote Sensing Symposium, Ukraine, pp. 378–381 (2011). https://doi.org/10.1109/MRRS.2011.6053679
40. Zaporozhets, A.: Experimental research of a computer system for the control of the fuel combustion process. In: Control of Fuel Combustion in Boilers. Studies in Systems, Decision and Control, vol. 287, pp. 61–87. Springer, Cham (2020). https://doi.org/10.1007/978-3-030-46299-4_4
41. Babak, V.P., Babak, S.V., Myslovych, M.V., Zaporozhets, A.O., Zvaritch, V.M.: Simulation and software for diagnostic systems. In: Diagnostic Systems For Energy Equipments. Studies in Systems, Decision and Control, vol. 281, pp. 71–90. Springer, Cham (2020). https://doi.org/10.1007/978-3-030-44443-3_3
42. Krasilnikov, A., Beregun, V., Harmash, O.: Analysis of estimation errors of the fifth and sixth order cumulants. In: 2019 IEEE 39th International Conference on Electronics and Nanotechnology (ELNANO), Ukraine, pp. 754–759 (2019). https://doi.org/10.1109/ELNANO.2019.8783910
43. Beregun, V., Harmash, O.: Application of cumulant coefficients for solving the problems of testing and diagnostics in control systems. In: 2018 IEEE 5th International Conference on Methods and Systems of Navigation and Motion Control (MSNMC), Ukraine, pp. 210–213 (2018). https://doi.org/10.1109/MSNMC.2018.8576176
44. Beregun, V.S., Krasilnikov, A.I.: Research of excess kurtosis sensitiveness of diagnostic signals for control of the condition of the electrotechnical equipment. Tekhnichna Elektrodynamika **4**, 79–85 (2017). https://doi.org/10.15407/techned2017.04.079
45. Antoni, J., Bonnardot, F., Raad, A., Badaoui, M.El.: Cyclostationary modelling of rotating machine vibration signals. Mech. Syst. Sign. Process. **18**(6), 1285–1314 (2004). https://doi.org/10.1016/S0888-3270(03)00088-8
46. Nataraj, C., Harsha, S.P.: The effect of bearing cage run-out on the nonlinear dynamics of a rotating shaft. Commun. Nonlinear Sci. Numer. Simul. **13**(4), 822–838 (2008). https://doi.org/10.1016/j.cnsns.2006.07.010

47. Huang, P., Pan, Z., Qi, X., Lei, J.: Bearing fault diagnosis based on EMD and PSD. In: 2010 8th World Congress on Intelligent Control and Automation, China, pp. 1300–1304 (2010). https://doi.org/10.1109/WCICA.2010.5554896
48. Myslovych, M.V., Sysak, R.M., Ostapchuk, L.B., Hertsyk, S.M.: Algorithms of operation and software of multilevel system for monitoring and technical diagnostics of electrical power facilities equipment. Tekhnichna Elektrodynamika **4**, 86–88 (2016). https://doi.org/10.15407/techned2016.04.086
49. Myslovych, M.V., Sysak, R.M.: On some peculiarities of design of intelligent multi-level systems for technical diagnostics of electric power facilities. Tekhnichna Elektrodynamika **1**, 78–85 (2015)
50. Zhang, Z., Wang, Y., Wang, K.: Fault diagnosis and prognosis using wavelet packet decomposition, Fourier transform and artificial neural network. J. Intell. Manuf. **24**, 1213–1227 (2013). https://doi.org/10.1007/s10845-012-0657-2
51. Sugumaran, V., Ramachandran, K.I.: Automatic rule learning using decision tree for fuzzy classifier in fault diagnosis of roller bearing. Mech. Syst. Sign. Process. **21**(5), 2237–2247 (2007). https://doi.org/10.1016/j.ymssp.2006.09.007
52. Marchenko, B.H., Myslovych, M.V.: Vibrodiagnostics of bearing units of electrical machines. Naukova dumka (1992)
53. Gyzhko, Yu., Myslovych, M.: Elements of the theory and practical application of systems for vibrodiagnostics of electrical machines moving parts. Tekhnichna Elektrodynamika **2**, 45–56 (2015)
54. Antoni, J.: Cyclostationarity by examples. Mech. Syst. Sign. Process. **23**(4), 987–1036 (2009). https://doi.org/10.1016/j.ymssp.2008.10.010
55. Makowski, M., Pietrzak, P., Pekoslawski, B., Napieralski, A.: Measurement synchronization in the vibration diagnostic system of high power electric machines. In: Proceedings of the 17th International Conference Mixed Design of Integrated Circuits and Systems—MIXDES 2010, Poland, pp. 566–569 (2010)
56. Napolitano, N: Cyclostationarity: new trends and applications. Signal Process. **120**, 385–408 (2016). https://doi.org/10.1016/j.sigpro.2015.09.011
57. Benesty, J., Chen, J., Huang, Y., Cohen, I.: Pearson correlation coefficient. In: Noise Reduction in Speech Processing. Springer Topics in Signal Processing, vol. 2, pp. 1–4. Springer, Berlin, Heidelberg (2009). https://doi.org/10.1007/978-3-642-00296-0_5
58. Zaporozhets, A.: Research of the process of fuel combustion in boilers. In: Control of Fuel Combustion in Boilers. Studies in Systems, Decision and Control, vol. 287, pp. 35–60. Springer, Cham (2020). https://doi.org/10.1007/978-3-030-46299-4_2
59. Kalsi, H.S.: Electronic instrumentation, 3rd edn., p. 829. Tata McGraw-Hill Education, New Delhi (2012)
60. Sait, A.S., Sharaf-Eideen, Y.I.: A review of gearbox condition monitoring based on vibration analysis techniques diagnostics and prognostics. In: Proulx, T. (eds.) Rotating Machinery, Structural Health Monitoring, Shock and Vibration, vol. 5, pp. 307–324. Conference Proceedings of the Society for Experimental Mechanics Series. Springer, New York, NY (2011). https://doi.org/10.1007/978-1-4419-9428-8_25
61. Baranski, M.: New vibration diagnostic method of PM generators and traction motors—detecting of vibrations caused by unbalance. In: 2014 IEEE International Energy Conference (ENERGYCON), Croatia, pp. 28–32 (2014). https://doi.org/10.1109/ENERGYCON.2014.6850401
62. Yan, R., Gao, R.X., Chen, X.: Wavelets for fault diagnosis of rotary machines: a review with applications. Signal Process. **96A**, 1–15 (2014). https://doi.org/10.1016/j.sigpro.2013.04.015
63. Tse, P.W., Peing, Y.H., Yam, R.: Wavelet analysis and envelope detection for rolling element bearing fault diagnosis—their effectiveness and flexibilities. J. Vib. Acoust. **123**(3), 303–310 (2001). https://doi.org/10.1115/1.1379745
64. Wang, Z., Chen, J., Dong, G., Zhou, Y.: Constrained independent component analysis and its application to machine fault diagnosis. Mech. Syst. Signal Process. **25**(7), 2501–2512 (2011). https://doi.org/10.1016/j.ymssp.2011.03.006

65. Kang, Y., Wang, C.-C., Chang, Y.-P.: Gear fault diagnosis in time domains by using Bayesian networks. In: Castillo, O., Melin, P., Ross, O.M., Sepúlveda Cruz, R., Pedrycz, W., Kacprzyk, J. (eds.) Theoretical Advances and Applications of Fuzzy Logic and Soft Computing. Advances in Soft Computing, vol. 42, pp. 741–751. Springer, Berlin, Heidelberg (2007). https://doi.org/10.1007/978-3-540-72434-6_75
66. Kudryavtseva, I.S., Naumenko, A.P., Odinets, A.I., Bardanov, V.E.: New diagnostic signs of the technical condition of piston compressors on the basis of characteristic function of the vibroacoustic signal. J. Phys.: Conf. Ser. **1260**(3), 032023 (2019). https://doi.org/10.1088/1742-6596/1260/3/032023

Chapter 5
Examples of Using Models and Measures on the Circle

Abstract Examples of using the developed models and measures on the circle for the study of cyclic signals in various subject areas are given. The object of study is the phase shift between cyclic signals. The limiting case of cyclic signals are periodic signals, in particular harmonious signals. The solutions to the problems of precision ultrasonic echo-pulse thickness measurement of products from materials with significant attenuation are considered. A high probability of detecting information signals against additive noise is achieved through the use of selective circular statistics - the resulting vector length. These statistics are determined during processing phase measurement data in a sliding mode. A method for processing the results of multi-scale phase measurements based on numerical systems of residual classes in phase range finders and direction finders is considered. The method is different in that it allows to control the correctness of eliminating the ambiguity of phase measurements. The features of statistical data processing in environmental monitoring systems based on unmanned aerial systems during the flight of objects of increased environmental hazard are analyzed. The given examples testify to the powerful methodological potential of using the developed models and circle measures for use in precision phase measuring systems.

Keywords Phase measuring system · Multi-scale phase measurements · Phase direction finder · Phase range finder · Residual class system · Ultrasonic echo-pulse thickness measurement

Developed in goniometry models and measures on a circle have found new application in the study of cyclic processes. Actually, the word "cycle" comes from the Greek "kyklos"—a circle, which is the most natural geometric figure for displaying angles. A cycle is understood as a set of interconnected phenomena or processes that form a cycle during a certain range of time—period. Process values separated by this time interval may not coincide exactly—the repeatability of the most characteristic features of the latter remains essential. Processes in which cycles can be distinguished are called cyclic.

V. P. Babak et al., *Models and Measures in Measurements and Monitoring*,
Studies in Systems, Decision and Control 360,
https://doi.org/10.1007/978-3-030-70783-5_5

Cyclic processes can be damped and undamped. An example of undamped cyclic process is the change of seasons of the year. Other examples are oscillations of a physical pendulum, damped oscillations of electric currents in a resonant circuit, processes of changes in blood pressure in a human's blood vessels, and the like.

Cyclic signals and their limiting case—periodic signals have found application in phasemetry. The main object of study in this field of measurement technology is the phase shift between two cyclic signals. Such measurements are known for a wide range of applied applications in radio engineering, non-destructive testing, medicine, radar and radio navigation, information security, etc. [1–7]. The use of the developed models and measures on the circle in phasemetry can significantly expand the capabilities of this measurement area.

The considered examples do not cover the whole variety of options for the application of models and measures on the circle for measurements, but they convincingly testify to their powerful methodological potential.

5.1 Phase Systems for Ultrasonic Echo-Pulse Thickness Measurement

During testing products made of materials with significant attenuation of ultrasonic vibrations, the problem of identifying ultrasonic pulse signals with a harmonious carrier signal against a background of significant noise arises. This problem has an effective solution through the determination of discrete phase characteristics of such signals and their further statistical analysis. This method is proposed and investigated in [8, 9].

The echo-pulse method of ultrasonic thickness measurement (UST) is based on the determination of the delay τ_d necessary for the propagation of the ultrasonic signal through the monitoring object. The latter, by the known velocity c of the propagation of a longitudinal ultrasonic wave in two directions between the surface and the bottom of the object, uniquely determines its thickness: $h = 0.5 \cdot c \cdot \tau_d$. Typically, the value of τ_d is estimated from the time interval between the bypasses of two bottom ultrasonic signals. This method is very sensitive to the presence of noise.

The determination of τ_d by the phase characteristics of the echo signals has certain advantages in noise immunity, and in combination with the statistical methods of processing these characteristics, it is possible to detect UST signals with a signal-to-noise ratio (s/n) close to one or less, and use them for precision measurement of the delay τ_d in structural materials with significant attenuation of ultrasonic waves.

The proposed method of ultrasonic thickness gauge is realized by forming an ultrasonic sounding pulse signal [10, 11], enter it into the control object, receive the signal after it is distributed in the object and reflection from the opposite side of the object, determine the phase characteristic of the reflected signals (echo signals), calculate the phase difference of the echo signals and a carrier signal, perform its sliding window processing, calculate the current values of r-statistics (selective-resulting length, SRL) for the data selected by the window, and time τ_d is found at its maximums. The thickness of the object is calculated by the formula

$$h = 0.5 \cdot \tau_{1,k} \cdot c / (k-1), \quad (5.1)$$

where $\tau_{1,k}$ is the time interval between the first and k-th bottom signals.

This method allows to determine the sequence of decaying bottom pulses in the presence of a noise-significant one, to isolate the delay $\tau_{1,k}$ between the first and k-th bottom signals and thereby reduce the RMS error of quantization of the time interval $\tau_d = \tau_{1,k} / k - 1$ by $\sqrt{k-1}$ times.

The methodology of this method is based on determining the current values of r-statistics. The procedure for obtaining and statistical evaluation of the results of phase measurements is shown in Fig. 5.1.

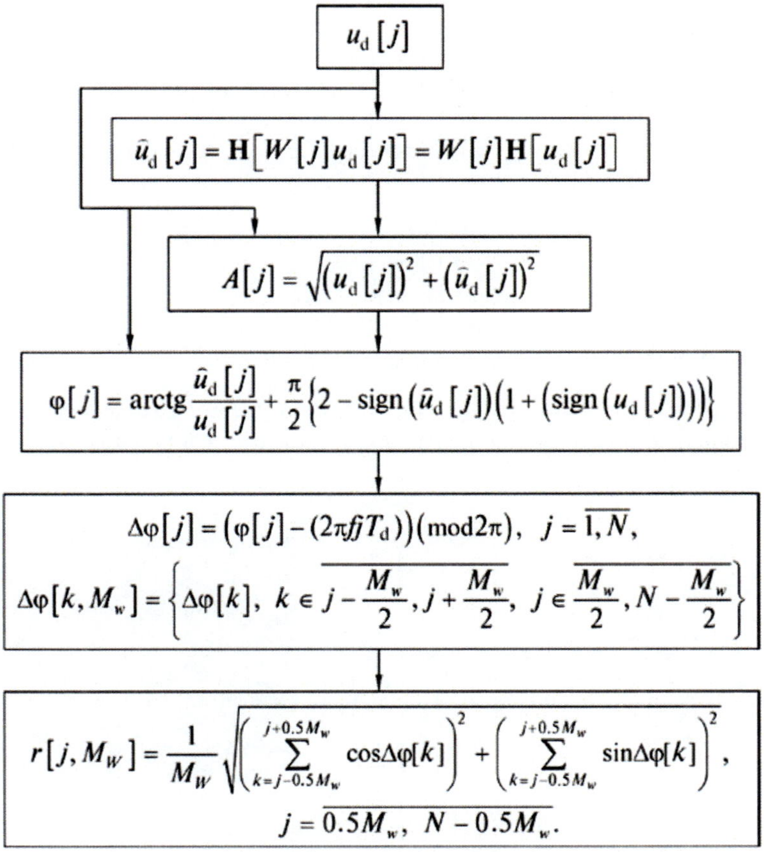

Fig. 5.1 Graphical representation of the r-statistic determination technique

An experimental study of the effectiveness of the proposed method was carried out as follows: an ultrasonic probe signal in the form of a radio pulse was emitted into a sample with a known thickness (or propagation time of an ultrasonic signal). The signal was analyzed, which is an additive mixture of a periodic (repetition period T_r) sequence of echo signals and the implementation of Gaussian noise $\xi(t)$ with zero mathematical expectation and dispersion σ^2. Its discretized prototype looks like

$$u_d[j] = \sum_{i=1}^{d} k_{EAP_i} \cdot u_{del}(j - \tau - (i-1)T_d) + \xi[j],$$

where k_{EAP_i}—electro-acoustic path coefficients for the *i*-th echo signal, $i = \overline{1, d}$; τ—delay of the first echo signal relative to the probe, $j = \overline{1, N}$; N the sample size, the value of which is obtained at discrete time moments jT_d; T_d—period of discretization.

The s/n ratio was determined by the $U_{max,i} / \sigma$ value, where $U_{max,i}$ is the maximum value of the envelope of the *i*-th echo signal.

For conducting experimental studies of this method of UST, a laboratory bench was developed, the structural diagram and general view of which are shown in Fig. 5.2a, b.

For generating the signal, a two-channel arbitrary waveform generator ANR-3122 was used with the following characteristics:

- frequency band—0.02 Hz … 10 MHz;
- digit capacity of the digital-to-analog converter—12 bits;
- maximum number of points per channel—128 K;
- maximum signal sampling frequency—80 MHz;
- duration of the front of the rectangular signal—20 ns.

Two types of converters of the combined type were used: Panametrics C308 with an operating frequency of 5 MHz and Panametrics C305 with a frequency of 2.25 MHz.

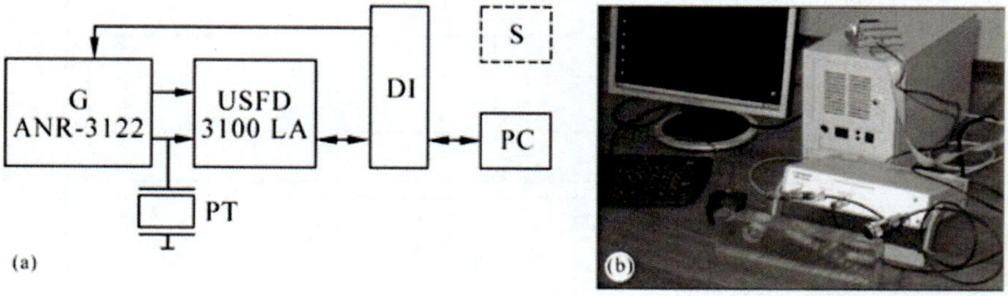

Fig. 5.2 Laboratory stand: **a** structure (G—signal generator, PT—piezoelectric transducer, USPC—ultrasonic flaw detector, PC—personal computer, DI—digital interface, S—software); **b** general view

An ultrasonic flaw detector USPC 3100 LA manufactured by Socomate (France) was used as a preliminary signal processing device for ultrasonic scanning, made structurally in the form of a PCI-card for a PC. The flaw detector provides amplification, filtering and analog-to-digital signal conversion (ADC), preliminary digital processing of experimental data and storing the result in random access memory for subsequent transmission by packets to a PC. Main technical characteristics of the USPC 3100 LA flaw detector:

- receiver frequency band—0.35–30 MHz;
- dynamic range of the signal amplitude—up to 105 dB;
- ADC sampling frequency—100 MHz;
- ADC resolution—10 bits.

The research was performed on 4 samples, the acoustic properties of the materials of the samples are given in Table 5.1.

The general view of the samples is shown in Fig. 5.3.

In all experiments, the sampling frequency is $f_{\mathrm{d}} = 100$ MHz. To sound the test object, a pulse signal with a rectangular bypass and a duration equal to 4 periods of the carrier signal was used.

Sample No. 1. The frequency of the fill signal is 2.3 MHz. The data obtained during the experiment and the calculation results are presented in Fig. 5.4.

From Fig. 5.4a, b it can be seen, that the analyzed signal is a sequence of bottom radio pulses in a mixture with Gaussian noise, which decay exponentially. The graph in Fig. 5.4c is a sawtooth function with a range of $[0, 2\pi)$. At time intervals within which bottom pulses are present, the graph takes the form of a curve that changes much more slowly (sections 1–7).

From a comparative analysis of envelope A (Fig. 5.4b) and statistics $r[j, M_W]$ (Fig. 5.4d), it can be concluded that the proposed method of phase ultrasonic scanning is more effective than the amplitude one, since it allows to detect more echo signals, i.e., the possibility of their detection at a lower s/n ratio. Indeed, according to the bypass signal $u_d[j]$, starting from $i = 5$, it is practically impossible to distinguish echo signals from the noise background. But according to the statistics graph $r[j, M_W]$ one can confidently detect 8 echo signals ($d = 8$, for the 8-th it can be written $U_{max,8} / \sigma_{\mathrm{n}} \approx 1.8$).

In order to determine the standard deviation (SD) for estimating the time interval τ, a series of $S = 20$ experiments was performed, with averaging of the obtained estimates.

The average value $\overline{\tau}_{kS}$ for the k-th interval for S experiments is determined by the formula $\overline{\tau}_{kS} = S^{-1} \sum_{s=1}^{S} \tau_{ks}$, where τ_{ks} is the delay value between k and $k+1$ pulses in the s-th experiment, $k = \overline{1, K}$; $s = \overline{1, S}$. RMS for $\overline{\tau}_{kS}$ was determined by the formula

$$\sigma_{KS} = \sqrt{(S-1)^{-1} \sum_{n=1}^{S} (\tau_{ks} - \overline{\tau}_{kS})^2}.$$

Table 5.1 Acoustic properties of sample materials

Material	Sample thickness *h*, mm	Signal propagation time, μs	Density x10^3, kg/m^3	Longitudinal wave velocity, 10^3 m/s	Young's modulus, hPa	Poisson's ratio	2.5 MHz attenuation coefficient, Np/m	Wave impedance Z, 10^6 Pa·s/m
Sample No 1 Carbon steel (steel grade 10 - quality structural carbon steel) CO-2A	56	18.9	7.856	5.925	210 (at 20 °C)	0.28	1.7	46.5
Sample No 2 Stainless steel (12X18H10T—stainless structural cryogenic steel: 0.12%—carbon, 18%—chromium, 10%—nickel)	17.8	6.2	7.920	5.742	198 (at 20 °C)	0.30	2.2	45.4
Sample No 3 Plexiglass CO1 (according GOST 14782-86)	25.4	19.3	1.190	2.674	3.1	0.35	30	3.2
Sample No 4 Porcelain ceramic (vitreous phase)	10.9	5.5	2.270	3.942	59	0.17	0.3	8.9

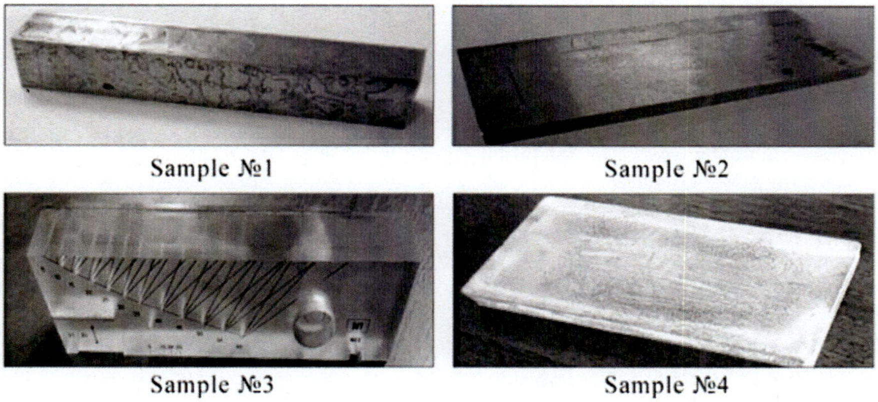

Fig. 5.3 General view of the studied samples

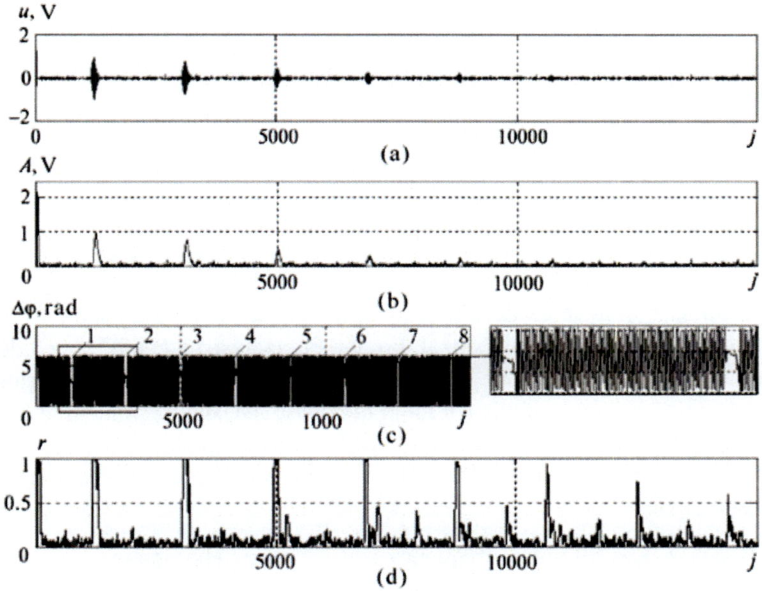

Fig. 5.4 Experimental and calculated data for sample No. 1: **a** a graph of the signal of UST; **b** the envelope of the signal; **c** a fragment of the graph of the difference in discrete phase characteristics of the UST signal and the carrier signal of a sinusoidal shape; **d** a graph of statistics $r[j, M_W]$ for a sliding window with aperture $M_w = 110$

For combining the results of a series of separate and independent measurements of the same magnitude, it is necessary to use the weighted average of these measurements to obtain the best estimate [12–14]:

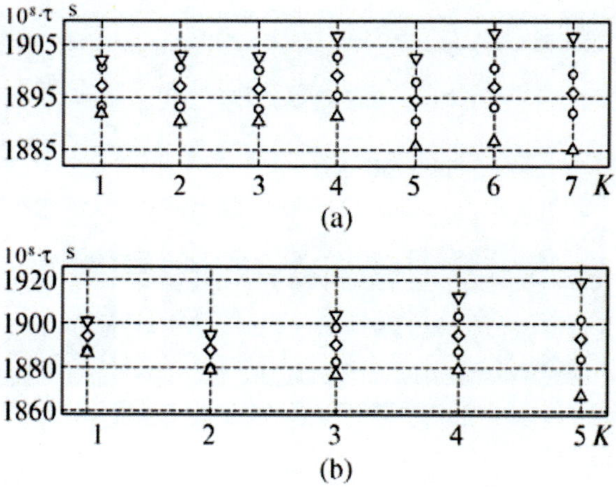

Fig. 5.5 Graphs of the average value of the signal propagation time $\overline{\tau}_{kS}$ and the RMS boundary for $\overline{\tau}_{kS}$ and $\overline{\tau}$, obtained from the r-statistics **a** and bypass signal **b**: rhombs—$\overline{\tau}_{kS}$ value; triangles—RMS boundaries for $\overline{\tau}_{kS}$; circle—RMS boundaries for $\overline{\tau}_{kS}$

$$\overline{\tau}_H = \sum_{k=1}^{K} \overline{\tau}_{kS} / \sigma_{KS}^2 \Big/ \sum_{k=1}^{K} \sigma_{KS}^{-2}. \tag{5.2}$$

RMS of this estimate is defined as

$$\sigma_{\overline{\tau}H} = \left(\sum_{k=1}^{K} \sigma_{KS}^{-2} \right)^{-0.5}. \tag{5.3}$$

Figure 5.5a, b shows the values $\overline{\tau}_{kS}$ and the RMS boundary for $\overline{\tau}_{kS}$ and $\overline{\tau}_N$, obtained from the r-statistics and bypass signal. From their analysis, it can be concluded that for s/n ratio $\gg$ 1, the amplitude method provides high accuracy in measuring time intervals. With a decrease in the s/n ratio, the phase measurement method becomes more efficient, and for values of s/n < 2 the amplitude method generally loses its ability to determine time intervals (the (5.6–5.8)th echo signals are not detected by the bypass signal).

Sample No. 2. The fill signal frequency is 2.23 MHz. The obtained experimental data and calculation results are presented in Fig. 5.6.

In this experiment, the envelope of the signal $u_d[j]$ starting from $i = 10$ practically does not allow echo signals to be distinguished from noise. But the graph of statistics $r[j, M_W]$ indicates that in this experiment $d = 18$ echo signals confidently appeared. For pulse number $i = 8$, the s/n ratio is ~1.

Sample No. 3. The fill signal frequency is 2.3 MHz. The obtained experimental data and calculation results are presented in Fig. 5.7.

From Fig. 5.7 a, b it can be seen that only one echo signal is detected behind the bypass signal, but the graph of $r[j, M_W]$ statistics indicates the possibility of detecting two pulses.

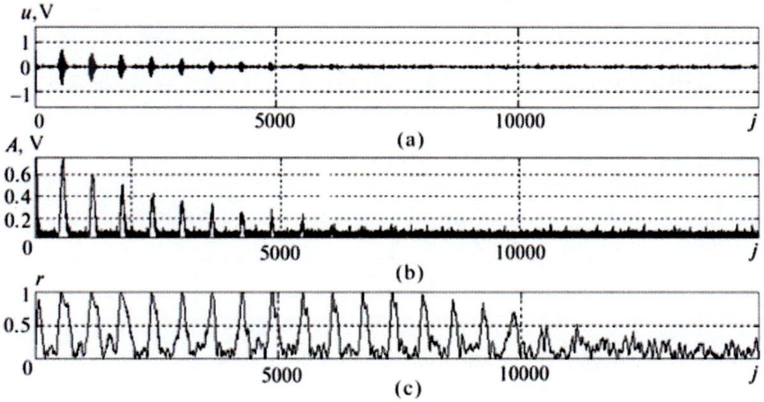

Fig. 5.6 Experimental data for sample No. 2: **a** $u_d[j]$ signal graph; **b** $A[j]$ envelope of signal graph; **c** $r[j, M_W]$ statistic graph

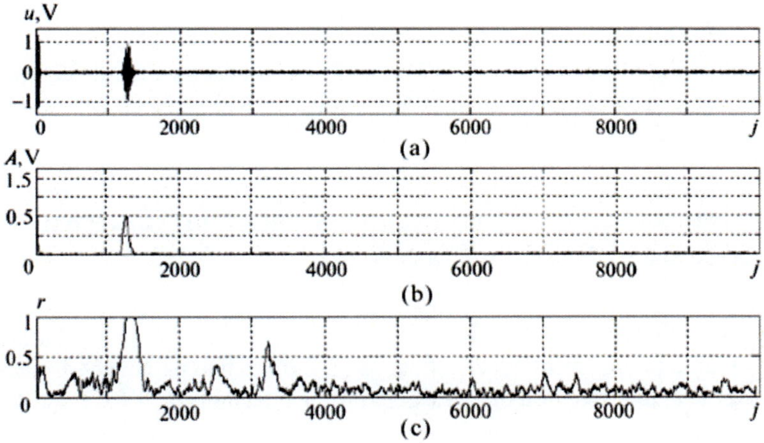

Fig. 5.7 Experimental data for sample No. 3: **a** UST signal graph; **b** $A[j]$ envelope of signal graph; **c** $r[j, M_W]$ statistic graph

Sample No. 4. The fill signal frequency is 2.3 MHz. The obtained experimental data and calculation results are presented in Fig. 5.8.

From a comparison of the graphs of the bypass signal and the $r[j, M_W]$ statistics, it can be concluded that in this case, the phase method confidently identifies 6 echo signals, while the amplitude method—4–5 echo signals. The s/n ratio for the last detected echo signals is 2.3. In general, the phase method for detecting UST signals with a harmonious carrier signal against the background of additive Gaussian noise with respect to s/m < 2 allows to obtain a larger amount of measurement information during the experiments compared to the amplitude method and, on this basis, increase the accuracy of measuring the thickness of products by ultrasonic echo pulse method.

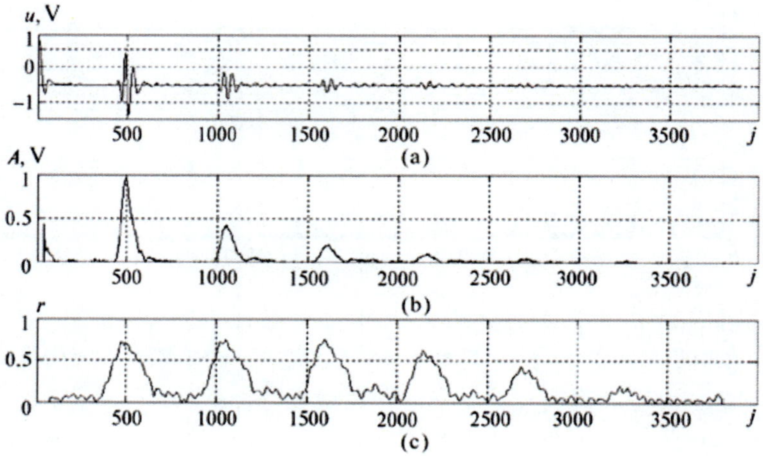

Fig. 5.8 Experimental data for sample No. 4: **a** $u_d[j]$ signal graph; **b** $A[j]$ envelope of signal graph; **c** $r[j, M_W]$ statistic graph

The considered method can also find application in ultrasonic defectometry in problems of detecting defects such as discontinuities with reduced dimensions.

5.2 Multiscale Phase Measurements Based on Numerical Systems of Residual Classes

The phase measurement method is based on the use of at least one pair of harmonious signals with a phase shift between them as an informative parameter. Usually this parameter is associated with the measured physical quantity $L \in [0, L_{\max}]$, where $L_{\max}$ is the maximum measured value, the dependence in the form

$$\varphi(L) \equiv kF(L) \bmod 2\pi. \tag{5.4}$$

Here $F(L)$ is a function of L, k is a dimensional coefficient. For example, in the task of measuring distance by the phase method [15, 16], this coefficient is $k = 2\pi/\lambda$, where λ is the wavelength in the medium of its propagation along the distance L, and $F(L) = L$. In this case, $\varphi(L)$ is comparable to kL by modulo 2π, and determines the nonlinear character of the transformation (5.4) in a significant dynamic range.

Homomorphic, with condition $kF(L) \geq 2\pi$, transformation (5.4) leads to a measurement uncertainty. An effective implementation of the phase method requires the unambiguous determination of the signals phase shifts of type $\Phi(L) = kF(L) = 2\pi\, n(L) + \varphi(L)$, $n = 0, 1, 2, \ldots$. This task is known as the solution (elimination, difference, docking of scales) of phase uncertainty and has solutions based on additional $i = 1, 2, \ldots$ phase measurements, for example, on several lower frequencies (that is, with lower k and, accordingly, lower measurement accuracy L) [17]. In this case, the measurements at each frequency form its own scale with the corresponding conversion coefficient, and the methods of overcoming the ambiguity are reduced to sequentially recalculating the measurement results from one scale to another (if the uniqueness condition is fulfilled for one of the scales: $k_i F(L_{\max}) < 2\pi$), or by enumerating all possible values n (by a significant limitation of n, which takes place in phase direction finders [18–20]) and the choice of the most probable value by a certain criterion.

At the same time, there is another possibility of n determining, based on using the features of the numerical system of residual classes (SRC). This idea is based on the similarity of the modular representation of the phase shifts of the signals (5.4) and for the recording of the numbers A in the SRC by residues in the form $a_i \equiv A \bmod p_i$, where the integers $p_i > 1$ are the SRC modules.

The SRC theory was most actively developed and used in the field of computer technology to build fault-tolerant high-speed computing tools [21–23]. [24] proposed the use of SRC for processing interferograms in optics. The authors suggested using this idea to eliminate ambiguity in phase rangefinders and phase direction finders [25, 26].

Definition and features of SRC. The essence of SRC is to represent the numbers A from the working interval $\left[0, A_{\mathrm{p}}\right)$ by the set of integral residuals $A_{\mathrm{SRC}} = (\alpha_1, \ldots, \alpha_m)$. The integers $a_i, i = \overline{1, m}$ are formed by dividing A by other integer coprime numbers $p_i, i = \overline{1, m}$—SRC modules, that is, $a_i \equiv A \bmod p_i$. The set of all residues for each module forms a ring of integers $\alpha_i \in [0, p_i)$ according to the corresponding module.

The restoration of the numbers A in the positional number system is based on the Chinese remainder theorem [27–29]: the restoration of A is possible in the case of one-to-one correspondence between A and A_{SRC}, which is achieved by the following conditions: (1) the modules of the system are mutually prime numbers; (2) the maximum renewable number satisfies the inequality $A_{\max} < A_{\mathrm{p}} = \prod_{i=1}^{m} p_i$. Under these conditions, there is an inverse transformation $A_{\mathrm{SRC}} \Rightarrow A$. The number A can be calculated, for example, by the algorithm

$$A = \sum_{i=1}^{m} a_i B_i \left(\bmod A_{\mathrm{p}}\right), \tag{5.5}$$

where $(B_1, \ldots B_i, \ldots B_m)$ is the system of orthonormal bases, which is calculated for the selected SRC modules, for example, according to the procedure described in [30].

Example 5.1 Let the number $A = 31$ is presented in the SRC according to the system of modules (5, 7): $A_{SRC} = (1, 3)$. The number A satisfies the condition $A \le A_{\max} = 34$; the system of orthonormal bases is equal $B = (21, 15)$. The condition of orthonormality is to fulfill for the elements of the basis a set of relations:

$$B_1 \bmod p_1 = 21 \bmod 5 = 1,\ B_1 \bmod p_2 = 21 \bmod 7 = 0,$$
$$B_2 \bmod p_1 = 15 \bmod 5 = 0,\ B_2 \bmod p_2 = 15 \bmod 7 = 1.$$

As a result of calculations by (5.5), it can be obtained the values: $A = (1 \cdot 21 + 3 \cdot 15)(\mathrm{mod} 5 \cdot 7) = 66 \bmod 35 = 31$.

The main theorems and statements about SRC are given in [31]. To justify the use of SRC in phasemetry, the following statement is important.

Proposition 1 If the number $A \in R$ and the module $p_i \in N$ have a common multiplier A, then

$$A \bmod p_i = a\left[\frac{A}{a} \bmod \left(\frac{p_i}{a}\right)\right]. \tag{5.6}$$

Proposition 1 gives formal reasons to perform modular operations, firstly, with real numbers, and secondly, with named numerical values of physical quantities. Let, for example, a physical quantity and its unit of measurement have numerical values A and p_i, and dimension [A]. Then, according to (5.6), it can be written

$$\{A[A]\} \bmod \{p_i[A]\} = (A \bmod p_i)[A], \tag{5.7}$$

From (5.6), another important conclusion for the practice of measurement follows: if a physical quantity varies in cycles that are not numerically determined by integers, then it is possible to reduce modular operations with such quantities to operations with integers. For example, let consider the fractional part of the total phase shift of the signals:

$$[\Phi(L)] \bmod 2\pi = 2\pi\left[\left(\Phi(L)/2\pi\right) \bmod 1\right] \text{ [rad]}. \tag{5.8}$$

In order to optimize computational costs, sometimes it is necessary to determine the remainder $\alpha_i = A \bmod p_i$ from the results of calculations for another module p_j. Such a recalculation according to (5.6) is performed as

$$\alpha_i = A \bmod p_i = \frac{p_i}{p_j}\left(\frac{Ap_j}{p_i} \bmod p_j\right). \tag{5.9}$$

If, for example, $p_j = 2^g$ is chosen, then the remainder α_i will be formed automatically as a result of performing Ap_j/p_i operations in the case where the digit grid of the calculator g is limited to binary bits.

An important feature of SRC is the possibility of organizing control (and even correction) of errors that occur during receiving balances and performing arithmetic operations with them. To do this, the base of the SRC is supplemented with an additional module $p_{m+1} > p_i, i \in \overline{1, m}$ (one or more). The new SRC has a full range of conversion of numbers $[0, A_p)$, $A_p = p_1 \ldots p_m p_{(m+1)} = A_p p_{m+1}$. A module system $(p_1, \ldots, p_i, \ldots, p_{m+1})$ that satisfies condition $p_1 < p_2 < \cdots < p_{m+1}$

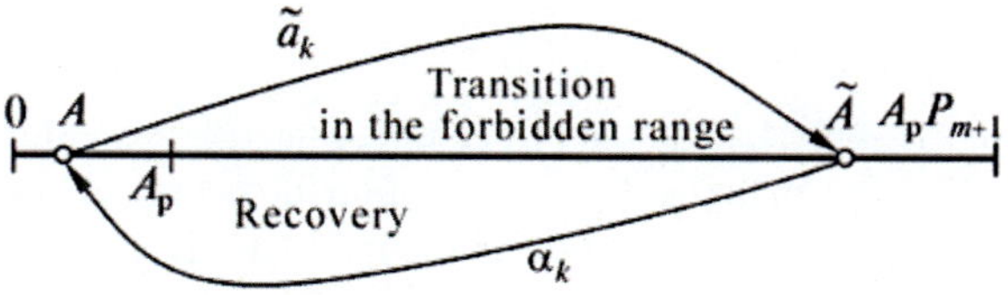

Fig. 5.9 Illustration of the process of creating and restoring an integer A

is called ordered, and the representation $A_{SRC,i} = (\alpha_1, \ldots, \alpha_{i-1}, \alpha_{i+1}, \ldots, \alpha_{m+1})$ obtained from A_{SRC} by removing the remainder α_i is called the projection of the number A by modulo p_i. An arbitrary error in one residue A_{SRC} is called single, in two residues—double, in several–multiple. The detection of errors in A_{SRC} is justified by the following statement.

Proposition 2 Let the modules of the ordered SRC $p_1, p_2, \ldots p_m, p_{m+1}$ are coprime numbers and let $A_{\mathrm{SRC}} = (\alpha_1, \alpha_2, \ldots \alpha_k, \ldots \alpha_{m+1})$ is the correct number. Then the number $\widetilde{A}_{\mathrm{SRC}} = (\alpha_1, \alpha_2, \ldots \widetilde{\alpha}_k \neq \alpha_k, \ldots \alpha_{m+1})$ with the distorted remainder $\widetilde{\alpha}_k$ is the wrong number.

So, the arbitrary distortion of one of the residues, which is the number $A < A_{\mathrm{p}}$ in the SRC, translates it, if it is restored, in the positional system to the interval $[A_{\mathrm{p}}, A'_p)$. The process of switching to the full range of the wrong number with the distortion of one remainder and returning the number to the working range after correcting the remainder is shown in Fig. 5.9.

Distortion of any remainder in the new representation of $(a_1, \ldots, a_{m+1})$ leads to the fact that the number $\widetilde{A} = \left(\sum_{i=1}^{m+1} a_i B'_i\right)(\bmod A_p)$ is restored, where $(B'_1, \ldots B'_i, \ldots B'_{m+1})$ is the new system of orthonormal bases, goes from the working range of $\left[0, A_{\mathrm{p}}\right)$ (from the range of the so-called regular digits) to the range of $\left[A_{\mathrm{p}}, A_{\mathrm{p}} p_{m+1}\right)$, is a sign of error.

Proposition 2 provides a theoretical basis for detecting and correcting errors in the data represented by SRC codes. The criterion for the absence of an error in A_{SRC} is the fulfillment of the condition

$$A = \left[\sum_{i=1}^{m+1} \alpha_i B_i\right](\bmod A_{\mathrm{p}}) < A_{\mathrm{p}}. \tag{5.10}$$

The safety check of A_{SRC} can be performed in another way—by comparing various projections of the reconstructed number. According to [32, 33], the introduction of only one additional module $p_{m+1} > p_m$ makes it possible to identify all single errors and about 95% of double errors.

Proposition 3 Let the correct number $A = (\alpha_1, \alpha_2, \ldots \alpha_k, \ldots \alpha_{m+1})$ be given in the SRC with ordered modules $p_1, p_2, \ldots p_m, p_{m+1}$, which are coprime numbers. The numbers $A_1, A_2, \ldots A_i \ldots A_{m+1}$ are reconstructed from the projections of $A_{SRC,i}$ on different bases of the SRC, that is,

$$A_1 = A_2 = \cdots = A_i = \cdots = A_{m+1} = A. \tag{5.11}$$

If $A > A_p$, and the projection $A_i < A_p$, this indicates the presence of an error in the i-th remainder. Correction of the detected errors requires additional calculations and can be performed using the methods given in [34–36].

Application of SRC for distance measurement in phase range finders. Let a harmonic signal (for example, of electromagnetic nature) propagate along the distance D in the forward and reverse directions, which leading to a phase shift of the signals $\Phi(D) = 4\pi D\lambda^{-1}$:

$$\begin{aligned} u_1(t) &= U_1 \sin(2\pi ft), \quad t \in [0, T_c]; \\ u_2(t, x) &= U_2 \sin[2\pi ft - \Phi(D)], t \in [0, T_c], \end{aligned} \tag{5.12}$$

where T_c, f—respectively, the observation time and signal frequency.

Phase shift is available for measurement:

$$\varphi(D) \equiv \Phi(D) \bmod 2\pi \equiv 4\pi D\lambda_i^{-1} \bmod 2\pi. \tag{5.13}$$

Let the distance $D \in (0, D_{\max})$ with a discrete step d_0 be determined unchanged on the observation time interval T_c. Let it will be organized measurements of the phase shifts of signals (5.12) at m frequencies $\{f_i, i = \overline{1, m}\}$ with wavelengths $\lambda_i \ll D_{\max}$ as follows. Working frequencies it will be chosen multiples of mutually prime numbers p_i, $i = \overline{1, m}$:

$$f_i = v/p_i d_0 = v/\lambda_i, \quad i = \overline{1, m}, \tag{5.14}$$

where v—wave propagation velocity in the medium.

The set of harmonic signals with frequencies (5.14) is a set with orthogonal components in the time interval $T_{opt} = d_0/v\prod_{i=1}^{m} p_i$, which makes it possible to apply a polyharmonic signal and perform phase measurements simultaneously at all frequencies.

The measurement result is a vector of phase shifts of the signals $\overline{\varphi_m} = (\varphi_1, \ldots \varphi_i, \ldots .\varphi_m)$; for each $\varphi_i(D)$ a comparison performs—$\varphi_i(D) \equiv 4\pi D\lambda_i^{-1}(\bmod 2\pi)$. The vector $\overline{\varphi_m}$ uniquely determines the distance D.

Comparison (5.13) according to (5.6) is presented as follows:

$$\frac{\varphi_i(D)}{2\pi} p_i \equiv \frac{2D}{\lambda_i} p_i (\bmod p_i). \tag{5.15}$$

Further, it will be chosen a discrete step for measuring phase shifts of signals

$$\Delta\varphi_i = 2\pi/p_i, \quad i = \overline{1, m}. \tag{5.16}$$

Then comparison (5.15) can be rewritten as

$$\frac{\varphi_i(D)}{\Delta\varphi_i} \equiv \frac{2D}{d_0}(\operatorname{mod} p_i). \tag{5.17}$$

Taking into account only the integer parts of the comparison (5.17), it will be written an equation for remainders:

$$\alpha_i(D) = \left[\varphi_i(D)/\Delta\varphi_i\right]^+. \tag{5.18}$$

The choice of the $\Delta\varphi_i$ values and the process of determining the residuals $\alpha_i(D)$ for the system of modules (5, 7) is illustrated in Fig. 5.10, for which $A = 24$ $A_{SRC} = (4, 3)$.

Based on the residues determined on all m frequencies, it will be compiled a system of comparisons:

$$\begin{cases} \alpha_1(D) \equiv A(D)(\operatorname{mod} p_1), \\ \ldots \\ \alpha_i(D) \equiv A(D)(\operatorname{mod} p_i), \\ \ldots \\ \alpha_m(D) \equiv A(D)(\operatorname{mod} p_m). \end{cases} \tag{5.19}$$

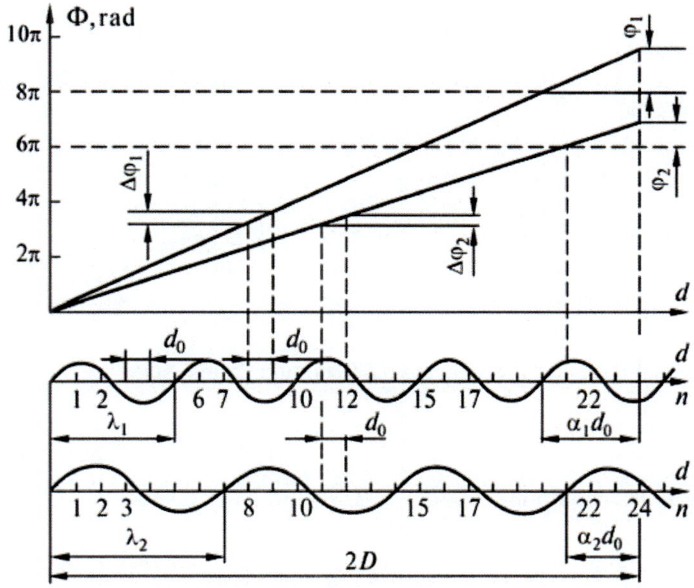

Fig. 5.10 Graphic illustration of the process of displaying the results of measuring the phase shifts of signals (distances) by integers in SRC with modules (5, 7)

In (5.19), the number $A(D)$ is equal to the number of segments of length $0.5d_0$, which fit at the measured distance, and $\alpha_i(D)$ are the residuals $A(D)$ in the modules p_i, $i = \overline{1, m}$. Thus, the initial conditions (5.14) and (5.16) imposed on the choice of frequencies and the discreteness of measuring phase shifts allow to display the numerical result of measuring $A(D)$ by the remainders of this number according to the system of modules $(p_1, \dots p_i, \dots p_m)$ and to reduce the ambiguity of the phase measurements to the task of restoring the number $A(D)$ with $A(D)_{\text{SRC}} = (\alpha_1(D), \dots \alpha_i(D), \dots \alpha_m(D))$.

With the proper choice of p_i and m, the result of measuring the distance has the form

$$D = \frac{d_0}{2}\left[\sum_{i=1}^{m}\left[\frac{\varphi_i(D)}{\Delta\varphi_i}\right]^{+} B_i\right](\text{mod}\, A_p). \tag{5.20}$$

Thus, the determination of the distance by expression (5.20) is similar to restoring the number A from its representation A_{SRC}, and there is a correspondence between the SRC data and the signal parameters:

$$n_i \Rightarrow f_i, \quad a_i \Rightarrow \varphi_i, \quad A \Rightarrow \Phi. \tag{5.21}$$

Example 5.2 The distance 1955.1 m is measured with a step of 1 m ($D = 1955.1$ m, $d_0 = 1$ m), the SRC modules are $p_1 = 13$, $p_2 = 17$, $p_3 = 19$. Electromagnet waves ($\nu = 3 \cdot 10^8$ m/s) are also used.

The operating range of the SRC is $A_p = p_1 p_2 p_3 = 13 \cdot 17 \cdot 19 = 4199$.

Checking the conditions for the unambiguous measurement of the distance $D < D\text{max}$ gives the result:

$$D\text{max} = 0.5 d_0 A\text{max} = 0.5 \cdot 4198 = 2099 \text{ m} > D.$$

The calculation of frequencies according to (5.14) gives the value

$$f_1 \approx 23.077 \text{ MHz}, \quad f_2 \approx 17.647 \text{ MHz}, \quad f_3 \approx 15.789 \text{ MHz}.$$

Orthogonal bases for the selected SZK module system are:

$$B_1 = 1938, \quad B_2 = 494, \quad B_3 = 1768.$$

Checking the conditions for the correct definition of bases is:

$$\left[\sum_{i=1}^{3} B_j\right](\text{mod}\, A_p) = [1938 + 494 + 1768](\text{mod} 4199) = [4200](\text{mod} 4199) = 1.$$

Table 5.2 Initial data

Parameters	f_1	f_2	f_3
$\Delta\varphi_i$, rad	0.4833	0.3696	0.3307
$\varphi_i(D)$, rad	4.9361	0.0691	4.9876

The calculated output data—$\Delta\varphi_i$, obtained by (5.16) and $\varphi_i(D)$ (5.13), are summarized in Table 5.2.

At the data processing stage, the residues $\alpha_i(D)$ (5.18) are firstly determined, and in the absence of measurement errors, it can be obtained:

$$\alpha_1(D) = 10, \quad \alpha_2(D) = 0, \quad \alpha_3(D) = 15.$$

The result of measuring the distance according to (5.20) is formed as

$$\begin{aligned} D &= 0.5 \cdot [10 \cdot 1938 + 0 \cdot 494 + 15 \cdot 1768](\text{mod}4199) \\ &= 0.5 \cdot 45900(\text{mod}4199) = 1955 \text{ m}. \end{aligned}$$

The phase method allows to refine the result obtained within one step. It should be noted that modular arithmetic requires error-free initial data for the correct recovery of numbers. Even minor errors in the measurement of phase shifts of signals can lead to distortion of the final result. Therefore, it is advisable to carry out the calculations in the extended SRC using the methods of error detection/correction.

Using of SRC for determining the azimuth in the phase direction finder. Phase direction finders [36] are used to determine the bearing—the angle between the direction to the source of radio signals with a harmonious carrier and one of the planes taken as the origin of the angular coordinates. In aviation and maritime navigation, bearing is understood as azimuth. The principle of operation of direction finders is based on the fact that the normal to the phase front of a plane wave propagating in a homogeneous medium coincides in space with the direction to the radiation source. Information on the orientation of the phase front of the wave in space is obtained by measuring and analyzing the phase shifts of signals received by spatially separated elements of linear antennas (Fig. 5.11).

The last should have wide radiation patterns and identical phase-frequency characteristics. For increasing the accuracy of bearing detection over a wide range of values, antenna systems with several bases are used and the bases' values are increased (the base is the distance between the phase centers of the antennas).

Received signals have form:

$$u(t, \alpha_x, l_i) = U_i \cos[2\pi ft - \varphi_i(\alpha_x, l_i)], \; t \in [0, T_c), \tag{5.22}$$

where U_i, f, $\varphi_i(\alpha_x, l_i)$ are the amplitude, frequency and initial phase of the signal at the output of the i-th element of the antenna ($i = 0, 1, 2$), $\varphi_{1,2} \in [0, 2\pi)$, $\varphi_0 = 0$;

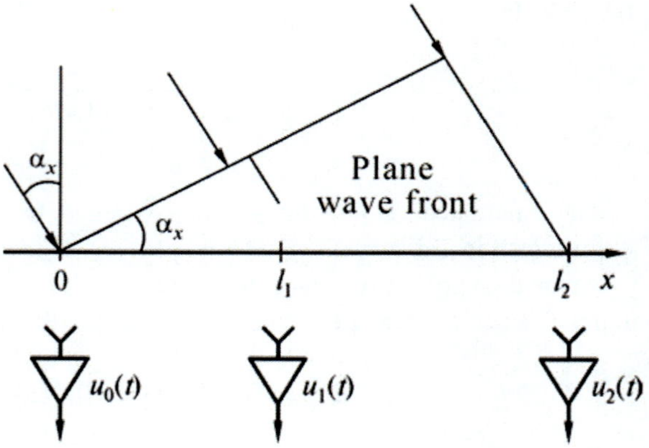

Fig. 5.11 Signal reception scheme of spatially separated elements of a direction finder antenna

l_i—i-th base of the antenna; T_c—signal observation time, $T_c > 1/f$; α_x—measured azimuth.

The linear antenna elements in the two-base phase direction finder (Fig. 5.11) are placed in space with relative to the reference element (with index $i = 0$) at a distance of l_1, l_2. It is allowed that the distance from the antenna to the signal source is much greater than l_2, which allows to consider the wave to be plane, $l_1 = p_1\Delta l, l_2 = p_2\Delta l$, where Δl is the quantum of the antenna bases, p_1, p_2 are integers.

The delay of the signals arriving at the first and second elements of the antenna relative to the zero element is

$$\tau_i(\alpha_x, l_i) = l_i \sin\alpha_x / \nu, i = \overline{1, 2}. \tag{5.23}$$

The total phase shifts of the signals between the zero and the i-th ($i = 1, 2$) antenna elements are analytically determined as

$$\Phi_i(\alpha_x, l_i) = 2\pi\, l_i \lambda^{-1} \sin\alpha_x, \quad i = \overline{1, 2}. \tag{5.24}$$

and their parts available for unambiguous measurement in $[0, 2\pi)$ are given by the formula

$$\varphi_i(\alpha_x, l_i) = 2\pi\Delta l p_i \lambda^{-1} \sin\alpha_x (\text{mod} 2\pi), \quad i = \overline{1, 2}. \tag{5.25}$$

The modular nature of the dependence of $\varphi_i(\alpha_x, l_i)$ on azimuth determines the possibility of eliminating the ambiguity of measurements α_x based on the use of modular arithmetic.

It can be obtained the conditions under which the task of eliminating the ambiguity of measurements of α_x in the phase direction finder is reduced to the task of restoring an integer of its representation in the SRC.

Proposition 4 In order for the results of measurements in a two-base phase direction finder, in the case of a single-beam propagation of a harmonic signal from a single radiation source, can be represented by residuals in the SRC with (p_1, p_2) modules, the following conditions must be met:

(1) the bases of the direction finder must be multiples (p_1, p_2);
(2) quanta of measurement of phase shifts of signals in the direction finder channels should be selected from condition $\Delta\varphi_{1(2)} = 2\pi / p_{2(1)}$.

This statement is not difficult to prove. It will be denoted the wavelength in the signal propagation medium λ, and the direction finder base $l_1 = p_1 \Delta l$, $l_2 = p_2 \Delta l$. Let the wave arrive at the direction finder at an angle α_x. The phase shifts in the interval $[0, 2\pi)$ in the direction finder channels are measured without errors. The total phase shift in the direction finder channels is determined by expression (5.24).

In order to present direction finder data in SRC, it is necessary to ensure equality of angle quanta at different bases. This condition is transformed into a condition for selecting various quanta of the distance measurement that the signal travels in various direction finder channels. Based on the logic of data presentation in SRC and in order to coordinate modular operations to determine the residuals and wavelength λ, it is necessary to fulfill the conditions

$$\lambda = \lambda_0 p_1 q_1, \quad \lambda = \lambda_0 p_2 q_2, \tag{5.26}$$

where λ_0 is part of the wavelength, $q_1, q_2, p_1, p_2 \in N$. It follows from (5.26) that $p_1 q_1 = p_2 q_2$. It can be turned this equality into an identity by setting $q_1 = p_2$, $q_2 = p_1$, then $\lambda_0 = \lambda / p_1 p_2$. So, the quantization of distances for different bases must be performed with steps

$$\Delta\lambda_1 = \lambda / p_2, \quad \Delta\lambda_2 = \lambda / p_1. \tag{5.27}$$

By quantizing the distance $l_{1(2)} \sin \alpha_x$ with quanta (5.27), it can be obtained the remainders

$$a_{1(2)}(\alpha_x, l_{1(2)}) \equiv \left[\frac{l_{1(2)} \sin \alpha_x}{\Delta\lambda_{1(2)}} \right]^{+} \bmod p_{2(1)} \equiv \left[\frac{\Delta l \sin \alpha_x}{\lambda_0} \right]^{+} \bmod p_{2(1)}. \tag{5.28}$$

Using statement 1, transform (5.28) can be formed to

$$a_{1(2)}(\alpha_x, l_{1(2)}) \equiv \left[\frac{p_{2(1)}}{2\pi} \left(\frac{2\pi\, l_1 \sin \alpha_x}{\lambda} \bmod 2\pi \right) \right]^{+}. \tag{5.29}$$

From (5.29), we have the part of $\Phi_{1(2)}(\alpha_x, l_{1(2)})$ available for measurement

$$\begin{gathered} 2\pi\, a_{1(2)}(\alpha_x, l_{1(2)}) / p_{2(1)} = \varphi_{1(2)}(\alpha_x, l_{1(2)}) = \\ = (2\pi\, l_{1(2)} \lambda^{-1} \sin \alpha_x) \bmod 2\pi = \Phi_{1(2)}(\alpha_x, l_{1(2)}) \bmod 2\pi \text{'} \end{gathered} \tag{5.30}$$

which was required to prove.

Given the above, it can be calculated the guiding sine, with an error up to the quantum of the phase shift

$$\sin\alpha_x = A\lambda/\Delta l p_1 p_2^2 = \lambda\left(\sum_{i=1}^{2} a_i(\alpha_x, l_i)B_i\right)(\operatorname{mod} A_p)/\Delta l p_1 p_2^2. \tag{5.31}$$

From Eq. (5.31) it can be obtained the azimuth value unique in a wide sector of angles, but with a significant quantization error. For increasing the accuracy of determining α_x by using the capabilities of precision measurement of phase shifts of signals, in (5.31), instead of the number A, it is necessary to substitute its updated (with a fractional part) value:

$$A_T = (\alpha_1(\alpha_x, l_1)B_1 + \alpha_2(\alpha_x, l_2)B_2) \bmod A_p$$
$$- \left[\frac{p_1}{2\pi}\left(\frac{2\pi\Delta l \sin\alpha_x}{\lambda}\right) \bmod 2\pi\right]^+ + \frac{p_1}{2\pi}\left(\frac{2\pi\Delta l \sin\alpha_x}{\lambda}\right) \bmod 2\pi. \tag{5.32}$$

Taking into account (5.31) and (5.32), the azimuth value is calculated as

$$\alpha_x = \arcsin\left(A_T\lambda/\Delta l p_1 p_2\right). \tag{5.33}$$

The sector of unambiguous determination of azimuth is limited by the angle

$$\alpha_{x,\max} = \arcsin\left(A_{\max}\lambda/\Delta l p_1 p_2\right). \tag{5.34}$$

Example 5.3 Let a plane harmonic electromagnetic wave arrive at a double-base linear antenna (Fig. 5.11) at an angle $\alpha_x = 60.75^\circ$. Let set the ratio $\Delta l/\lambda = 1.1$ and let the antenna bases relate as $l_1/l_2 = 11/13$. It is necessary to determine the azimuth α_x from the results of measuring the phase shifts of the direction finder signals, applying for phase correction polysemy SRC. There is no error in measuring phase shifts of signals.

Based on the initial data, it takes $p_1 = 11$, $p_2 = 13$, therefore $A_p = 11\cdot 13 = 143$. The correctness of the task is confirmed by the fulfillment of condition $\alpha_x < \alpha_{x,\max}$:

$$\alpha_{x,\max} = \arcsin\left(\frac{11\cdot 13 - 1}{11\cdot 13\cdot 1.1}\right)\cdot\frac{180}{\pi} \approx 64.2^\circ > \alpha_x = 60.75^\circ.$$

The solution to the problem is carried out in two stages. At the first, data for calculations are prepared: the orthogonal basis is $B_1 = 66$, $B_2 = 78$. The expected phase shifts of the signals calculated from (5.24), (5.25) are summarized in Table 5.3.

At the second stage, using the obtained $\varphi_i(\alpha_x)$, α_x is determined.

Table 5.3 Signal phase shifts

Phase shift	Basis l_1	Basis l_2
$\Phi_i(\alpha_x)$, rad	66.3328	78.3933
$\varphi_i(\alpha_x)$, rad	3.50100	2.9951

Further, it can be calculated the residuals by expression (5.30) and obtained the following result:

$$a_1(\alpha_x, l_1) = \left[13 \cdot 3.501/2\pi\right]^+ = 7, a_2(\alpha_x, l_2) = \left[11 \cdot 2.9951/2\pi\right]^+ = 5.$$

Then it can be found the number A by the Formula (5.5):

$$A = (7 \cdot 66 + 5 \cdot 78) \bmod (143) = 137,$$

and its specified value is according to the Formula (5.32):

$$A_{\mathrm{T}} = 137 - 5 + 5.2435 = 137.2435.$$

The azimuth is determined by (5.33):

$$\alpha_x = \arcsin\left(\frac{137.2435}{1.1 \cdot 11 \cdot 13}\right) \cdot \frac{180}{\pi} = 60.7499°,$$

which corresponds to the initial data of Example 5.3.

In direction finders based on SRC, a sign of distortion of one residue is obtaining a result in the field of complex numbers.

Table 5.4 presents a comparative analysis of the use of SRC for multiscale phase measurements in range finders and direction finders.

Thus, the possibility of presenting the measurement data in phase range finders and direction finders in SRC and the solution on this basis of the ambiguity of

Table 5.4 Results of phase measurements in range finders and direction finders

No.	Characteristic	Phase range finder	Phase direction finder
1	Measured parameter L, unit	Distance, meter	Flat angle, rad
2	Nature of dependence $\varphi(L)$	Linear	Nonlinear
3	Number of working frequencies	m	1
4	Number of measuring channels	1	$m + 1$
5	Approximate number of whole phase cycles n, which is determined uniquely	~100 … 1000	~10 … 40
6	Required number of RCS modules	3 … 7	2 … 4

phase measurements is proved. The combination of SRC and the phase measurement method provides the last with a unique property—the ability to detect and correct gross measurement errors. The search for distorted residues and their shutdown/correction are implemented by increasing the antenna elements (in the direction finder) or operating frequencies (in the range finders) and by complicating the algorithm for processing the measurement data. This helps to maintain the operability of such measuring instruments in conditions of a significant decrease in the s/n ratio.

5.3 Statistical Data Processing in Environmental Monitoring Systems Based on Unmanned Aerial Systems

For solving the applied problems of monitoring environmental parameters, it is especially important [37–39]:

- assessment of the spatial distribution of harmful emissions in order to determine the most probable directions of their distribution with reference to geographical coordinates;
- comparative analysis of the development in space and time of negative processes of anthropogenic nature;
- analysis of the processes of entry, accumulation and distribution of harmful substances from pollution sources, taking into account the landscape and physical and geographical features of the location of the pollution sources;
- prediction of the most likely scenarios for the development of processes of the impact of harmful substances on the environment.

The solution of these tasks allows to create new and improve existing models for the spread of the effects of harmful emissions, industrial incidents and disasters, and make informed management decisions to minimize the consequences of accidents.

Obtaining measurement information to solve such problems should be carried out using computerized information-measuring systems (CIMS) based on unmanned aerial systems (UAS). Such an implementation of the hardware component of the environmental control system allows to maintain the system's operability in a wide range of changes in meteorological conditions, time of year and day; conduct chemical and radiation reconnaissance; quickly reconfigure the system to control various harmful substances; change in accordance with the task the list of points or control routes; establish, in the presence of a subsystem for monitoring the current time, causal relationships between various events in the control zone; control environmental parameters and characteristics in hazardous conditions for humans; develop and refine models of the processes of atmospheric transport of radionuclides and other harmful substances and their deposition on the earth's surface for the assessment and prediction of environmental pollution by harmful emissions of energy facilities; receive reliable experimental data.

In CIMS based on the UAS, it is usually assumed that the sensors move in space in the area of pollution sources along certain closed paths. The last may change due to changes in the wind rose, other meteorological conditions or control tasks. The current position of an unmanned aerial vehicle (UAV) with sensors in space and time is periodically recorded with sufficient accuracy, for example, using the GPS system.

The primary measurement information obtained in this way, for example, the exposure dose rate or the background radiation level, which is determined using Geiger-Muller counters, can be used to construct histograms of the distribution of such data, function gradient fields, lines of the same pollution levels (contours) of the space with various harmful emissions and the like.

In the process of processing the primary measurement data, another important task arises—their analysis in the polar coordinate system, which allows to study and predict the most dangerous directions of the spread of harmful emissions by the methods of statistical analysis of angular observations [40–48].

Let consider the task of analyzing the primary data of environmental control using CIMS based on the UAS. Let the UAV move in the vicinity of the control object at a considerable distance from it and a constant height along a closed path, which is a second-order curve. This trajectory covers a source of harmful emissions, the dimensions of which are much smaller than the size of the trajectory. This allows to consider the source of atmospheric pollution as a point source, whose coordinates (x_0, y_0) on the earth's surface are known. Sensor data (results of measuring the parameter s) are obtained with regular, that is, periodic in time sampling with a determinate period T_d, and with reference in time and space, that is, as a function of $s(x, y, t, \bar{p})$, where $\bar{p}$ is the vector of meteorological parameters.

It is necessary to construct a circular distribution in the space of the parameter s.

The solution to this measurement problem has certain features characteristic of angular measurements. To simplify the problem, it can be assumed that the conditions for the measurement experiment are unchanged (stationary), which allows to consider the experimental results as independent of the arguments t and $\bar{p}$.

The process of accumulation of primary measurement information occurs during the movement of the UAS along a closed path. For an example in Fig. 5.12, in the Cartesian coordinate system xOy, the external curve 2 is such path. The origin

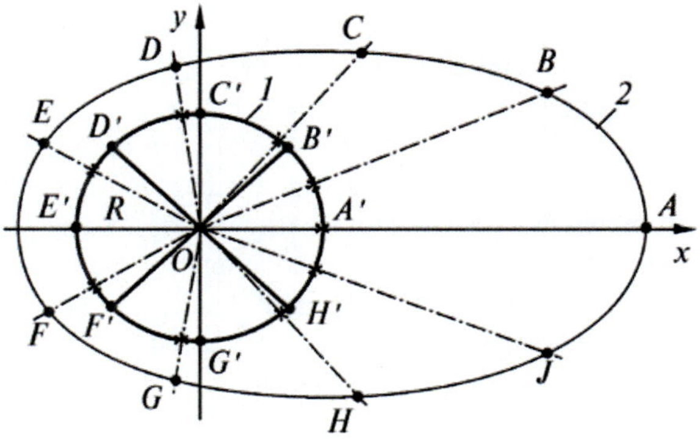

Fig. 5.12 Example of mapping observation results with an arbitrary second-order curve (curve 2) onto a circle (curve 1)

of the coordinate system $x_0 = 0$, $y_0 = 0$ is tied to the emission source. A set of $(A, B, C, D, E, F, G, H, J)$ points with known coordinates at which the controlled parameter is determined are indicated on the flight path. In Fig. 5.12, these points are distributed uniformly along the external curve (i.e., with the same distances between the points along the flight path), which, with subject to regular sampling, is possible in the case of wind absence, that is, when $V = 0$. In the case of projecting these points on a circle of radius R (curve 1), it can be got a set of points unevenly distributed around the circle (on the Fig. 5.12 they are indicated by asterisks). Such a transformation of data from curve 2 to curve 1 in each case should be performed taking into account the coordinates of the points on these curves and the physics of the formation of fields of harmful emissions or radiation.

In the presence of wind or other destabilizing meteorological factors, the angular unevenness of the obtained experimental data will increase. Such irregularity is not artificial, but naturally arises as a consequence of the method of obtaining primary measurement information using UAVs, therefore, in principle, it cannot be eliminated by choosing the mode of its collection.

Processing measurement data in the case of uneven discretization on a circle is inconvenient to perform. With this in mind, it is advisable to bind the experimental data to the set of points received by dividing the circle into equal class intervals. In Fig. 5.12 such points are designated as $(A', B', C', D', E', F', G', H')$. They are obtained by dividing the central angle 2π into 8 equal class intervals by the value of $\pi/4$. The parameter value at these points can be obtained by interpolating the data distributed unevenly around the circle. In the general case, the number of class intervals may not coincide with the set of points on curve 2.

With a significant number of split points, it is sufficient to apply linear interpolation. The objective conditions of the measurement experiment can lead to a decrease in the number of interpolation nodes. The result of this, for example, is the absence of a radio channel. In this case, the measurement information together with the corresponding spatial coordinates is accumulated on board the UAV. With a limited amount of random access memory (RAM) and a significant total flight time, this necessitates an increase in the sampling interval and the solution of the approximation problem presented on the circle of sparse data. In the case of reducing the split points and the need for more accurate reproduction of the dependence $s(l)$, interpolation is used by higher order polynomials.

An example of data approximation and the transition from non-uniform to uniform sampling (in angle) is shown in Fig. 5.13. On this Fig. 5.13, a circle of radius R is represented by a scan of length $2\pi R$.

From the thus obtained distribution in the space of values of the parameter s, it is necessary to determine the angular statistics, which evaluate the probabilistic characteristics of the angular distribution of this parameter in the vicinity of the source of emissions. The methodology for the development of such measurements in CIMS based on the UAS and obtaining angular statistics of the distribution of controlled quantities in space is presented by the structure of the experimental data processing algorithm (Fig. 5.14).

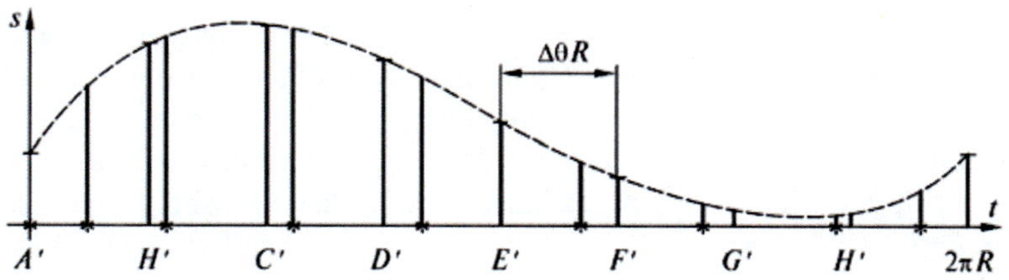

Fig. 5.13 Illustration of the transition to uniform data discretization (on a circle) using interpolations of unevenly sampled input data

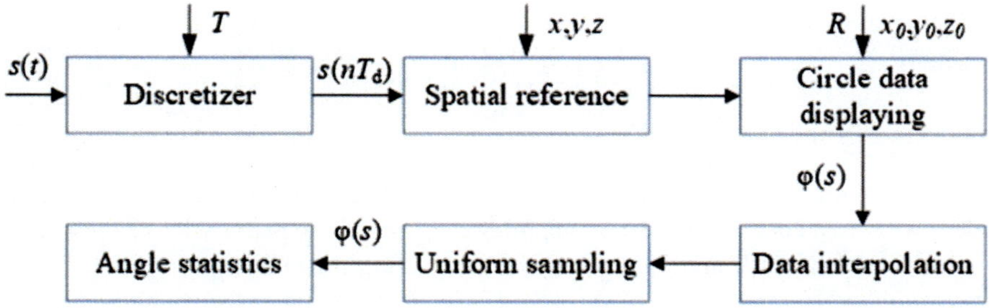

Fig. 5.14 The structure of the algorithm for the formation and processing of angular data in CIMS based on UAS

Let consider in more detail the questions of obtaining angular statistics of the distribution in the space of the parameter s. For this purpose, it was presented the obtained data in a cylindrical coordinate system (ρ, θ, s). The value of the parameter s is reflected on the polyline (or curve after interpolation of the experimental data) on the cylindrical surface, as shown in Fig. 5.15. For convenience, the radius at the base of the cylinder is $R = 1$, which does not affect on the ratio of angles and the resulting

Fig. 5.15 Graphical representation of the distribution of the parameter $s(l)$ in a cylindrical coordinate system

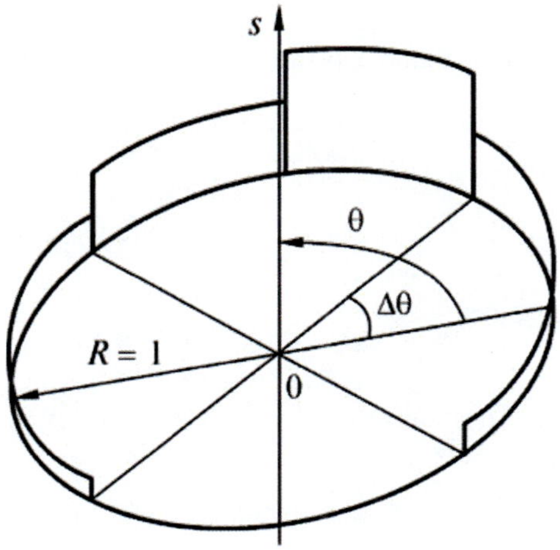

angular statistics. This simplifies the analysis, simplifies the analytical results and is generally accepted in the theory of statistical analysis of angular observations.

Let justify the transition from the sample of values of the parameter $s[j]$, $j = \overline{1, m}$ to the sample of angles as follows. Suppose that according to the results of preliminary processing, it can be obtained j class-intervals with centers at points $\theta[j] = 2\pi(j - 0.5)/m$, $j = \overline{1, m}$. To transfer the probabilistic properties of the parameter s to a sample of angles, it is necessary to choose the frequency of the angles falling into the j-th class interval, which is proportional to the corresponding relative value

$$s^*[j] = \left[s[j] \left(\sum_{j=1}^{m} s[j] \right)^{-1} K \right]^{+}, j = \overline{1, m}. \tag{5.35}$$

In expression (5.35), the $s^*[j]$ value is interpreted as the number of obtained values of the angle in the j-th class interval. For matching the numerical values of $s^*[j]$ with their contents and obtaining integer values of the quantity $s^*[j]$, the dimensionless coefficient K and the operation of extracting the integer part of the number are introduced into formula (5.35). The greater K, the smaller the rounding error. It is advisable to choose the value of K such that the rounding error does not exceed ~10^{-3}, which allows to neglect it in comparison with the relative measurement error of the controlled parameter. The values of $s^*[j]/K$, $j = \overline{1, m}$ are the columns of the circular histogram that are responsible for the empirical values of the probability density of the measured parameter. Formula (5.35) is obtained for stepwise interpolation of the distribution of the parameter s. During interpolating polynomials of higher orders, the frequency values can be obtained by the formula

$$s^*[j] = \left[K \int_{2\pi m^{-1}(j-1)}^{2\pi m^{-1} j} s(l)dl \Big/ \int_{0}^{2\pi} s(l)dl \right]^{+}, j = \overline{1, m}. \tag{5.36}$$

According to the results of such a transformation, it can be written m sets of angles

$$\begin{aligned} &\left(\theta_1[1], \ldots \theta_{s^*[1]}[1]\right), \\ &\ldots \\ &\left(\theta_1[j], \ldots \theta_{s^*[j]}[j]\right), \\ &\ldots \\ &\left(\theta_1[m], \ldots \theta_{s^*[m]}[m]\right). \end{aligned} \tag{5.37}$$

These sets together form statistics of angles Θ of volume

$$M = \sum_{j=1}^{m} s^*[j]. \tag{5.38}$$

For the obtained Θ-statistics, the sine and cosine moments of the n-th order with respect to the zero direction are determined

$$C_n = M^{-1} \sum_{j=1}^{m} s^*[j] \cos(n\theta[j]),\ S_n = M^{-1} \sum_{j=1}^{m} s^*[j] \sin(n\theta[j]). \tag{5.39}$$

In formulas (5.39), it is important to use weight processing of angles with coefficients $s^*[j]$. Such a procedure is essential and provides a probabilistic representation of the distribution in space of various physical quantities with a single dimensionless angular measure. This makes it possible to unify the processing of heterogeneous physical quantities obtained from UAV flight information and perform their angular analysis.

Through certain sine and cosine moments (5.39) of the Θ-statistics, it can be estimated other accepted in the theory of statistical analysis of angular observations, selective circular characteristics: circular average, length of the resulting vector, circular dispersion and standard deviation, circular median and mode, etc.

References

1. Lupenko, S., Lutsyk, N., Lapusta, Y.: Cyclic linear random process as a mathematical model of cyclic signals. Acta Mechanica et Automatica **9**(4), 219–224 (2015). https://doi.org/10.1515/ama-2015-0035
2. Lunden, J., Kassam, S.A., Koivunen, V.: Nonparametric cyclic correlation based detection for cognitive radio systems. In: 2008 3rd International Conference on Cognitive Radio Oriented Wireless Networks and Communications (CrownCom 2008), pp. 1–6. Singapore. https://doi.org/10.1109/crowncom.2008.4562527
3. Karpenko, O., Kuznetsov, A., Sai, V., Stasev, Yu.: Discrete signals with multi-level correlation function. Telecommun. Radio Eng. **71**(1), 91–98 (2012). https://doi.org/10.1615/TelecomRadEng.v71.i1.100
4. San Emeterio, J.L., Rodriguez-Hernandez, M.A.: Wavelet cycle spinning denoising of NDE ultrasonic signals using a random selection of shifts. J. Nondestr. Eval. **34**, 270 (2015). https://doi.org/10.1007/s10921-014-0270-8
5. Shahidan, S., Pulin, R., Bunnori, N.M., Holford, K.M.: Damage classification in reinforced concrete beam by acoustic emission signal analysis. Constr. Build. Mater. **45**, 78–86 (2013). https://doi.org/10.1016/j.conbuildmat.2013.03.095
6. Li, D., Ruan, T., Yuan, J.: Inspection of reinforced concrete interface delamination using ultrasonic guided wave non-destructive test technique. Sci. China Technol. Sci. **55**, 2893–2901 (2012). https://doi.org/10.1007/s11431-012-4882-x
7. Yu, Y., Guan, J.: Investigation of signal features of pulsed eddy current testing technique by experiments. Insight Non-Destruct. Test. Condition Monitor. **55**(9), 487–491 (2013). https://doi.org/10.1784/insi.2012.55.9.487

8. Derhunov, O., Kuts, Y., Shengur, S., Monchenko, O., Oliinyk, Y.: Improvement of ultrasonic testing method for materials with significant attenuation. Eastern-Europe. J. Enterprise Technol. **1**, **9**(91), 54–61 (2018). https://doi.org/10.15587/1729-4061.2018.122858
9. Blyznjuk, E.D., Eremenko, V.S., Kuts, YuV, Bystraya, I.N., Monchenko, E.V., Tsapenko, V.K.: Phase signal detector for ultrasonic nondestructive testing. Tech. Diagnost. Non-Destruct. Test. **2**, 21–24 (2011)
10. Fang, N., Xi, D., Xu, J., Ambati, M., Srituravanich, W.: Ultrasonic metamaterials with negative modulus. Nat. Mater. **5**, 452–456 (2006). https://doi.org/10.1038/nmat1644
11. Pantea, C., Rickel, D.G., Migliori, A.: Digital ultrasonic pulse-echo overlap system and algorithm for unambiguous determination of pulse transit time. Rev. Sci. Instrum. **76**, 114902 (2005). https://doi.org/10.1063/1.2130715
12. Babak, V., Eremenko, V., Zaporozhets, A.: Research of diagnostic parameters of composite materials using Johnson distribution. Int. J. Comput. **18**(4), 483–494 (2019)
13. Eremenko, V., Zaporozhets, A., Isaenko, V., Babikova, K.: Application of wavelet transform for determining diagnostic signs. In: Eremenko, V., Zaporozhets, A., Isaenko, V., Babikova, K. (eds.) CEUR Workshop Proceedings, vol. 2387, pp. 202–214. http://ceur-ws.org/Vol-2387/20190202.pdf
14. Zaporozhets, A., Eremenko, V., Isaenko, V., Babikova, K.: Approach for creating reference signals for detecting defects in diagnosing of composite materials. In: Shakhovska, N., Medykovskyy, M. (eds.) Advances in Intelligent Systems and Computing IV. CCSIT 2019. Advances in Intelligent Systems and Computing, vol. 1080, pp. 154–172. Springer, Cham (2020). https://doi.org/10.1007/978-3-030-33695-0_12
15. Coddington, I., Swann, W.C., Nenadovic, L., Newbury, N.R.: Rapid and precise absolute distance measurements at long range. Nat. Photon. **3**, 351–356 (2009). https://doi.org/10.1038/nphoton.2009.94
16. Kuts, Y.V., Yeremenko, V.S., Monchenko, E.V., Protasov, A.G.: Ultrasound method of multilayer material thickness measurement. In: AIP Conference Proceedings, 1096, 1115 (2009). https://doi.org/10.1063/1.3114079
17. Payaro, M., Wiesel, A., Yuan, J., Lagunas, M.A.: On the capacity of linear vector Gaussian channels with magnitude knowledge and phase uncertainty. In: 2006 IEEE International Conference on Acoustics Speech and Signal Processing Proceedings, pp. IV–IV. France (2006). https://doi.org/10.1109/icassp.2006.1661031
18. Dubrovin, A.V.: Potential direction-finding accuracy of systems with antenna arrays configured as a set of an arbitrary number of rings. J. Commun. Technol. Electron. **51**, 252–254 (2006). https://doi.org/10.1134/S1064226906030028
19. Henault, S., Antar, Y.M.M., Rajan, S., Inkol, R., Wang, S.: Impact of experimental calibration on the performance of conventional direction finders. In: 2009 Canadian Conference on Electrical and Computer Engineering, pp. 1123–1128. Canada (2009). https://doi.org/10.1109/ccece.2009.5090302
20. Anikin, A.S., Denisov, V.P.: Estimation of the small sized radio direction finder errors in case of scattered signals. In: 2016 17th International Conference of Young Specialists on Micro/Nanotechnologies and Electron Devices (EDM), pp. 61–63. Russia (2016). https://doi.org/10.1109/EDM.2016.7538692
21. Bogatyrev, V.A.: Exchange of duplicated computing complexes in fault-tolerant systems. Autom. Control Comput. Sci. **45**, 268–276 (2011). https://doi.org/10.3103/S014641161105004X
22. Sand, M., Potyra, S., Sieh, V.: Deterministic high-speed simulation of complex systems including fault-injection. In: 2009 IEEE/IFIP International Conference on Dependable Systems & Networks, pp. 211–216. Portugal (2009). https://doi.org/10.1109/dsn.2009.5270335
23. Bosilca, G., Delmas, R., Dongarra, J., Langou, J.: Algorithm-based fault tolerance applied to high performance computing. J. Parallel Distrib. Comput. **69**(4), 410–416 (2009). https://doi.org/10.1016/j.jpdc.2008.12.002

24. Euillades, L.D., Euillades, P.A., Pepe, A., Blanco, M.H., Baron, J.H.: On the generation of late ERS deformation time series through small doppler and baseline subsets differential SAR interferograms. IEEE Geosci. Remote Sens. Lett. **8**(2), 238–242 (2011). https://doi.org/10.1109/LGRS.2010.2060466
25. Kuts, YuV: Measurement of cumulative phase shifts. Tekhnichna elektrodynamika **5**, 67–72 (2001)
26. Kuts, V.Y., Kuts, Y.V.: Modular arithmetic application to calculate the azimuth for phase direction finder. Vistnyk NTUU KPI Seria – Radiotekhnika Radioaparatobuduvannia, vol. 64, pp. 23–32 (2016)
27. Xu, G.: On solving a generalized Chinese remainder theorem in the presence of remainder errors. In: Akbary, A., Gun, S. (eds.) Geometry, Algebra, Number Theory, and Their Information Technology Applications. GANITA 2016. Springer Proceedings in Mathematics & Statistics, vol. 251, pp. 461–476. Springer, Cham (2018). https://doi.org/10.1007/978-3-319-97379-1_21
28. Kaya, K., Selcuk, A.A.: Robust threshold schemes based on the Chinese remainder theorem. In: Vaudenay, S. (eds.) Progress in Cryptology—AFRICACRYPT 2008. AFRICACRYPT 2008. Lecture Notes in Computer Science, vol. 5023, pp. 94–108. Springer, Berlin, Heidelberg (2008). https://doi.org/10.1007/978-3-540-68164-9_7
29. Wang, C., Yin, Q.Y., Wang, W.J.: An efficient ranging method based on Chinese remainder theorem for RIPS measurement. Sci. China Inform. Sci. **53**, 1233–1241 (2010). https://doi.org/10.1007/s11432-010-0105-x
30. Kasianchuk, M.N., Nykolaychuk, Y.N., Yakymenko, I.Z.: Theory and methods of constructing of modules system of the perfect modified form of the system of residual classes. J. Autom. Inform. Sci. **48**(8), 56–63 (2016). https://doi.org/10.1615/JAutomatInfScien.v48.i8.60
31. Omondi, A., Premkumar, B.: Residue Number Systems. Theory and Implementation, p. 296. Imperial College Press, London (2007)
32. Eremenko, V., Zaporozhets, A., Isaenko, V., Babikova, K.: Application of wavelet transform for determining diagnostic signs. In: CEUR Workshop Proceedings, vol. 2387, pp. 202–214. http://ceur-ws.org/Vol-2387/20190202.pdf
33. Babak, V.P., Babak, S.V., Myslovych, M.V., Zaporozhets, A.O., Zvaritch, V.M.: Methods and models for information data analysis. In: Diagnostic Systems for Energy Equipments. Studies in Systems, Decision and Control, vol. 281, pp. 23–70. Springer, Cham (2020). https://doi.org/10.1007/978-3-030-44443-3_2
34. Morrassi, A., Vestroni, F.: Dynamic Methods for Damage Detection in Structures. Springer, Wien (2008). https://doi.org/10.1007/978-3-211-78777-9
35. Kuts, Y.V., Lysenko, Y.Y., Dugin, A.L., Zakrevskii, A.F.: Analysis of an Eddy-current transducer with impulsive excitation in the nondestructive testing of cylindrical objects. materials science, pp. 431–437 (2016). https://doi.org/10.1007/s11003-016-9975-4
36. Nataraj, C., Harsha, S.P.: The effect of bearing cage run-out on the nonlinear dynamics of a rotating shaft. Commun. Nonlinear Sci. Numer. Simul. **13**(4), 822–838 (2008). https://doi.org/10.1016/j.cnsns.2006.07.010
37. Yan, A.-M., Kerschen, G., De Boe, P., Golinval, J.-C.: Structural damage diagnosis under varying environmental conditions—part I: a linear analysis. Mech. Syst. Signal Process. **19**(4), 847–864 (2005). https://doi.org/10.1016/j.ymssp.2004.12.002
38. Kussul, N., Shelestov, A., Skakun, S.: Grid and sensor web technologies for environmental monitoring. Earth Sci. Inf. **2**, 37–51 (2009). https://doi.org/10.1007/s12145-009-0024-9
39. Kurzhanski, A.B., Khapalov, A.Y.: Mathematical problems motivated by environmental monitoring. IFAC Proc. Vols. **23**(8), Part 5, 529–534 (1990). https://doi.org/10.1016/s1474-6670(17)51788-7
40. Babak, S., Babak, V., Zaporozhets, A., Sverdlova, A.: Method of statistical spline functions for solving problems of data approximation and prediction of objects state. In: CEUR Workshop Proceedings, vol. 2353, pp. 810–821. http://ceur-ws.org/Vol-2353/paper64.pdf
41. Zaporozhets, A., Babak, V., Isaienko, V., Babikova, K.: Analysis of the air pollution monitoring system in Ukraine. In: Systems, Decision and Control in Energy I. Studies in Systems, Decision and Control, vol. 298. Springer, Cham (2020). https://doi.org/10.1007/978-3-030-48583-2_6

42. Liukkonen, M., Heikkinen, M., Hitunen, T., Halikka, E., Kuivalainen, R., Hiltunen, Y.: Artificial neural networks for analysis of process states in fluidized bed combustion. Energy **36**(1), 339–347 (2011). https://doi.org/10.1016/j.energy.2010.10.033
43. Babak, S., Myslovych, M., Sysak, R.: Module structure of UAV-based computerized systems for remote environment monitoring of energy facilities. In: 2016 17th International Conference Computational Problems of Electrical Engineering (CPEE), pp. 1–3. Poland (2016). https://doi.org/10.1109/cpee.2016.7738752
44. Babak, S., Myslovych, M.: Practical application peculiarities of autonomous diagnostic complexes for thermal control of overhead power lines. Techn. Electrodyn. **1**, 73–80 (2016). https://doi.org/10.15407/techned2016.01.073
45. Zaporozhets, A., Kovtun, S., Dekusha, O.: System for monitoring the technical state of heating networks based on UAVs. In: Shakhovska N., Medykovskyy M.O. (eds.) Advances in Intelligent Systems and Computing IV. CSIT 2019. Advances in Intelligent Systems and Computing. Springer, Cham, pp. 935–950 (2020). https://doi.org/10.1007/978-3-030-33695-0_61
46. Zaporozhets, A.O.: Correlation analysis between the components of energy balance and pollutant emissions. Water Air Soil Pollut. **232,** 114 (2021). https://doi.org/10.1007/s11270-021-05048-9
47. Zaporozhets, A.O., Khaidurov, V.V.: Mathematical models of inverse problems for finding the main characteristics of air pollution sources. Water Air Soil Pollut. **231**, 563 (2020). https://doi.org/10.1007/s11270-020-04933-z
48. Zaporozhets, A.: Review of quadrocopters for energy and ecological monitoring. In: Babak, V.P., Isaenko, V.M., Zaporozhets, A. (eds.) Systems, Decision and Control in Energy I, Studies in Systems, Decision and Control. Springer, Cham, pp. 15–36 (2020). https://doi.org/10.1007/978-3-030-48583-2_2

Chapter 6
Models and Measures for Standardless Measurements of the Composite Materials Characteristics

Abstract This chapter is devoted to the application of virtual measures in methods of standardless diagnostics of composites, which allows to simulate various types of defects or damage levels of a product without the use of physical standards, which reduces the time, technical and economic costs associated with the manufacture of reference samples. A method for constructing virtual measures (standards) that can be used to train and configure diagnostic systems, assess the reliability of diagnostic methods is considered. The method allows to synthesize a model of the information signal of the flaw detector, which takes into account the deterministic and random components of the characteristics of real signals. The use of neural networks in standardless diagnostic systems allows not only to recognize the information signals received during control, but also to store information about the patterns and the relationship of the characteristics of the information signal and the state of the control object, provides the ability to correctly classify information signals corresponding to possible defects that were not encountered during training network.

Keywords Standardless diagnostics · Virtual measure · Mathematical model · Information signal · Diagnostic systems · Neural networks · Network architecture

Virtual measure for evaluating the characteristics of composite materials can be considered as an image of a signal or signal field obtained using a simulation model built on the basis of a priori information about the regularity of changes in the characteristics of an information signal with a corresponding change in the state of the research object (RO). Knowledge about the nature of changes in information signals can be obtained either experimentally using standard samples (physical measures) or based on mathematical modeling—constructing a functional dependence of changes in informative signs on state changes. Virtual measures allow to create a library of images of information signals, which allows to abandon material measures (standard samples, which reflect the state of the RO) during the training and configuration of diagnostic systems.

V. P. Babak et al., *Models and Measures in Measurements and Monitoring*,
Studies in Systems, Decision and Control 360,
https://doi.org/10.1007/978-3-030-70783-5_6

The methods for creating virtual measures include the ability of some neural network classifiers to conduct cluster analysis and create new classes or images that were not reproduced as reference ones during training.

Virtual measures also allow to simulate probabilistic measures in the form of stochastic distortions of information signals or signal fields and various kinds of noise (for example, quantization noise during studying the characteristics of measuring systems, or structural noise inherent in various kinds of media).

The creation of virtual measures allows to solve the problem of standardless diagnostics, when the creation of physical standards that reproduce various types of defects and the state of the RO is difficult or impossible.

6.1 Method for Creating Virtual Measures for Information Signals

The process of diagnosing complex objects is characterized by a large influence of random factors caused by changes in the properties of objects that arise due to the complexity of the manufacturing, installation and operation processes, a large number of types of possible defects that cannot be formalized, imperfection of control methods and diagnostic equipment, and other factors.

The reliability of the control in this case is determined not only by the physical methods used to obtain information about the technical condition of the object, but also by the mathematical models underlying the diagnostic methods, and information processing methods with the aim of creating spaces of diagnostic features and making diagnostic decisions.

In diagnostic problems, the presence of an adequate model by which the synthesis of information signals (virtual measures) occurs, which is characteristic of objects with various degrees of damage or types of defects, is of great importance, since it allows to solve several problems simultaneously. Firstly, the existence of such a model allows to construct many virtual measures that correspond to the possible states of the diagnostic object, which means that they can be used to train and configure the diagnostic system without the physical production of reference samples. Secondly, the model of information signals can be used to select the threshold value of the sensitivity of diagnostic systems, assess their legal capacity and adjust the main parameters to determine the reliability of control and classification and the like.

It will be considering the method of synthesis of virtual measures using a mathematical model that takes into account the deterministic and random components of the characteristics of real signals [1]. This method allows to synthesize any required number of various information signals, which reduces the time, technical and economic costs associated with the manufacture of reference samples.

In diagnostic systems, a wide class of information signals, there are pulsed or radio-pulsed signals, which characterized not only by the amplitude, duration, frequency, and phase of the carrier, but also by their shape. For example, such information signals are inherent in acoustic, electromagnetic and other non-destructive testing methods. Changing the shape of pulsed signals in many cases is the most informative and noise-tolerant characteristic, so it is advisable to consider the construction of a mathematical model of the signal, which allows to simulate just this change.

Modeling a virtual measure in the form of an information signal as a whole, rather than its individual parameters, allows to development classifiers that operate with the entire set of signal characteristics and, thus, provide the construction of more adequate crucial diagnostic rules.

As an example, to build a simulation model, it was used information signals from a system that implements control of composite products and coatings by the low-speed impact method [2–5]. The objects under investigation are composite panels with damage from impacts with different kinetic energies from 2.3 to 5.1 kJ [6].

Since information signals are characterized by a large number of parameters, the inclusion of which can significantly complicate the construction of a simulation model, one of the main steps is to reduce the dimension of the feature space, that is, the choice of the most informative parameters that have maximum sensitivity to changes in the properties of the diagnostic object. Having determined the most informative signs, by changing them, it is possible to simulate an information signal in order to create a library of virtual standards for setting up diagnostic systems, developing diagnostic methods, verifying the reliability of systems, their validation, and the like. The main reasons for reducing the dimension of the feature space is the fact that the use in the diagnosis of signs, the value of which is greatly influenced by the action of random factors, can lead to a decrease in the reliability of control and making the wrong decision. Reducing the number of features can reduce computational complexity and increase the universality of the classifier. A formal approach to the selection of such signs does not exist. However, at their choice, general requirements can be formulated, such as: suitability for measurement, information content and noise immunity.

During solving the problem of evaluating the information content of signs, it is necessary to take into account the influence of random factors, such as the presence of interference in the measuring channels, random errors of the sensors, spatial heterogeneity of objects, and the like.

Thus, from the vector of signs of dimension M, $A \in \{a_0, a_1, \ldots, a_{M-1}\}$ it is necessary to choose the most informative ones, that is, to obtain a new vector of signs of dimension M', $B \in \{b_0, b_1, \ldots, b_{M-1}\}$ (the set of signs that must be selected during the selection process), moreover $M' < M$. Then the selection problem actually reduces to mapping one set to another: $A \to B$.

The proposed method is a way of distinguishing features characterized by large distances between classes and small within classes.

One of the methods for selecting diagnostic features is an analysis of variance. It is carried out to assess the degree of change in informative signs, under the influence of certain variable factors (multivariate analysis) or factor (one-factor analysis). In general, the task of dispersion analysis is to isolate from the total variance the variance due to the influence of the factor (factor dispersion) and due to the action of unaccounted factors (residual variance) [2, 7]. Then the level of influence of the factor is distinguished by the coefficient of significance:

$$\beta_x = C_x / C_y, \tag{6.1}$$

where C_x—factorial variance; C_y—total variance.

The degree of factor influence a on a change in the value of a diagnostic sign is determined by comparing it with a certain threshold level β_{th}. Signs that have a large degree of factor influence on a change in their values, form a resulting vector, which can be further used in diagnostic and classification tasks.

Since a simulation model of the information signal as a whole is constructed, and not its individual components, it is advisable to use its spectral representation. The use of the most frequent Fourier transform complicates the simulation procedure, since it reflects an array of real numbers (discrete information signal) into a complex-valued array. Therefore, for the spectral conversion of information signals, it is advisable to use the discrete orthogonal functions of Hartley, Chebyshev, Laguerre and others, which allow to obtain real values of spectral coefficients [2, 8–10].

The calculated pulse spectra are damped. This made it possible to limit the number of analyzed coefficients to the first n_1, which have more than 99% of the total energy of the studied information signals. For the schedule in each basis, it can be distinguish a set of coefficients that meet two requirements: they are the largest in magnitude compared to others, and they allow to separate the values of the coefficients of spectral decomposition, the values of which differ significantly for each class.

Quantitatively, this conclusion can be justified using the coefficient obtained from Bessel inequality, which characterizes the energy contribution of the first n_1 components of the decomposition to the total energy of the information signal:

$$K_{n,n_1} = \sum_{j=0}^{n_1-1} |a_j|^2 / \sum_{k=0}^{n-1} |a_k|^2, \tag{6.2}$$

where a_k—the value of spectral factor; n—total amount of spectral components.

Thus, if it selects the n_1 first decomposition coefficients for diagnostic features, they will characterize the decomposition components, which make the largest contribution to the energy of the information signal. It also reduces computational costs and improves the efficiency of working with the received signals.

Some of the expansion coefficients are more subject to the influence of random factors than others, that is, they have a large intra-group dispersion. In addition, the spectral decomposition coefficients change differently when the degree of defectiveness of the product changes, that is, they have different values of factor dispersion. It does not make sense to take into account all the obtained n_1 spectral components, since this will lead to an unjustified complication of the simulation model and a corresponding increase in computational costs.

For the obtained values of the decomposition coefficients a_k, $k = \overline{0, n_1}$, in order to identify the most informative coefficients and reduce the space of signs, the dispersion analysis is carried out. In this case, for each of the first coefficients, the total scattering C_y (scattering within one group of coefficients characterizing one class), intragroup scattering C_z and intergroup scattering (scattering between groups of coefficients characterizing different classes) C_x of values of the investigated attribute is determined. A measure of the degree of defectiveness of a product on a change in the value of informative features is the significance coefficient β_x.

For forming a set of diagnostic features, the following a_k, $k = \overline{0, n_1}$ coefficients should be used, the value of β_{xk} for which exceeds a certain threshold level. In Fig. 6.1, as an example, diagrams are given for assessing the effect of changes in the degree of defectiveness on the spectral components of a pulsed information signal. Columns corresponding to the decomposition coefficients for which the β_x coefficient exceeds the level of 0.95, highlighted in solid color. An analysis of the above diagrams shows that the five spectral decomposition coefficients (a_0, a_2, a_4, a_5, a_8) are the most informative.

Thus, the procedure for ranking diagnostic features is carried out in two stages: the selection of the coefficients of the largest number of spectral decomposition coefficients n_1 of the coefficients that have the highest energy, and the selection of of those coefficients (n_1'), the value of which most depends on the degree of defectiveness of the sample and less depends from the influence of random factors. That is, from a space of dimension M, a subspace with dimension M' is selected which

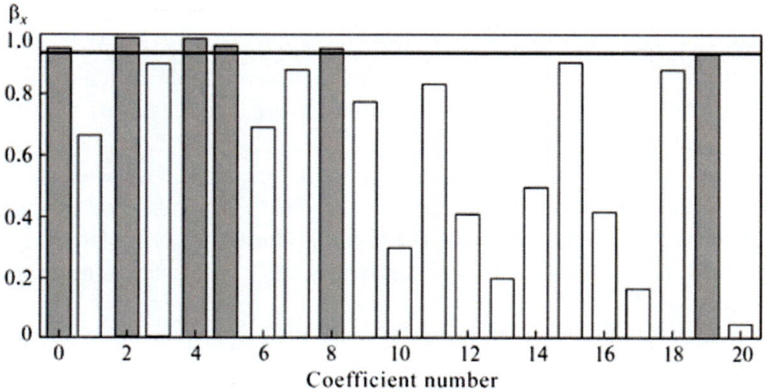

Fig. 6.1 Assessing the effect of defectiveness on the coefficients of spectral decomposition

makes it possible to approximate, with a given accuracy, a given type of information signals corresponding to certain measuring transducers. This is analytically described by the expression:

$$\left|X_i(\mathrm{Z}) - \underset{i}{\overset{*}{X}}(\mathrm{Z})\right| \leq \alpha, \tag{6.3}$$

where $X_i(\mathrm{Z})$—physical information signal from a space of dimension M; $\underset{i}{\overset{*}{X}}(\mathrm{Z})$—approximated signal based on the subspace M'; α—permissible error (discrepancy) between signals; $Z = \{z_0, z_1, \ldots, z_j\}$—area of signal detection area $X(Z)$, $j \in \overline{0, N-1}$; N—the number of samples of a discrete signal $X(Z)$.

This approach makes it possible to significantly reduce the number of spectral decomposition coefficients for analysis and simulation of an information signal. In this problem, the dimension of the feature space will be reduced to $\mathrm{N} = 5$.

For determining the values of the corresponding decomposition coefficients characteristic of information signals, which corresponding to the states of objects for which these signals were not investigated, that is, to build virtual reference signals, it is necessary to obtain a function that approximates the distribution of the values of each of the spectral decomposition coefficients depending on the state of the object under study. Such a function can be determined by interpolating the known values of the expansion coefficients, for example, power polynomials or splines. Further, for each spectral component, it is necessary to select the desired value of the degree of defectiveness x of the object, determine the value of the spectral components from the established functional dependencies, and perform the inverse transformation [11–13].

In interpolation problems, spline interpolation is more efficient than polynomial interpolation because it gives reliable results even with lower degrees of polynomials. Also, during using it, the Runge phenomenon does not occur, which occurs when using polynomial interpolation, especially applying polynomials of high orders (degrees). The systems of linear equations that need to be solved to construct splines are very well conditioned, which makes it possible to obtain polynomial coefficients with high accuracy. As a result, even with very large sample sizes N, the computational scheme does not lose stability. The construction of a table of spline coefficients requires $O(N)$ operations, and the calculation of the spline value at a given point is—$O(\log_2 N)$.

As an example, Fig. 6.2 shows the interpolation functions constructed using cubic Hermite splines for the first two spectral components depending on the degree of defectiveness. As can be seen from the Fig. 6.2, during applying Hermite's cubic splines, there are no oscillations and negative values of the spectral decomposition coefficients.

For evaluating the effectiveness of the considered algorithms, a comparison was made of simulated signals and physical signals obtained during the control of the object. In this case, a prototype composite panel was used with a section that was damaged by a 2.9 kJ impact. Figure 6.3 shows the real signal from the site with

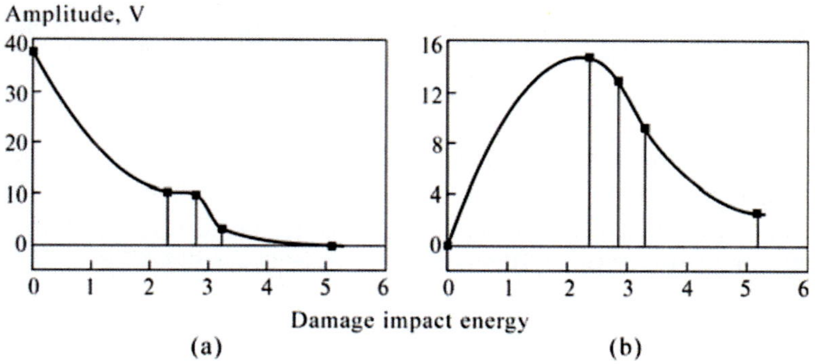

Fig. 6.2 Approximation by Hermite cubic splines of the 1st (**a**) and 2nd (**b**) components of the spectral decomposition of the information signal

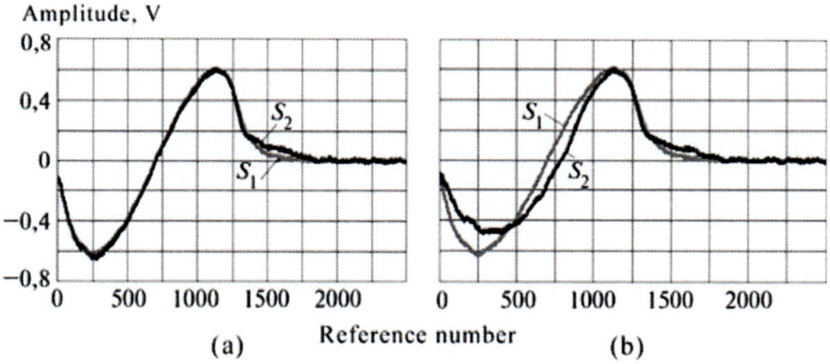

Fig. 6.3 Real S_1 and simulated S_2 information signals

damage caused by an impact with an energy of 2.9 kJ—(curve S_1) and the simulated signal (curve S_2): with the help of Hermite cubic splines (Fig. 6.3a) and quadratic splines (Fig. 6.3b).

The numerical estimation of the discrepancy between the simulated and real signals was carried out by calculating the mean square error between these signals: for approximation by Hermite cubic splines, this error is 2.53×10^{-3}, for approximation by quadratic splines—3.97×10^{-2}.

Substituting the desired value of the sample defectiveness degree, the corresponding value of each coefficient of spectral decomposition is obtained. By restoring the signal with the obtained coefficients, it can be obtained a simulated corresponding information signal—a virtual measure. Such modeling allows to minimize the number of samples, which reproduces the corresponding state of the studied object, which must be used during setting up diagnostic systems, and to optimize the conduct of experimental studies on a physical object.

During diagnosing objects, information signals are characterized by a deterministic and random component. The random component describes such factors as the presence of interference in the measuring channels, random errors of the transducers, spatial heterogeneity of the objects of study, and the like. So, to build an adequate simulation model of such signals, it is necessary to take into account both components.

Unfortunately, models based on physical equations describing the transformation of information signals depending on the state of the object have a number of disadvantages that do not allow them to be used in the calculations and the formation of the space of diagnostic signs. Therefore, it is advisable to build stochastic models of information signals, which allow you to take into account their random changes in time, apply statistical processing methods, expanding the space of diagnostic features and increasing the reliability of diagnosis.

For selecting a method for constructing a simulation model of the information signal of a diagnostic system, it is necessary to conduct a statistical assessment of the obtained values of the coefficients of spectral decomposition. In the case when it is possible to accept the hypothesis of a Gaussian distribution of features, it suffices to determine the following estimates:

- expected value;

$$M\{U\} = \overline{U} = \frac{1}{B}\sum_{i=1}^{B} u_i, \tag{6.4}$$

where u_i—research sample; B—sample volume;

- standard deviation

$$\sigma(U) = \sigma = \sqrt{\frac{1}{B-1}\sum_{i=1}^{B}(u_i - \overline{u})^2}. \tag{6.5}$$

A simulation model of the information signal, taking into account its deterministic and random component, can be represented as [14–18]:

$$S_i(Z) = \sum_{j=0}^{n-1}\left[a_{i,j} + \eta_j\right] g_j(Z),\ i = \overline{0, L-1}, \tag{6.6}$$

where $a_{i,j}$—deterministic component of the signal, which is found according to the algorithm described previously through the functional dependence of the distribution of the values of the coefficients of spectral decomposition depending on the state of the object; η_j—random component that is based on the eigenvalues and eigenvectors of the covariance matrix of the coefficients of the signal spectral decomposition; $g_j(Z)$—selected basis of the orthogonal functions of the discrete argument; L—volume of the worked out sample of the information signal; n—number of components in the signal spectrum.

The determined component of the signal is as follows. It will be considering the vector $X(\mathrm{Z}) = (X_0, X_1, \ldots, X_\gamma)$, the elements of which are obtained as a result of uniform sampling of the sensor signal $X(t)$. Then it can be finding the vector $Y(\mathrm{Z}) = (Y_0, Y_1, \ldots, Y_\gamma), Y_\gamma = \mathbf{M}\big[X_\gamma\big]$ is the mathematical expectation of the vector $X(Z)$, $\gamma = \overline{0, N-1}$, N is the dimension of this vector. Then the values are determined:

$$a_{i,j} = \big\langle Y(\mathrm{Z})_i, g_j(\mathrm{Z})\big\rangle, \quad i = \overline{0, L-1}, \quad j = \overline{0, n-1}, \tag{6.7}$$

where $a_{i,j}$ is the j-th coefficient of spectral decomposition of the i-th implementation of the information signal; L is the sample dimension.

For the case under consideration, the first set of diagnostic features of the model characterizing the deterministic component is formed from 5 components of the decomposition ($n = 5$) of the information signal according to the discrete argument chosen by the basis of the Hartley orthogonal functions (the number of samples of the discrete signal $X(Z)$ is $N = 2500$). Modeling a certain degree of damage to the sample occurs by changing the values of the necessary components of the signal spectrum to values characteristic of information signals received at objects with an appropriate degree of damage.

The second set of diagnostic features characterizing the random component of the model is determined based on the Karhunen–Loève transform. The Karhunen–Loève transformation has fundamental importance, since it leads to the construction of uncorrelated features. Thus, we have the equation:

$$\eta_j = \sum_{k=0}^{n-1} \xi_k \phi_k(j), \; j = \overline{0, n-1}, \tag{6.8}$$

where $\xi_k = \sum_{j=0}^{n-1} \eta_j \phi_k(j)$ are the expansion coefficients, which are independent Gaussian random variables with variances $D_{\xi_k} = \sigma_k^2, \; k = \overline{0, n-1}$; $\big\{\phi_k(j), k, j = \overline{0, n-1}\big\}$ is the orthogonal basis whose $\phi_k(j)$ elements are eigenvectors of the covariance matrix R of the real signal.

Dispersions $D_{\xi_k} = \sigma_k^2, \; k = \overline{0, n-1}$ are equal to the eigenvalues $\lambda_k, k = \overline{0, n-1}$ of the covariance matrix R, corresponding to the eigenvectors $\phi_k(j), k = \overline{0, n-1}$. The elements of the matrix R have the form:

$$r_{i,j} = \frac{\sum_{k=0}^{n-1} \big(v_{k,i} - m_i\big)\big(v_{k,j} - m_j\big)}{n-1}, \tag{6.9}$$

where $v_{i,j}$ are the elements of the matrix V of spectral decomposition coefficients of information signals $X(Z)$; m_i are the elements of the matrix M of the mathematical expectation of each coefficient of spectral decomposition of the information signal.

Matrices V and M are formed as follows:

$$V = \begin{pmatrix} v_{0,0} & v_{0,1} & \cdots & v_{0,n-1} \\ v_{1,0} & v_{1,1} & \cdots & v_{1,n-1} \\ \vdots & \vdots & \ddots & \vdots \\ v_{B-1,0} & v_{B-1,1} & \cdots & v_{B-1,n-1} \end{pmatrix}, \quad M = \{m_0, m_1, \ldots, m_{n-1}\}, \tag{6.10}$$

where B is the number of information signals implementations; n is the number of spectral decomposition coefficients of one implementation, $m_i = \sum_{k=0}^{B-1} v_{k,i}/B$.

The total energy of the vector $\eta = \{\eta_0, \eta_1, \ldots, \eta_{n-1}\}$ is defined as:

$$\sum_{i=0}^{n-1} R_{ii} = \sum_{k=0}^{n-1} \lambda_k. \tag{6.11}$$

The sets of eigenvalues λ_k and eigenvectors $\phi_k(j)$ uniquely characterize the covariance matrix R, and hence the vector η, since the second set of signs is advisable to choose $n_2 = n_1 = 5$ eigenvalues and the corresponding eigenvectors of the covariance matrix of vector η.

Modeling of information signals is performed according to the algorithm shown in Fig. 6.4.

The selection of the coefficients of the orthogonal decomposition a_{k_1}, $k_1 = \overline{0, n_1 - 1}$ of the eigenvalues λ_{k_2} and eigenvectors $\phi_{k_2}(j)$, $k_2 = \overline{0, n_2 - 1}$, $j = \overline{0, n - 1}$ was carried out using realizations of the estimates of these characteristics obtained in the analysis of real information signals.

Since each component of the spectral decomposition is characterized by a different scattering value depending on the degree of damage to the object and serial number, in the simulation modeling scheme, respectively, for each coefficient, there are eigenvalues λ_k and eigenvectors $\phi_k(j)$ different in value. Thus, each of the spectral components will be in to varying degrees of exposure to random factors on its deterministic component, which happens during analyzing real information signals, received during the diagnosis of products. So, it can be noted that the described approach allows to create a simulation model that most fully describes the real information signal.

Using the described simulation algorithm, models of the implementation of information signals corresponding to various degrees of defectiveness of real samples were generated (250 realizations for leather of 5 samples). Figure 6.5 shows the real and simulated signals that take place during testing a defect-free sample and samples with different degrees of damage, where curve S_1 is a real signal, and curve S_2 is a simulated signal.

For comparing the physical and simulated signals, the value of the mean-square error between their values was determined: for a defect-free sample, this error was 3.6×10^{-3}; for samples with impact damage with energies of 2.3 kJ, 2.8 kJ, 3.2 kJ,

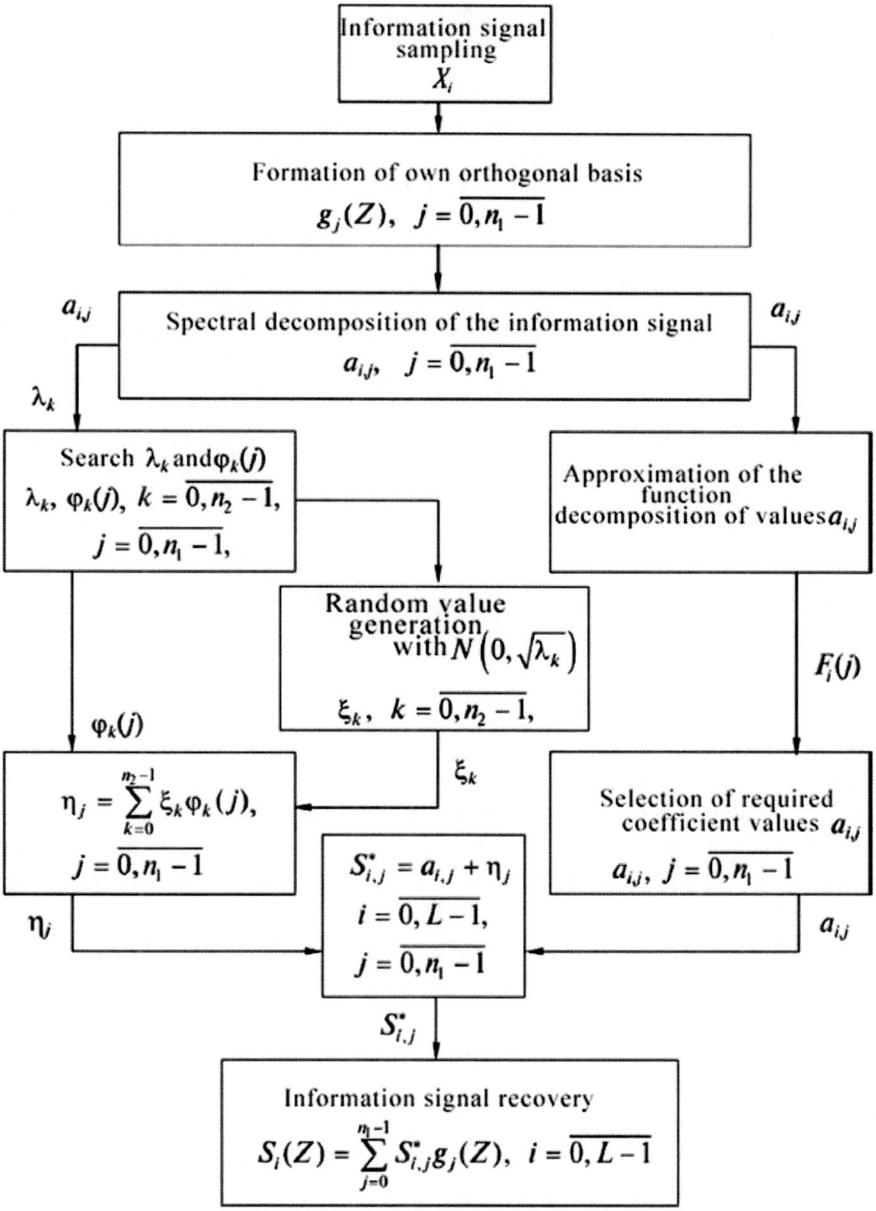

Fig. 6.4 Algorithm for simulation of information signals

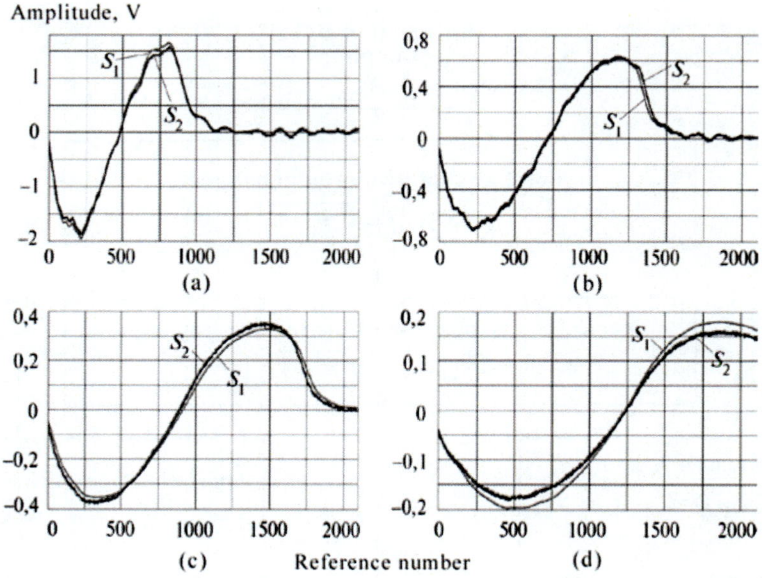

Fig. 6.5 Physical and simulated information signals: **a** sample without damage and samples with damage by impact **b** 2.3 kJ; **c** 3.2 kJ; **d** 5.1 kJ

5.1 kJ, the value of the relative error, respectively, is 2.4×10^{-3}, 2.0×10^{-3}, 2.6×10^{-3}, 1.8×10^{-3}.

Thus, it can be concluding that the obtained simulation models correspond in parameters and characteristics to real information signals, and can be used later in the formation of the training sample and the creation of a library of virtual standards that can be used to configure diagnostic systems (especially with multi-parameter diagnostics), and also for the purpose of forming a control sample to verify the reliability of the system classifier and its validation with a limited number, or the presence of physical reference samples. In addition, simulation models of information signals can be used during the training of classifiers built on the basis of artificial neural networks [19–22].

6.2 Neural Network Technologies in Standardless Flaw Detection of Composite Materials

Due to the large range of composite materials and their defects, as well as the complexity of manufacturing standard samples for all types of possible designs with all possible defects, an important task in the control of composites is the implementation of the principles of standardless flaw detection. The principle of standardless flaw detection is that control is carried out without the use of standard samples with normalized characteristics of defects, or their minimum number is used.

Standardless flaw detection is possible in two versions:

1. There are samples of the studied composite materials that do not contain defects, and the flaw detector is set up using standard samples, and using methods of standardless flaw detection, samples of informative parameters from the reference sample are compared with the samples of the studied sample and a decision is made according to certain criteria.
2. In the absence of samples of the studied materials, a sample of informative parameters from the defect-free zone of the studied product is taken as a training one, for example, according to the 3-point method.

Standardless flaw detection is based on the determination of differences in the informative parameter in one area of the test product relative to another. To construct the decisive rules of standardless flaw detection in the classical version, statistical criteria for testing hypotheses are used, with the help of which the statistical significance of the differences between the compared samples of informative parameters is estimated. Therefore, the construction and application of standardless decision rules requires the accumulation of statistical material and its processing.

Decisive rules based on the verification of statistical hypotheses use statistical criteria that are divided into criteria of agreement and criteria of homogeneity, parametric and nonparametric depending on the available a priori information about the RO, their distribution laws, etc [23]. The advantage of these methods is that the hypothesis is accepted or rejected at a given level of significance according to a certain rule (criterion) for testing a hypothesis, and the value of statistics obtained on the basis of the experiment data, and the specific on this criterion is compared with the critical value, which is determined for the known laws of statistical distributions criterion for a given level of significance. These methods make it possible to simultaneously take into account signs of various nature, since they are characterized by dimensionless quantities—the probabilities of their occurrence under various conditions of the RO [24–28]. The disadvantages of using statistical hypothesis testing methods for constructing decision rules are their significant complications during using several informative features, since it is necessary to analyze multidimensional distribution laws.

In addition to these statistical methods, it is advisable to use neural network classifiers in combination with simulation methods for information signals to construct decision rules for standardless flaw detection, which can significantly increase the reliability of the control and resolution during determining the type of defect or its size.

Using modern diagnostic methods and tools, it can be getting a large number of informative parameters characterizing the state of the diagnostic object. Consequently, there is a large dimension of the space of diagnostic features, by which decision rules are built. Thus, the problem arises of selecting informative features by which the control and formation of the rules for making a diagnostic decision will be carried out, as well as the task of developing methods and algorithms that allow efficient processing of data in a multidimensional space of signs and do not require significant hardware resources. The use of statistical methods in this case is difficult, because it is necessary to analyze multidimensional probability distribution functions, which significantly increases the computational cost of implementing the corresponding algorithm. Also, additional difficulties arise in the development and implementation of the appropriate software for information-diagnostic systems. It is necessary to use sophisticated methods of processing information signals and constructing decision rules based on statistical criteria that lead to the formation of complex decision rules; therefore, during using standardless flaw detection, it is necessary to apply such methods and algorithms that allow to efficiently process data in a multidimensional space of diagnostic signs, and accumulate the information received about possible defects, determine patterns in changing the values of information signal parameters in the dependence of the degree of imperfection of diagnosing object, expand the base classes of defects and refine it in the course of work without a full re-education system.

The most acceptable in this case are classification methods based on artificial neural networks. In particular, direct-connected networks are a universal and effective means of approximating functions, since they generate a large number of regression models, which allows them to be used to classify objects in a multidimensional space of diagnostic features. In diagnostic tasks, a trained neural network not only knows how to recognize (classify) information signals received during monitoring, but also stores information about the patterns and interconnection of the characteristics of the information signal and the state of the monitoring object, and can also correctly classify information signals corresponding to possible defects that were not encountered during network training.

The choice of architecture is a very important step in the development of a neural network classifier as part of diagnostic systems. It is necessary to clearly know which class of problems this diagnostic system can solve, since various neural network architectures are designed to solve specific groups of problems.

Since the fields of application of the most well-known paradigms intersect, various neural network architectures can be used to solve a specific problem, and the results can be significantly different. So, to choose the best architecture, you need to conduct detailed research.

Among the most popular and effective basic paradigms of neural networks that are used in classification and cluster analysis problems, the following networks can be distinguished: multilayer perceptron, radial basis neural networks, neural networks and Kohonen maps, networks of adaptive resonance theory. In addition, the so-called hybrid neural networks can be used to solve the problems posed, which can combine and contain the concepts of some basic paradigms (types of neurons, network architecture, training methods, etc.).

Table 6.1 shows the most common neural network architectures and related learning algorithms (the list is not exhaustive).

A network of a certain architecture requires an appropriate learning algorithm and is designed to solve a limited class of problems. In addition to the considered learning algorithms, the following algorithms are also common: Adaline and Madaline [29],

Table 6.1 Neural network architecture

Paradigm	Learning rule	Architecture	Learning algorithm	Task
With teacher	Error correction	Single and multilayer perceptron	Perceptron learning algorithms Back distribution	Image classification Function approximation Forecasting
	Boltzmann	Recurrent	Boltzmann learning algorithm	Image classification
	Hebb	Multilayer direct distribution	Linear discriminant analysis	Data analysis Image classification
Without teacher	Hebb	Direct distribution	Principal component analysis	Data analysis Data compression
		Hopfield network	Associative memory training	Associative memory
	Competition	LVQ network	Competition Vector quantization	Categorization Data compression
		SOM	SOM	Categorization Data analysis
		ART networks	ART-1, ART-2, Fuzzy-ART	Categorization Image classification
Mixed	Error correction and competition	RBF network	RBF learning algorithm	Image classification Function approximation Forecasting

linear discriminant analysis, Sammon's projections, analysis of the main components [30–33], etc.

For selecting the architecture of a neural network that can be used in problems of standardless flaw detection, the necessary conditions are: the ability to learn without a teacher, the ability to solve problems of data categorization (cluster analysis) and classification. From Table 6.1 it can be seen that among the architectures considered, neural networks of adaptive resonance theory (ART) satisfy these requirements.

The algorithm for constructing a classifier based on neural networks is as follows:

1. Work with data:
   - prepare a sample of training examples specific to this task;
   - break down the entire set of data into a training, control and test set.

2. Pretreatment:

   - select a system of features specific to this task, and perform data conversion accordingly for feeding to the network input. As a result, it is desirable to obtain a linearly separate space of the data set;
   - select the encoding of the initial values.

3. Design, training and quality assessment of the network:

   - choose the network topology: the number of layers, the number of neurons in the layers, and the like;
   - select the function of neurons activation;
   - choose a network learning algorithm;
   - evaluate the quality of the network based on a test set or other criterion, optimize the architecture;
   - choose the best option for the network and evaluate the reliability of its work.

4. Application and diagnostics:

   - explore the degree of influence of various factors on decision making;
   - make sure that the network provides the necessary reliability of anomaly detection and classification;
   - if necessary, return to stage 2, changing the way objects are presented or changing the database;
   - practically use the network to solve the problem.

The generalized block diagram of the diagnostic unit implements a standardless method, based on the neural network classifier is presented in Fig. 6.6. Such a unit is implemented in the form of specialized diagnostic system software.

It should be noted that the learning process of the diagnostic unit can be carried out not only experimentally on standard or real (controlled) samples, but also by means of simulation modeling of information signals corresponding to various types of possible defects. In simulation modeling, models of information signals are used, which are observed during the control of a particular product by a certain type of primary transducer.

Models of information signals differ from each other depending on the transducers used, control methods, primary processing of information signals and the like.

Standardless product diagnostics by neural networks of adaptive resonance theory. In the process of solving the problem of standardless diagnostics, a dilemma arises: how to make sure that the memory of the neural network remains plastic, capable of accepting new data and constructing new classes corresponding to new types of defects, while maintaining stability, which ensures that information about already known classes will not be destroyed and not destroyed in the process of functioning.

The networks and algorithms of the adaptive resonance theory [34–36] make it possible to maintain the plasticity necessary for studying new classes of objects, while at the same time preventing the change of previously memorized classes.

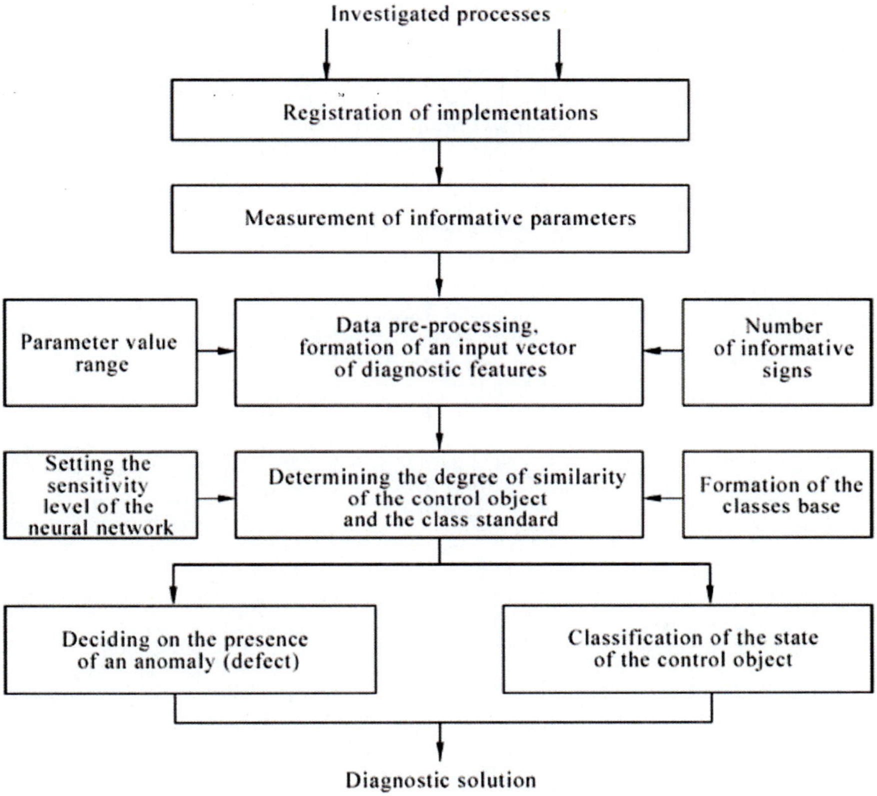

Fig. 6.6 Block diagram of the diagnostic unit

ART networks contain several paradigms, each of which is determined by the form of input data and the way they are processed. ART-1 is designed to process binary input vectors, and ART-2 and Fuzzy-ART networks can classify both binary and continuous data vectors [37–40].

ART networks are a vector classifier and operate according to the algorithm shown in Fig. 6.7. The input vector is classified depending on which of the reference vectors it has previously memorized by the network. The ART network expresses the decision to classify the input vector in the form of excitation of one of the neurons of the recognition layer.

If the input vector does not match any of the stored images, a new category is created (a new neuron is selected and a new vector is remembered), which corresponds to the input vector. If it is determined that the input vector is similar to one of the previously stored vectors by a certain similarity criterion, the reference vector in the memory of the neural network will change (learn) under the influence of the new input vector in such a way as to become more similar to this input vector.

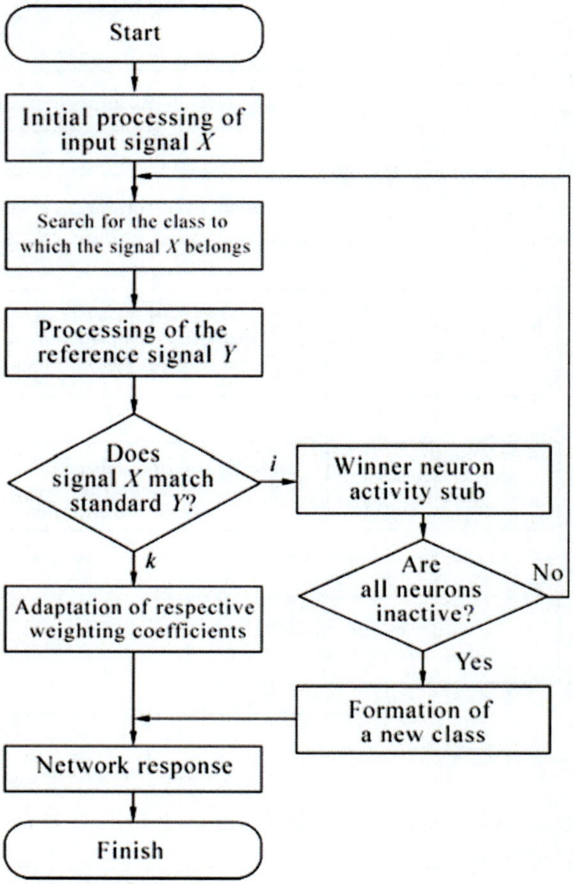

Fig. 6.7 Algorithm of operation of neural networks of the ART family

The memorized reference vector will not vary unless the current input vector is not similar to it. Thus, the stability-plasticity dilemma is solved. A new vector can create additional classification categories, however, a new input vector corresponding to new types of defects cannot force an existing memory to change or erase it.

Classifier based on the modified neural network ART-2. The ART-2 neural network was designed to analyze continuous input signals. Grossberg and Carpenter in [37–39] described several possible architectures of the ART-2 network (Fig. 6.8).

The basic equations describing the operation of the comparison layer F_1 network are:

$$p_i = u_i + \sum_j g(y_j)v_{j,i},\ q_i = p_i/(e + |p|),\ u_i = z_i/(e + |z|), \tag{6.12}$$

$$z_i = f(t_i) + b \cdot f(q_i),\ s_i = x_i + a \cdot u_i,\ t_i = s_i/(e + |s|), \tag{6.13}$$

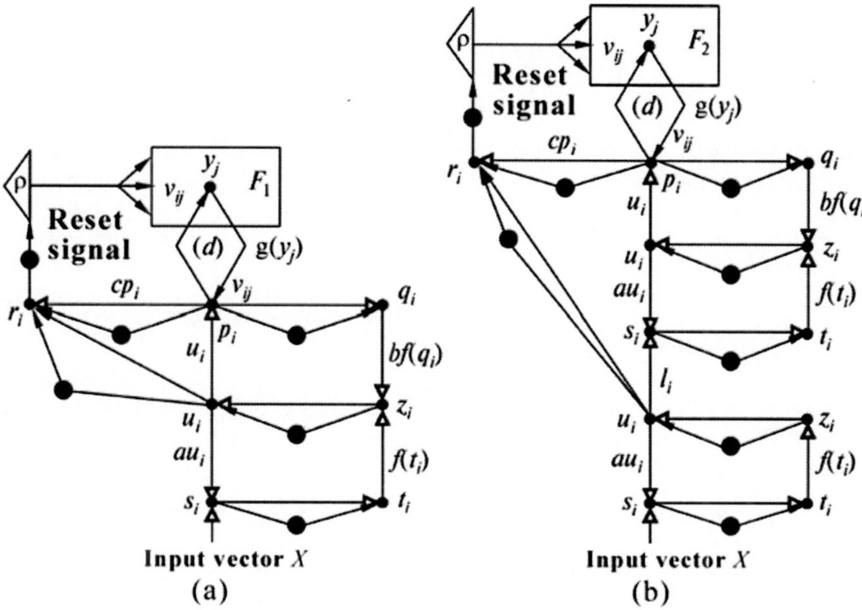

Fig. 6.8 Types of architectures of ART-2 neural networks: white arrows—specific operations in the F_1 and F_2 layers of the ART-2 network; black arrows—operations of network controls; black circles—operations of vector norm search

where |•| is L_2-norm of a vector in Euclidean space; y_j is output of the j-th neuron of the recognition layer F_2; $v_{j,I}$ is elements of the matrix of weight coefficients V; a and b are coefficients that are found experimentally; e is parameter characterizing the relationship between the operating time of the neurons of the layers F_1 and F_2, $0 < e \leq\leq 1$; $f(x)$ is nonlinear signal function of activation of neurons, can be continuously differentiated:

$$f(x) = \begin{cases} 2 \cdot \theta \cdot x^3 / \left(x^2 + \theta^2\right) & at\ 0 \leq x \leq \theta; \\ x & at\ x \geq \theta, \end{cases} \tag{6.14}$$

or piecewise linear:

$$f(x) = \begin{cases} 0\ at\ 0 \leq x \leq \theta; \\ x\ at\ x \geq \theta. \end{cases} \tag{6.15}$$

The basic equations describing the operation of the F_2 recognition layer of the ART-2 network:

$$T_j = \sum_i p_i w_{i,j},\ T_k = \max\left\{T_j : j = \overline{1, m}\right\}, \tag{6.16}$$

$$g(y_k) = \begin{cases} d \ at \ T_k = \max(T_j), \\ 0 \ in \ other \ cases, \end{cases} \quad (6.17)$$

where $w_{i,j}$ are elements of the matrix of weight coefficients W; d is coefficient that are found experimentally.

Thus, the vector p_i will arrive at the comparison device:

$$p_i = \begin{cases} u_i \text{ for inactive neurons of the } F_2 \text{ layer;} \\ u_i + d \cdot v_{k,i} \text{ with an active neuron} - \text{winner } k. \end{cases} \quad (6.18)$$

The comparison device activates the suppression signal if the condition is not met: $\rho/(e + |r|) \geq 1$. The sensitivity coefficient of the classifier ρ is selected in the interval [0, 1], and $r = (r_1, r_2, \ldots, r_n)$ is vector which characterizing the degree of difference between the input vector X and the reference vector W_k in the network memory $r_i = \frac{u_i + c \cdot p_i}{e + |u| + |c \cdot p|}$, c is weight coefficient, which is selected from the inequality $c \cdot d/(1 - d) \leq 1$.

In the case of the correct classification of the input vector, the suppression signal does not start, and the weights of the matrices W and V are modified as follows:

$$v_{j,i}^{new} = v_{j,i}^{old} + \Delta v_{j,i}, \ \Delta v_{j,i} = g(y_j)\left[p_i - v_{j,i}\right] = d \cdot \left(p_i - v_{k,i}\right), \quad (6.19)$$

$$w_{i,j}^{new} = w_{i,j}^{old} + \Delta w_{i,j}, \ \Delta w_{i,j} = g(y_j)\left[p_i - w_{i,j}\right] = d \cdot \left(p_i - w_{i,k}\right), \quad (6.20)$$

where $v_{j,i}^{new}$ and $v_{j,i}^{old}$ are weight coefficients of matrix V, respectively, before and after modification; $w_{i,j}^{new}$ and $w_{i,j}^{old}$ are weight coefficients of matrix W, respectively, before and after modification.

At the beginning of the operation of the ART-2 network and during the formation of a new neuron (in the case of the formation of a new class), the values of the corresponding weight coefficients are initialized with the initial values

$$v_{j,i} = 0, \ w_{i,j} \leq 1/(1 - d)\sqrt{N}, \ i = \overline{1, N}, \ j = \overline{1, m}, \quad (6.21)$$

where N is the dimension of the input vector X; m is the number of neurons in the recognition layer F_2 (the number of stored classes).

The ART-2 neural network is not sensitive to the order of presentation of the input vectors, it can work with both binary and continuous signals, it has high performance, and high reliability of data classification. In addition, ART-2 has the ability to independently correct classification errors after a certain number of cycles of re-presenting the training sample.

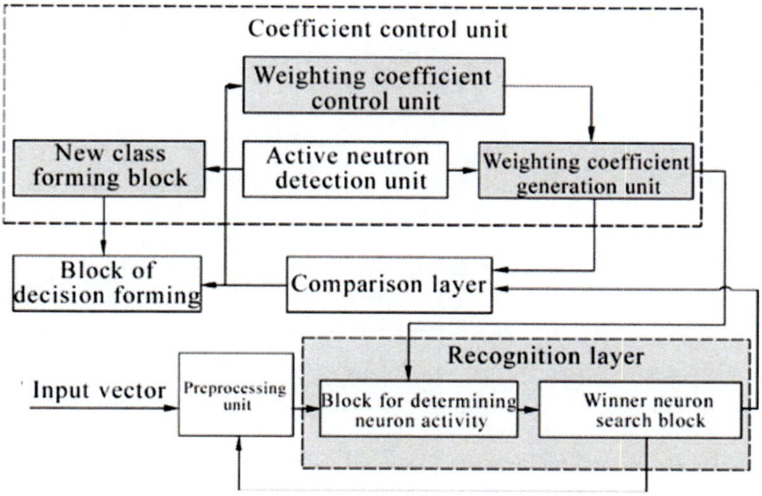

Fig. 6.9 Block diagram of the ART-2 modified neural network

Figure 6.9 shows a block diagram of a modified ART-2 network, where blocks are identified that have been modified and supplemented with respect to the classical network.

Using the developed ART-2 network and its operation algorithm, a high reliability of monitoring the condition of products made of composite materials was obtained in comparison with the classical implementation. In the Table 6.2 for comparison, the results of the classic and modified architectures ART-2 are shown. For analysis, it was used a sample of 2500 implementations of information signals from a flaw detector that implements the low-speed impact method obtained by monitoring cellular panels.

In general, it should be noted that the modified and classical ART-2 networks have the following advantages: stability of stored information and the ability to dynamically expand own knowledge base; high resolution for data classification;

Table 6.2 Results of work of architectural network

Network architecture	Index				
	Sampling analysis time, ms	Amount of memory for saving the network, kB	Reliability of control at the value of sensitivity coefficient ρ		
			0.965	0.975	0.985
Classical ART-2	1740	203	0.98	0.99	0.98
ART-2 with one weighting coefficient matrix	1080	127	0.99	0.99	0.98
ART-2 with two comparison criteria	1610	203	0.99	1.00	1.00

Fig. 6.10 Fuzzy-ART neural network architecture

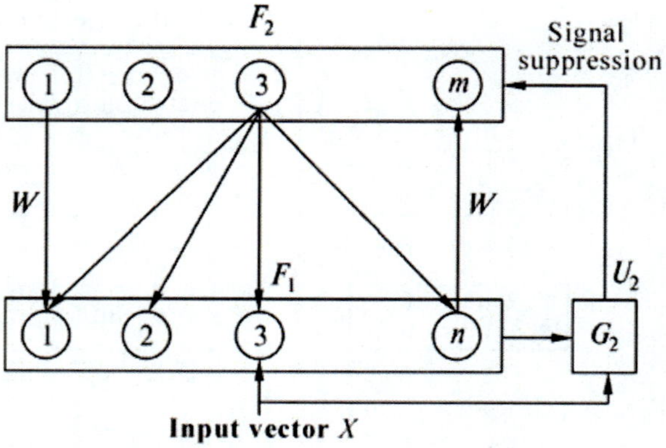

increased noise immunity; invariance with respect to the presentation of input vectors; the existence of the ability to change the speed of learning a network; upon repeated presentation of the training sample, the neural network is capable of correcting errors itself that were made at the previous stage of training; ability to work with continuous signals.

Classifier based on the modified neural network Fuzzy-ART. The expanded architecture of the ART-1 network through the introduction of fuzzy elements is called Fuzzy-ART [39] or a fuzzy ART network. Fuzzy operators are used for:

- definition of class k (determination of the winner neuron);
- calculating the degree of similarity of the input and reference vectors;
- adaptation of the weights of the neurons of the network.

Figure 6.10 shows the basic architecture of the Fuzzy-ART network.

It should be noted a characteristic feature of the Fuzzy-ART network: both matrices of weighting coefficients (W and V) are combined into one matrix W. The components of the input vectors X are real numbers, normalized in the interval [0, 1]. The recognition layer F_2 contains neurons for storing information about the studied class nomenclature. At the beginning of the neural network learning process, each class j is set to an inactive initial state. Between the input and the recognition layer is a matrix of weight coefficients W.

All its elements (weighting coefficients) are first initialized to units, i.e., $w_{ij} = 1$ for $i = 1, 2, \ldots, n$ and $j = 1, 2, \ldots, m$ (the index i corresponds to the element of the input vector, and j is neuron (class number) recognition layer). The following parameters are also set on the Fuzzy-ART network:

- selection parameter $\alpha > 0$, which determines the choice of class at the time of classification;
- correction coefficient $\eta \in [0, 1]$, significantly affects the learning speed of the neural network;
- sensitivity coefficient of the classifier or the level of similarity of the input and reference vectors $\rho \in [0, 1]$, affects the process of formation of classes.

For determining the class k to which the input vector X belongs, first the degree of activation of neurons L is determined, while the input vector X belongs to the class $y_j = \frac{|X \cap W_j|}{\alpha + |W_j|}$, for which the degree of activation is maximum: $y_k = \max_j(y_j)$.

In cases where two classes have the same maximum value of the degree of activation, the class is selected by the reference signal with the lowest index. This ensures the fact that the neurons of the recognition layer F_2 are allocated for each class of input signals in the sequence 1, 2, …, n.

After the first recognition phase, a comparison phase occurs. It consists in comparing the current input vector with a prototype of a particular class k. When this criterion is met, the process of adaptation (modification) of the weight coefficients is activated. Otherwise, the search continues for an alternative class in the recognition or selection layer of a new neuron to form a new class of input signals. The correction of the vector of weights occurs according to the rule

$$W_j(t+1) = \eta\big[X \cap W_j(t)\big] + 1(1-\eta),$$

where t is current stage number.

The classic Fuzzy-ART network has some drawbacks that make it unsuitable for solving the problem of standardless diagnostics. This problem can be solved by replacing certain components of the network. Figure 6.11 shows a block diagram of a modified Fuzzy-ART network, where blocks that have been added or changed compared to the classic network implementation are highlighted.

The classical architecture of this network is sensitive to the order of presentation of input vectors during operation. For solving this drawback, a modification of the classical architecture and the algorithm of the functioning of the Fuzzy-ART network is proposed. For this, the fuzzy logic operator OR $\cap$ was additionally used:

$$X \cup Y = \max(X, Y), \tag{6.22}$$

that is, for a certain G, the equality will be fulfilled:

$$x_i \cup y_i = \max(x_i, y_i) = y_i. \tag{6.23}$$

In the modified Fuzzy-ART network, the following expression is defined in the comparison layer in the second phase:

$$\rho^* = \frac{|(X \cup W_k) - (X \cap W_k)|}{|X|}. \tag{6.24}$$

By the expression (6.24), the degree of similarity of the input vector X and the reference vector W_k in the neural network database is determined. At $w_{k,i} = x_i$ (for all $i = \overline{1,n}$), $\rho^* = 0$ and it will increase in proportion to the increase in the difference

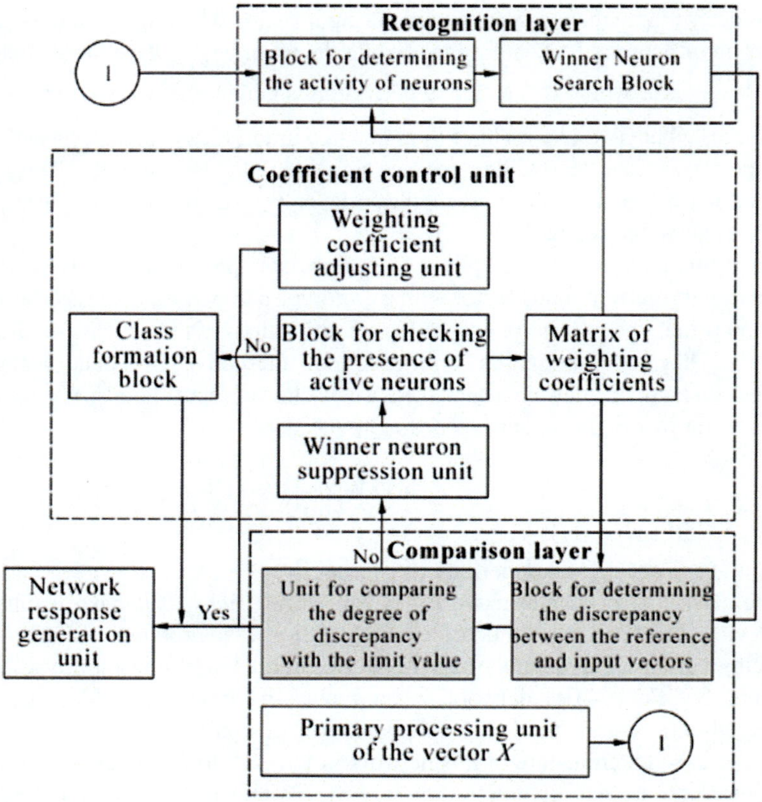

Fig. 6.11 Fuzzy-Art modified network flowchart

between the two vectors. Element G_2 activates the suppression signal if the condition $\rho^* \leq 1 - \rho$ is not satisfied, where ρ is the sensitivity coefficient of the classifier.

This approach provides Fuzzy-ART independence from the order of presentation of the input vectors, and such a neural network can be used to solve the problems of standardless diagnostics.

Table 6.3 shows the results of a study of the main characteristics of the modified neural network Fuzzy-ART.

Table 6.4 shows a comparative analysis of the ART-2 and Fuzzy-ART networks.

Table 6.3 Main characteristics of the modified neural network

Network architecture	Index				
	Sampling analysis time, ms	Amount of memory for saving the network, kB	Reliability of control at the value of sensitivity coefficient ρ		
			0.90	0.92	0.93
Fuzzy-ART	1530	115	0.99	1.00	0.98

Table 6.4 Results of a comparative analysis of some networks

Index	Neural network type		
	ART-2	Fuzzy-ART	Hybrid network
Training speed, s	1.6	1.5	5.7
Reliability of control	0.98–0.99	0.98–0.99	0.96–0.98
Ability to find new (abnormal) objects	Yes	Yes	Yes
Sample classification time, ms	1610	1530	2210
Amount of memory for saving the network, kB	203	115	219

From Table 6.4 it can be seen that classifiers based on the modified ART-2 and Fuzzy-ART networks provide high control reliability, speed, require a fairly small amount of physical memory for storage, and also dynamically expand their knowledge base about possible classes in the process (without stopping the process diagnostics) and without a complete retraining of the network, in fact provides the ability to conduct standardless diagnostics [40–44].

Results of the study of classifiers built on the basis of neural networks of the ART family. It was studied both the classical architectures of the ART-2 and Fuzzy-ART networks and the modified architectures that were described in the previous paragraphs. As diagnostic signs, it was used a change in the shape of the information signal of the flaw detector, which implements the low-speed impact method. As a result, the optimal level of the sensitivity coefficient ρ was determined for each of the architectures of ART networks. Also, to study the reliability of the control by the indicated neural networks, the implementations of the flaw detector information signals corresponding to the damage of the sample section with the energy of the damaging impact of 0.81, 1.38, 3.08, 4.21 and 4.59 kJ were simulated. The simulation of these signals was carried out using the method described in paragraph 1. The simulated signals are presented in Fig. 6.12.

The results of the developed ART networks are shown in Tables 6.5 and 6.6, which indicate the number of signals assigned to each class during the diagnosis of cell panels.

Also, the dependence of the reliability of monitoring the state of cellular panels using neural networks ART-2 and Fuzzy-ART on the value of the sensitivity coefficient of the network ρ was investigated.

As can be seen from the Tables 6.5 and 6.6, for certain values of the sensitivity coefficient ρ, the network created one common class for several types of defects or vice versa for a single type of defect the network created several classes. In the tables, the dash (–) indicates situations where the network could not classify the signals presented (for example, in Table 6.5 for defect sections 5–9 at $\rho = 0.96$ or $\rho = 0.995$). Analyzing the results, it can be concluded, that for ART-2 neural network with a changed architecture, it is necessary to select a value of the network sensitivity coefficient in the range $0.975 \leq \rho \leq 0.986$, and for the modified Fuzzy-ART neural

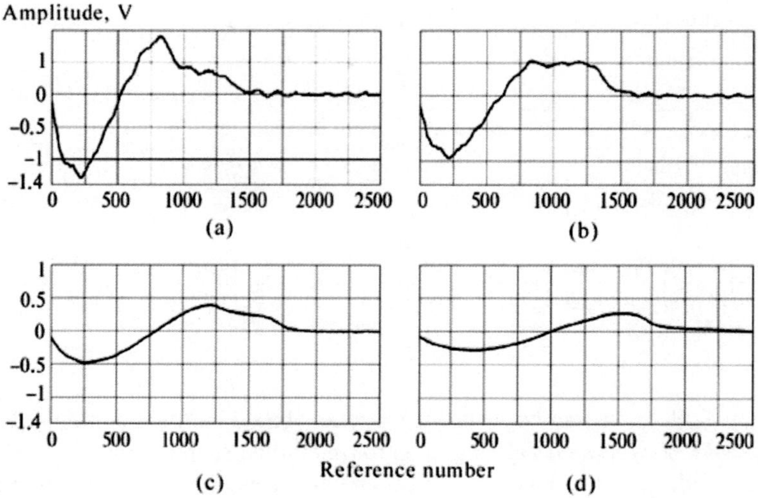

Fig. 6.12 Simulated information signals corresponding to defects as a result of a damaging impact with an energy of 0.81 kJ (**a**), 1.38 kJ (**b**), 3.08 kJ (**c**), 4.21 kJ (**d**)

Table 6.5 Sensitivity coefficients of ART-2 network

Area type	Network sensitivity coefficient, ρ					
	0.96	0.975	0.98	0.985	0.99	0.995
No defect	100	100	100	100	100	86
						14
Defect 1 (2.23 kJ)	200	100	100	100	100	100
Defect 2 (2.81 kJ)		100	100	100	98	50
					2	50
Defect 3 (3.24 kJ)	100	100	100	100	99	92
					1	8
Defect 4 (5.11 kJ)	100	100	100	100	100	100
Defect 5 (0.81 kJ)	–	100	100	100	–	–
Defect 6 (1.38 kJ)	–	100	100	100	–	–
Defect 7 (3.08 kJ)	–	100	100	100	–	–
Defect 8 (4.21 kJ)	–	99	96	74	–	–
			4	40		
Defect 9 (4.59 kJ)	–	101	8		–	–
			92	86		

Table 6.6 Sensitivity coefficients of Fuzzy-ART network

Area type	Network sensitivity coefficient, ρ				
	0.88	0.90	0.92	0.93	0.94
No defect	100	100	100	98	88
				2	10
					2
Defect 1 (2.297 kJ)	200	100	100	100	100
Defect 2 (2.812 kJ)		100	100	97	42
				3	58
Defect 3 (3.240 kJ)	100	100	100	100	100
Defect 4 (5.109 kJ)	100	100	100	100	100
Defect 5 (0.804 kJ)	–	100	100	–	–
Defect 6 (1.378 kJ)	–	100	100	–	–
Defect 7 (3.082 kJ)	–	100	100	–	–
Defect 8 (4.212 kJ)	–	75	92	–	–
		28	8		
Defect 9 (4.586 kJ)	–		91	–	–
		97	9		

network, the optimal values of the network sensitivity coefficient are in the range $0.90 \le \rho \le 0.92$.

The results of the study of the reliability of control D of cellular panels during using the modified ART-2 network are shown in Fig. 6.13, and the Fuzzy-ART networks—in Fig. 6.14. The reliability of the control was determined taking into account the errors of the first α and second β kind.

For studying the sensitivity and resolution of the classifier based on neural networks ART-2 and Fuzzy-ART, synthesized information signals were used in the experiments. 100 realizations of information signals corresponding to damage to a sample area with a damage energy of 0.5, 0.6, 2.0, 2.5, 3.0 and 4.2 kJ were simulated. The simulated signals are presented in Fig. 6.15.

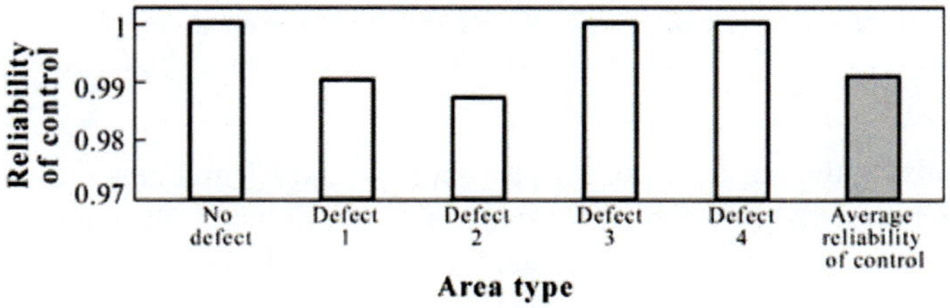

Fig. 6.13 Reliability of control using a classifier based on the ART-2 network

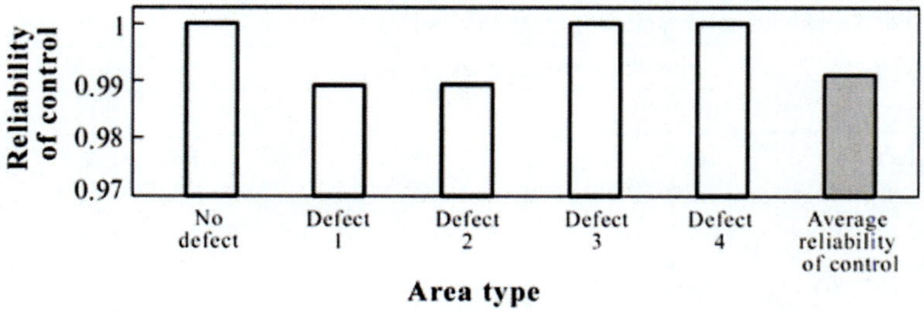

Fig. 6.14 Reliability of control using a classifier based on the Fuzzy-ART network

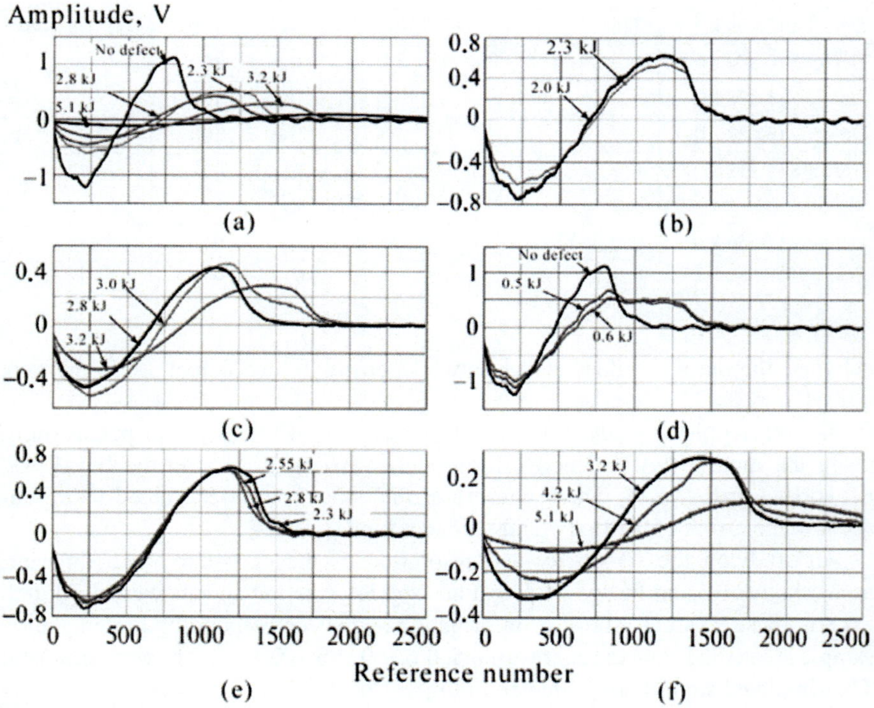

Fig. 6.15 Realization of information signals corresponding to areas with varying degrees of damage, and synthesized signals: **a** 2.3, 2.8, 3.2, 5.1 kJ and without defect **b** 2.0 and 2.3 kJ; **c** 2.8, 3.0 and 3.2 kJ; **d** 0.5, 0.6 kJ and without defect **e** 2.3, 2.55 and 2.8 kJ; **f** 3.2, 4.2 and 5.1 kJ

For studying the sensitivity of the classifier based on neural networks ART-2 and Fuzzy-ART to defects in composite materials (determining the minimum level of damage to a sample that can be detected and classified), information signals were used (Fig. 6.15a–f), which are typical for areas with different degree of damage.

Table 6.7 Results of the study of the reliability of control

Area with damage caused by hit with energy *A*										
(No defect) (kJ)	0.5	0.6	2.0	2.3	2.5	2.8	3.0	3.2	4.2	5.1
100	0	100	100	100	100	100	100	100	100	100

The control results are given in Table 6.7, which lists the number of correctly classified signals for each type of defect during the diagnosis of cell panels. It can be seen from the obtained results that the developed classifier allows to reliably identify defects corresponding to damages caused by impacts with an energy of 0.5 kJ, and also distinguishes defects caused with close values of the energy of a damaging impact (for example, 2.0 and 2.3 kJ; 2.3, 2.5 and 2.8 kJ; 2.8, 3.0 and 3.2 kJ; 3.2, 4.2 and 5.1 kJ). The minimum degree of damage to samples of cellular panels, which can be reliably determined using the developed classifier based on neural networks of adaptive resonance theory, corresponds to a defect caused with a damage energy of 0.6 kJ.

For studying the resolution of the classifier, 5 main sections of the test samples without defect and areas with damage with an energy of 2.23, 2.81, 3.24 and 5.11 kJ were selected. In the experiment, the energy intervals of the damaging impact were determined for each of the given sections of the honeycomb, in which the signals will be correctly classified.

For solving this problem, the information of the flaw detector was synthesized corresponding to areas with damage caused by impact with an energy from 0 to 5.11 kJ with a step of 0.01 kJ. Further, the received signals were fed to the input of the classifier to determine the class to which they belong. Thus, the resolution of the classifier was investigated and the energy intervals of damaging impact were determined for the indicated sections of the cellular panel. The class boundaries obtained using the ART-2 and Fuzzy-ART (F-ART) networks are shown in Table. 6.8.

From the above results it is seen that the developed classifiers based on the ART-2 and Fuzzy-ART neural networks with a modified architecture have high reliability indicators for diagnostics and can automatically expand their own knowledge base about possible defects of the studied objects, and they are also characterized by the reliability of diagnostics—97–99%

The results of a study of the neural network classifier for standardless diagnostics are also given in [45–51].

Table 6.8 Intervals of damaging impact energy

Network type	Area type									
	No defect		Defect 1 (2.30 kJ)		Defect 2 (2.81 kJ)		Defect 3 (3.24 kJ)		Defect 4 (5.11 kJ)	
	from, kJ	to, kJ	from, kJ	to, kJ	from, kJ	to, kJ	from, kJ	to, kJ	from, kJ	to, kJ
F-ART	0	0.54	2.09	2.55	2.56	2.95	3.17	3.834	4.96	5.11
ART-2	0	0.53	2.05	2.56	2.57	2.94	3.19	3.95	4.98	5.11

References

1. Babak, V., Zaporozhets, A., Kuts, Y., Scherbak, L., Eremenko, V.: Application of material measure in measurements: theoretical aspects. In: Zaporozhets A., Artemchuk V. (eds.) Systems, Decision and Control in Energy II. Studies in Systems, Decision and Control. Springer, Cham, pp. 261–269 (2021). https://doi.org/10.1007/978-3-030-69189-9_15
2. Babak, V., Eremenko, V., Zaporozhets, A.: Research of diagnostic parameters of composite materials using Johnson distribution. Int. J. Comput. **18**(4), 483–494 (2019)
3. Zaporozhets, A., Eremenko, V., Isaenko, V., Babikova, K.: Approach for creating reference signals for detecting defects in diagnosing of composite materials. In: Shakhovska, N., Medykovskyy, M. (eds.) Advances in Intelligent Systems and Computing IV. CCSIT 2019. Advances in Intelligent Systems and Computing, vol. 1080, pp. 154–172. Springer, Cham (2020). https://doi.org/10.1007/978-3-030-33695-0_12
4. Eremenko, V., Zaporozhets, A., Isaenko, V., Babikova, K.: Application of wavelet transform for determining diagnostic signs. In: Eremenko, V., Zaporozhets, A., Isaenko, V., Babikova, K. (eds.) CEUR Workshop Proceedings, vol. 2387, pp. 202–214. https://ceur-ws.org/Vol-2387/20190202.pdf
5. Zaporozhets, A., Eremenko, V., Serhiienko, R., Ivanov, S.: Methods and hardware for diagnosing thermal power equipment based on smart grid technology. In: Shakhovska, N., Medykovskyy, M. (eds.) Advances in Intelligent Systems and Computing III. CSIT 2018. Advances in Intelligent Systems and Computing, vol. 871, pp. 476–489. Springer, Cham (2019). https://doi.org/10.1007/978-3-030-01069-0_34
6. Eremenko, V., Zaporozhets, A., Babak, V., Isaienko, V., Babikova, K.: Using hilbert transform in diagnostic of composite materials by impedance method. Period. Polytech. Electr. Eng. Comput. Sci. **64**(4), 334–342 (2020). https://doi.org/10.3311/PPee.15066
7. Wu, Z., Liu, K., Wang, Y.: Validation and evaluation of damage identification using probability-based diagnostic imaging on a stiffened composite panel. J. Intell. Mater. Syst. Struct. **26**(16), 2181–2195 (2015). https://doi.org/10.1177/1045389X14549873
8. Patra, A., Unbehauen, H.: Identification of a class of nonlinear continuous-time systems using Hartley modulating functions. Int. J. Control **62**(6), 1431–1451 (1995). https://doi.org/10.1080/00207179508921607
9. Unbehauen, H., Rao, P.: Identification of continuous-time systems: a tutorial. IFAC Proc. Vols. **30**(11), 973–999 (1997). https://doi.org/10.1016/S1474-6670(17)42970-3
10. Liu, Z., Liu, Z., Peng, Y.: Dimension reduction of Karhunen-Loeve expansion for simulation of stochastic processes. J. Sound Vib. **408**, 168–189 (2017). https://doi.org/10.1016/j.jsv.2017.07.016
11. Khan, M., Panigrahi, S.: Non-destructive testing. In: Principles of Engineering Physics, vol. 1, pp. 785–804. Cambridge University Press, Cambridge (2016). https://doi.org/10.1017/9781316876947.013
12. Czichos, H., Saito, T., Smith, L.E.: Springer Handbook of Metrology and Testing. Springer, Berlin, Heidelberg (2011). https://doi.org/10.1007/978-3-642-16641-9
13. Makitalo, M., Foi, A.: Optimal inversion of the generalized Anscombe transformation for Poisson-Gaussian noise. IEEE Trans. Image Process. **22**(1), 91–103 (2013). https://doi.org/10.1109/TIP.2012.2202675
14. Zaporozhets, A., Eremenko, V., Serhiienko, R., Ivanov, S.: Development of an intelligent system for diagnosing the technical condition of the heat power equipment. In: 2018 IEEE 13th International Scientific and Technical Conference on Computer Sciences and Information Technologies (CSIT), pp. 48–51. Ukraine (2018). https://doi.org/10.1109/STC-CSIT.2018.8526742
15. Babak, V.P., Babak, S.V., Myslovych, M.V., Zaporozhets, A.O., Zvaritch, V.M.: Principles of construction of systems for diagnosing the energy equipment. In: Diagnostic Systems for Energy Equipments. Studies in Systems, Decision and Control, vol. 281, pp. 1–22. Springer, Cham (2020). https://doi.org/10.1007/978-3-030-44443-3_1

16. Chaki, S., Harizi, W., Bourse, G., Ourak, M.: Multi-technique approach for non destructive diagnostic of structural composite materials using bulk ultrasonic waves, guided waves, acoustic emission and infrared thermography. Compos. A Appl. Sci. Manuf. **78**, 358–361 (2015). https://doi.org/10.1016/j.compositesa.2015.08.033
17. Katunin, A., Danczak, M., Kostka, P.: Automated identification and classification of internal defects in composite structures using computed tomography and 3D wavelet analysis. Arch. Civil Mech. Eng. **15**(2), 436–448 (2015). https://doi.org/10.1016/j.acme.2014.08.002
18. Arkadov, G.V., Getman, A.F., Rodionov, A.N.: 7—Optimisation of non-destructive testing. In: Woodhead Publishing Series in Energy, Probabilistic Safety Assessment for Optimum Nuclear Power Plant Life Management (PLiM), pp. 216–253. Woodhead Publishing (2012). https://doi.org/10.1533/9780857093998.2.216
19. Eremenko, V.S., Pereidenko, A.V.: Software of information-measurement system for standardless diagnostic of composite materials. Softw. Eng. **2**(3), 65–76 (2012). https://doi.org/10.5923/j.se.20120203.04
20. Eremenko, V.S., Pereidenko, A.V., Rogankov, V.O.: System of standardless diagnostic of cell panels based on fuzzy-ART neural network. In: 2011 Microwaves, Radar and Remote Sensing Symposium, pp. 181–183. Ukraine (2011). https://doi.org/10.1109/MRRS.2011.6053630
21. Sambath, S., Pagaraj, P., Selvakumar, N.: Automatic defect classification in ultrasonic NDT using artificial intelligence. J. Nondestr. Eval. **30**, 20–28 (2011). https://doi.org/10.1007/s10921-010-0086-0
22. De Albuquerque, V.H.C., Tavares, J.M.R.S., Durao, L.M.P.: Evaluation of delamination damage on composite plates using an artificial neural network for the radiographic image analysis. J. Compos. Mater. **44**(9), 1139–1159 (2010). https://doi.org/10.1177/0021998309351244
23. Zaporozhets, A.O.: Correlation analysis between the components of energy balance and pollutant emissions. Water Air Soil Pollut. **232**, 114 (2021). https://doi.org/10.1007/s11270-021-05048-9
24. Duchene, P., Chaki, S., Ayadi, A., Krawczak, P.: A review of non-destructive techniques used for mechanical damage assessment in polymer composites. J. Mater. Sci. **53**, 7915–7938 (2018). https://doi.org/10.1007/s10853-018-2045-6
25. D'Orazio, T., Leo, M., Distance, A., Guaragnella, C., Pianese, V., Cavaccini, G.: Automatic ultrasonic inspection for internal defect detection in composite materials. NDT&E Int. **41**(2), 145–154 (2008). https://doi.org/10.1016/j.ndteint.2007.08.001
26. Potapov, A.I., Makhov, V.E.: Methods for nondestructive testing and diagnostics of durability of articles made of polymer composite materials. Russ. J. Nondestr. Test. **54**, 151–163 (2018). https://doi.org/10.1134/S1061830918030087
27. Gholizadeh, S.: A review of non-destructive testing methods of composite materials. Proc. Struct. Integr. **1**, 50–57 (2016). https://doi.org/10.1016/j.prostr.2016.02.008
28. Astaneh, A.V., Guddati, M.N.: Dispersion analysis of composite acousto-elastic waveguides. Compos. B Eng. **130**, 200–216 (2017). https://doi.org/10.1016/j.compositesb.2017.07.040
29. Azzam-Jai, A., Ouassaid, M.: Adaptive adaline neural PQ strategy-based multipurpose PV interfaced shunt active power filter. In: 2019 8th International Conference on Systems and Control (ICSC), pp. 71–76. Morocco (2019). https://doi.org/10.1109/ICSC47195.2019.8950643
30. Kuts, Y.V., Yeremenko, V.S., Monchenko, E.V., Protasov, A.G.: Ultrasound method of multilayer material thickness measurement. AIP Conf. Proc. **1096**(1), 1115–1120 (2009). https://doi.org/10.1063/1.3114079
31. Izenman, A.J.: Linear discriminant analysis. In: Modern Multivariate Statistical Techniques. Springer Texts in Statistics, pp. 237–280. Springer, New York, NY (2013). https://doi.org/10.1007/978-0-387-78189-1_8
32. Wang, H., Ding, C., Huang, H.: Multi-label linear discriminant analysis. In: Daniilidis, K., Maragos, P., Paragios, N. (eds.) Computer Vision—ECCV 2010. ECCV 2010. Lecture Notes in Computer Science, vol. 6316, pp. 126–139. Springer, Berlin, Heidelberg (2010). https://doi.org/10.1007/978-3-642-15567-3_10

33. Cai, T., Liu, W.: A direct estimation approach to sparse linear discriminant analysis. J. Am. Stat. Assoc. **106**(496), 1566–1577 (2011). https://doi.org/10.1198/jasa.2011.tm11199
34. Freeman J.A., Skapura, D.M.: Neural networks. Algorithms, applications, and programming techniques, 401 p. Addison-Wesley Publishing Co., Massachusetts (1991)
35. Srinivasan, D., Choy, M.C., Cheu, R.L.: Neural networks for real-time traffic signal control. IEEE Trans. Intell. Transp. Syst. **7**(3), 261–272 (2006). https://doi.org/10.1109/TITS.2006.874716
36. Vavilov, V.P., Nesteruk, D.A.: Active thermal inspection of composite materials with the use of neural networks. Russ. J. Nondestr. Test. **47**, 655–662 (2011). https://doi.org/10.1134/S1061830911100160
37. Carpenter, G.A., Grossberg, S., Rosen, D.B.: Fuzzy ART: fast stable learning and categorization of analog patterns by an adaptive resonance system. Neural Netw. **4**(6), 759–771 (1991). https://doi.org/10.1016/0893-6080(91)90056-B
38. Carpenter, G.A., Grossberg, S., Rosen, D.B.: ART 2-A: an adaptive resonance algorithm for rapid category learning and recognition. Neural Netw. **4**(4), 493–504 (1991). https://doi.org/10.1016/0893-6080(91)90045-7
39. Carpenter, G.A., Grossberg, S.: A massively parallel architecture for a self-organizing neural pattern recognition machine. Comput. Vis. Graph. Image Process. **37**(1), 54–115 (1987). https://doi.org/10.1016/S0734-189X(87)80014-2
40. Babak, V.P., Babak, S.V., Myslovych, M.V., Zaporozhets, A.O., Zvaritch, V.M.: Technicalf. In: Diagnostic Systems For Energy Equipments. Studies in Systems, Decision and Control, vol. 281, pp. 91–133. Springer, Cham (2020). https://doi.org/10.1007/978-3-030-44443-3_4
41. Garcia, D., Palazzetti, R., Trendafilova, I., Fiorini, C., Zucchelli, A.: Vibration-based delamination diagnosis and modelling for composite laminate plates. Compos. Struct. **130**, 155–162 (2015). https://doi.org/10.1016/j.compstruct.2015.04.021
42. Eremenko, V.S., Pereidenko, A.V., Suslov, E.F., Lutsenko, T.V.: Composite materials nondestructive testing system based on low frequency acoustic methods. In: 2013 IEEE 7th International Conference on Intelligent Data Acquisition and Advanced Computing Systems (IDAACS), pp. 61–65. Germany (2013). https://doi.org/10.1109/IDAACS.2013.6662640
43. Ponomarev, A.V., Ponomareva, O.V.: Digital technologies in non-destructive testing. IOP Conf. Ser. J. Phys. Conf. Ser. 1327, 012038 (2019). https://doi.org/10.1088/1742-6596/1327/1/012038
44. Yin, A., Wang, X., Glorieux, C., Yang, Q., Dong, F., He, F., Wang, Y., Sermeus, J., der Donck, T.V., Shu, X.: Texture in steel plates revealed by laser ultrasonic surface acoustic waves velocity dispersion analysis. Ultrasonics **78**, 30–39 (2017). https://doi.org/10.1016/j.ultras.2017.02.016
45. Meng, M., Chua, Y.J., Wouterson, E., Ong, C.P.K.: Ultrasonic signal classification and imaging system for composite materials via deep convolutional neural networks. Neurocomputing **257**, 128–135 (2017). https://doi.org/10.1016/j.neucom.2016.11.066
46. Addin, A.O., Sapuan, S.M., Othman, M.M.: Prediction and detection of failures in laminated composite materials using neural networks—a review. Polym. Polym. Compos. **14**(4), 433–442 (2006). https://doi.org/10.1177/096739110601400409
47. Farhana, N.I.E., Abdul Majid, M.S., Paulraj, M.P., Ahmadhilmi, E., Fakhzan, M.N., Gibson, A.G.: A novel vibration based non-destructive testing for predicting glass fibre/matrix volume fraction in composites using a neural network model. Compos. Struct. **144**, 96–107 (2016). https://doi.org/10.1016/j.compstruct.2016.02.066
48. Wang, C.J., Fang, H., Wang, H.: ESammon: a computationaly enhanced sammon mapping based on data density. In: 2016 International Conference on Computing, Networking and Communications (ICNC), USA, pp. 1–5. https://doi.org/10.1109/ICCNC.2016.7440696
49. Tan, A.-H., Subagdja, B., Wang, D., Meng, L.: Self-organizing neural networks for universal learning and multimodal memory encoding. Neural Netw. **120**, 58–73 (2019). https://doi.org/10.1016/j.neunet.2019.08.020
50. Tymoshchuk, P., Shatnyi, S.: Hardware Implementation Design of Parallelized Fuzzy Adaptive Resonance Theory Neural Network. In: 2019 IEEE XVth International Conference on the Perspective Technologies and Methods in MEMS Design (MEMSTECH), pp. 61–66. Ukraine (2019). https://doi.org/10.1109/MEMSTECH.2019.8817406

51. Keshin, G.A., Ilham, S., Ozkan, C.: The Fuzzy ART algorithm: Aa categorization method for supplier evaluation and selection. Expert Syst. Appl. **37**(2), 1235–1240 (2010). https://doi.org/10.1016/j.eswa.2009.06.004

Chapter 7
Monitoring the Air Pollution with UAVs

Abstract Monitoring of air pollution using unnamed aerial systems (UAS)s is considered. The specifics and content of the information support of the measuring system using UASs are described. The mathematical model of the local atmospheric pollution field in the form of a random field is substantiated. A model of the Wiener field is considered, which describes the Brownian motion of the pollutants movements in the air. A model of the vector random field is proposed for the case of the formation of a local pollution field, the characteristics of which are estimated in the framework of the correlation theory. As an example, the analysis of the structure of a multifunctional measuring system using UASs for remote monitoring of air pollution is considered and the results of experimental studies of monitoring air pollution by radionuclides are presented, which confirms the prospect of further use of UASs for monitoring air pollution.

Keywords Remote monitoring · Sources of air pollution · Characteristics of local atmospheric pollution fields · Monitoring information support · Unmanned aerial vehicle · Measuring system

7.1 Statistical Approach for Studying the Structure of Fields of Meteorological Elements from Air Pollution Sources

The scientific and technical direction of environmental studies of energy facilities combines a wide range of important and relevant world-class problems. The real situation of the technical equipment of energy facilities in Ukraine, which basically worked out their technical resources, requires the creation of new monitoring systems for remote environmental monitoring. This requires the use of modern achievements of science and technology, the latest information and measurement technologies, and their effectiveness depends on the development and practical use of hardware and software measuring complexes and systems [1–3].

V. P. Babak et al., *Models and Measures in Measurements and Monitoring*,
Studies in Systems, Decision and Control 360,
https://doi.org/10.1007/978-3-030-70783-5_7

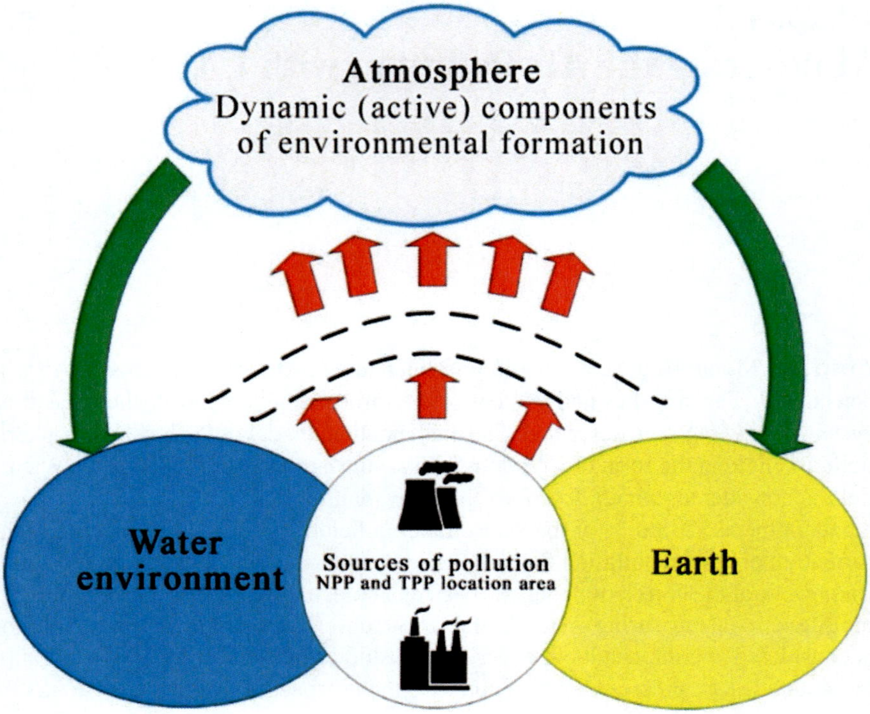

Fig. 7.1 Illustrative scheme for the spread of pollution in areas of NPP and TPP

The state of the environment in the surface layer of the earth (including in the areas where various industrial facilities and human activity are located) is determined mainly by three main factors: the state of the atmosphere, the surface of the earth (soil or water surface), the presence and amount of green plantations (Fig. 7.1).

The determining factor, due to its active movement in space, is the atmosphere. Consider the basic issues of the formation of the atmosphere state in the areas of nuclear power plants (NPPs) and thermal power plants (TPPs). The meteorological fields form in the atmosphere.

It should be noted that due to the active movement in the space of these fields, various (including harmful and dangerous) substances are being generated resulting from the operation of NPPs and TPPs. Depending on the characteristics (wind speed and direction, atmospheric pressure, temperature, etc.) of meteorological fields, as well as on the quantity, volume and presence of certain pollutants coming from NPPs and TPPs, they are transferred in certain directions and to certain distances followed by falling onto green spaces and the surface of the earth (soil, water surface) of harmful and unsafe substances [4–7].

Currently, there are a number of regulatory documents that regulate the monitoring of NPP equipment according to certain characteristics, which are essentially measures. The implementation of such monitoring is carried out by stationary and mobile information-measuring tools. Recently, numerous methods and tools have appeared that, based on the latest technologies, allow to apply remote multifunctional monitoring of the environment and various equipment of NPPs in hard-to-reach and dangerous places for humans. This is of particular importance during performing work to eliminate the consequences of accidents and disasters at NPPs.

Given the availability of such advanced technical equipment, new opportunities appear for remote multifunction monitoring at a qualitatively new level, including the use of unmanned aerial vehicles (UAVs) [5, 8–14].

Based on this, it became possible to monitor the equipment of NPPs, which works as normal, in addition to existing methods and tools through monitoring the physical processes that characterize the environment and the technical condition of the equipment of NPPs, namely the concentration of aerosols containing various radionuclides. It is possible to remotely measure them in the upper atmosphere of the environment of NPPs, as well as to study the distribution of aerosol concentrations along the plane and height [15–18].

Significantly increases the need to expand the range of tasks that monitoring solves in the event of emergency situations or in the disaster mode of NPP. In these cases, the main role is transferred to UAVs, which must be equipped with appropriate remote control equipment to ensure:

- measuring the level of radiation and the concentration of radionuclides in the upper atmosphere and in layers above the emergency facilities of NPPs;
- remote measuring of the temperature state of emergency facilities of NPPs.

The main source of environmental pollution of TPPs during their operation both in the standard and in emergency mode is the furnace-boiler equipment (burners, combustion chambers, boilers, etc.) [19, 20]. His work is accompanied by the emission into the atmosphere of the combustion products of organic fuel (coal, fuel oil, gas). The negative impact on the environment is also carried out by the emission of exhaust steam, part of which enters the environment after the operation of steam turbines and cooling in cooling towers [21–23].

In addition to the above components that pollute the environment during the operation of TPP equipment, local harmful components arise that are formed as a result of the operation of equipment for the preparation of fuel (coal) and chemical treatment of water. These are, first of all, noise-vibration and chemical components. As a result of the operation of generators, transformers, and other converting and distribution equipment, there are also separate local harmful components associated with electromagnetic, noise-vibration, and thermal processes.

In the event of emergency situations (damage of furnace-boiler equipment, smoke filter systems, etc.), a sharp increase in the environmental impact of the combustion products of TPPs is possible. Local pollution also occurs during accidents of powerful transformers, which may be accompanied by the release of transformer oil and fires.

The use of specialized computerized information-measuring systems (IMS) in addition to the existing methods and means of environmental monitoring and TPP equipment makes it possible (including using UAVs) to carry out remote multifunction monitoring, which allows to measure the degree of hazardous products concentration of combustion (CO, fraction ash, smoke, etc.) [4, 5, 24]. The use of UAVs provides the receipt of these data remotely, in different layers of the atmosphere of the environment of TPPs.

The multifunctionality of modern means of remote monitoring of TPP facilities, in addition to the above, provides for the expansion of its capabilities through the use of environmental video surveillance equipment and TPP equipment, measurement and analysis of the composition of substances in the environment, etc.

Let dwell in more detail on some features of the formation of meteorological fields in the atmosphere, since they are the active component that acts as a carrier of pollutants from NPPs and TPPs. These issues are discussed in sufficient detail in numerous scientific papers on meteorology, therefore, in this paper it will be dwelling on the characteristic features of the formation of meteorological fields to further substantiate the models and methods of their study from the point of view of the carrier of pollutants from NPPs and TPPs [25–27].

The accumulated experience of research in various fields of science and technology makes it possible to justify the direction of research on the state and basic characteristics of the environment (including meteorological fields). The main areas of such research include:

- theoretical research;
- simulation (modeling), including performing computational (virtual) experiments;
- experimental, including a wide range of field measuring experiments.

A characteristic feature of the atmosphere is the disordered turbulent nature of the movements that occur in it. Fields of meteorological elements are very variable. The dependence of the instantaneous field values on spatial coordinates, as well as the time course of these values, becomes very complex, and during registering these values under the same conditions, they turn out to be different each time. Because of this, such a description of the fields is impossible in which it is possible to set their instantaneous values at each point in space at each moment in time by a deterministic model.

A probabilistic or statistical approach to the study of the field structure of meteorological elements is natural, in which each field is considered as a random field, and methods of the theory of random functions are used to describe it.

The basis of this approach is the refusal to consider the features of individual instantaneous field values and the transition to the consideration of some averaged properties of the statistical set of field realizations corresponding to a certain set of fixed external conditions.

In the experimental determination of the statistical characteristics of a random field, it is assumed that there is a certain set of its realizations, which correspond to the same research conditions, or, for a homogeneous ergodic field, the presence of one field realization in a sufficiently large space-time range.

Statistical set of implementations. The formation of such a set of homogeneous implementations of the studied meteorological field is necessary with statistical material in the process of conducting a full-scale measuring experiment to assess the state and characteristics of the field.

Meteorological fields physically do not allow mass repetition under the same external conditions. Researchers do not have at their disposal a statistical ensemble of realizations of the studied fields, despite the vicissitudes of the indicated conditions.

In meteorology, a single process is usually divided into parts that are conventionally taken for different implementations, that is, observations made in different spatial areas or at different points in time are used as realizations of a random field. At the same time, during implementation, corresponding to the same external conditions, observations are made that were performed in similar, in a certain sense, spatial areas or time intervals that can be used for joint statistical processing.

In the theory of random functions, situations that correspond to the same external conditions are called those in which the laws of the distribution of a random field are preserved. In practice, usually these distribution laws are unknown, therefore, the selection of similar situations is performed on the basis of everyday meteorological experience and the results of previous studies.

Based on these requirements and the physical nature of meteorological processes, the following points can be noted that should be considered during combining experimental data into one statistical population.

At choosing time moments corresponding to similar situations, one should proceed from the presence of daily and annual variations of meteorological elements. The presence of the diurnal course leads to the fact that the same can be considered moments concerning one particular time of day. Due to the annual course, it is impossible to consider the moments corresponding to different seasons of the year as corresponding to similar situations. Strictly speaking, only sales received on the same day and hour annually should be considered similar. However, this turns out to be practically meaningless, since in this case it is possible to operate with only a small set of realizations, averaging over which will be insufficient to reliably obtain statistical characteristics. Therefore, in practice, they are usually combined into one set of implementations, which does not refer to one day, but to a certain interval of the year, for example, a month or a season. That is, all available implementations that correspond to a specific time of day and season, obtained from observations that have been carried out over a number of years, are combined into one aggregate. In order for the implementations to be independent, a sufficient time interval between observations should be chosen. For example, it is known that during the day the air pressure changes slightly, therefore, there is a large dependence between its values at different points in time. This dependence remains noticeable over the next two days, therefore, at selecting a realizations set of the pressure field, observations are usually used with an interval of at least three days.

In addition, for accepting the daily and annual variations at combining various implementations into one statistical set, an additional classification of empirical material can be carried out according to some special features. So, during researching the wind field, implementations corresponding to different circulation conditions are separated, for example, separately identifying jet flows or classifying realizations by the magnitude of wind speed, etc. Separation by types of circulation is sometimes also used at researching the pressure field (geopotential).

At combining similar spatial volumes, that is, implementations obtained in different geographical points, it is assumed that these points should belong to the same climatic regions.

Field uniformity and isotropy. At studying the spatial structure of meteorological fields, it is important to observe the conditions of uniformity and isotropy of the field, which imposes certain restrictions on the spatial extent of the studied field. It is assumed that in a real turbulent flow, which, generally, is not homogeneous and isotropic, one can single out such a scale within which the conditions of homogeneity and isotropy should be approximately satisfied. Such fields are called locally homogeneous and isotropic.

Depending on the scale of the studied fields in meteorology, the microstructure, mesostructure and macrostructure are divided.

The microstructure describes the features of the fields in the intervals from fractions of millimeters to hundreds of meters. In this region, there is local uniformity and isotropy in three dimensions.

The statistical mesostructure describes the features of the fields in the range from a kilometer to tens of kilometers. In this area, the difference between vertical and horizontal directions is clearly manifested. Homogeneity and isotropy are approximately performed only in the horizontal direction.

Variability and interconnections at spatial scales of the order of hundreds or more kilometers are described by a statistical macrostructure. Macroprocesses associated with atmospheric formations of a synoptic and even global nature, the physical nature of which differs significantly from the nature of disordered small-scale turbulent pulsations.

In many cases, it is convenient to consider macro processes as random and describe them by analogy with small-scale processes as a kind of macro-turbulent exchange. However, this analogy is formal.

In this region, the conditions of homogeneity and isotropy are approximately satisfied only in the horizontal plane. In the field of meso and macroturbulence, it can be spoken about the homogeneity and isotropy of only deviations of meteorological elements from the climatic norm, while at the same time, climatic norms themselves at these scales can experience significant differences. At the same time, it is practically impossible to use the ergodic property, which greatly facilitates the study of a homogeneous field and allows one to determine statistical characteristics from one implementation of sufficient duration. Indeed, the mathematical expectation of the meteorological field depends on the coordinates, therefore, one calculation cannot be used to calculate the mathematical expectations, but a sufficiently large number of realizations must be used. In addition, in empirical studies of the structure of large-scale meteorological fields, data are used from observations obtained at weather stations, which in a spatial region are combined into one statistical population. Usually there is little such data, that is, implementation is important only in a small number of discrete points, and, therefore, averaging over one implementation will not be effective (justified in the sense of a point estimate).

During studying the structure of a field, such statistical characteristics as mathematical expectation, correlation or spectral functions, and probability distribution functions that are necessary for solving monitoring problems are evaluated.

7.2 Air Pollution Research Information

In the future, by the concept of the environment of energy facilities or simply the environment, it is necessary to understand a limited (local) region of space, which consists of the underlying earth's surface, the surface atmospheric zone up to a height of 20 km and surrounds energy facilities, including NPPs, TPPs, in a radius up to 100 km.

Airspace from the environment—the atmosphere—is an integrated environment within which a significant number of spatio-temporal fields of various physical nature are formed and propagated.

Let dwell only on physical atmospheric fields. Atmospheric meteorological fields or simply meteorological fields, including meteorological fields of pressure (geopotential), temperature and wind are considered as natural fields formed by processes on the earth, in the atmosphere and in space.

As a result of the interconnection and interaction of the processes of functioning of energy facilities with meteorological fields, concentration fields of atmospheric pollution impurities or simply atmospheric pollution fields are formed, which reflect the dynamics of the processes of transport and diffusion of pollution impurities of energy objects in time and space by meteorological fields.

In the general case, the objects of research are the processes of observation, measurement, control, identification (recognition), diagnostics, and forecasting the dynamics in time and space of the characteristics and parameters of atmospheric pollution fields.

In modern information technologies, field studies in integrated form, the above processes are called monitoring processes or simply monitoring.

For implementing the monitoring process, IMS uses both hardware and software systems for various purposes, surveillance tools, and information technology, vehicles, including appropriate information support [28–30].

The term "information support" also requires a certain interpretation and concretization, since information support is an integral product of a significant number of components that are part of the so-called soft equipment of the systems, objects, and complexes under study.

The fundamentals of creating information support are based on the results of the following two fundamental directions of research.

The first direction. Investigation of the processes of interconnection and interaction of naturally created atmospheric meteorological fields with the functioning processes of industrial and vital objects, primarily NPPs, TPPs and the formation of atmospheric pollution fields on their basis, the dynamics of the state and characteristics of which are the results of remote monitoring, and more generally case by environmental monitoring results.

The second direction. The use of scientific, technical, technological, industrial, economic achievements and resources to create modern measuring instruments—a hardware-software unmanned aircraft measuring system of characteristics, parameters of the studied environmental fields, which allows to use of various types of sensor subsystems with resolution and accuracy of measurement of the contribution of individual impurity components environmental pollution, ensures the safety of a full-scale measurement experiment in hazardous places and during accidents of NPPs and TPPs for given spatial areas of the environment and time intervals of the remote control process.

The composition of information support during the creation of systems includes various components, namely:

- physical and mathematical models that are homomorphic reflect the formation of atmospheric fields of environmental pollution in time and space, the functioning of computerized remote control systems during performing a sequence of n necessary functions;
- bases of design and operational parameters, characteristics of measurement systems, unmanned aerial systems, subsystems, modules that are used to create systems;
- databases of tests, measurements, control and diagnostics during operation of systems, results of their processing;
- algorithmic support for computer simulation problems, including computational problems of computer modeling of various options for the structure of the created system;
- algorithmic support for the operation of various subsystems of the created system.

Atmospheric pollution fields are physical fields, while the chemical composition of pollution impurities is diverse, for example, the nomenclature of the radionuclides composition. The characteristics and concentration parameters of the studied impurities of pollution, their dynamics in time and space are the measurement data by various measuring tools, systems, complexes. The effectiveness of using such tools is determined, first of all, by the capabilities of their technical sensor modules, like primary measuring transducers.

Along with the functioning of existing monitoring systems, information support should serve as the basis for creating a scientific and technical basis for the development of perspective systems. If it can be tried to introduce a certain classification of information support of technical systems, it can be taken various classification features as a basis. So, the following classification features characterize both current and future options for the performance of specified functions by monitoring systems, for example:

- functional-current informational support, which changes insignificantly during the development, implementation and operation of the current model of the system;
- innovative information support, which reflects both the results of the operation of existing systems and new developments in various fields of science and technology;
- a priori support of systems at the first stages of the life cycle, namely at the design and manufacturing stages, during design characteristics and parameters of the system are mainly used;
- a posteriori software is mainly used at the stages of operation and, partially, testing the system and the content of the adjusted software, the use of which makes it possible to adjust the calculated characteristics and obtain real system characteristics, including characteristics of the accuracy of measurement results, reliability, residual life in the statistical evaluation of real data measurements, control, diagnostics and forecast.

Other signs of the classification of information support reflect in more detail the monitoring process:

- theoretical, algorithmic, and software, or scientific and technical, is used at all, especially in the first stages, in the development of physical and mathematical models, the rationale for a variant of the structure of the systems, determination of the design characteristics of the system;
- production and technological support or design and technological at the stages of preparation for production, production of experimental and serial samples of the system;
- normative and technical support is used at all stages, but its main role is in testing, certification, transfer to operation and at the stage of operation, repair and modernization of the system.

The step-by-step creation and use of information support of IMS is schematically shown in Fig. 7.2.

The weight and significance of each of these types of information support is different at different stages of the life cycle of monitoring systems. Regulatory, technical and design and technological information support consists of a generally accepted regulatory framework, standards, guidelines, existing databases, and sanitary standards. Their main function is to provide information support for existing systems that implement well-known, generally accepted monitoring principles and methods. In a certain sense, they can be considered as a basic component of information support [31–34].

The main subjects of research are models, methods and means of remote monitoring of the dynamics in space and time of the environmental characteristics of energy facilities.

The definition of physical and mathematical models of the studied pollution fields are based on the following. Despite the fact that such models are copies of the studied

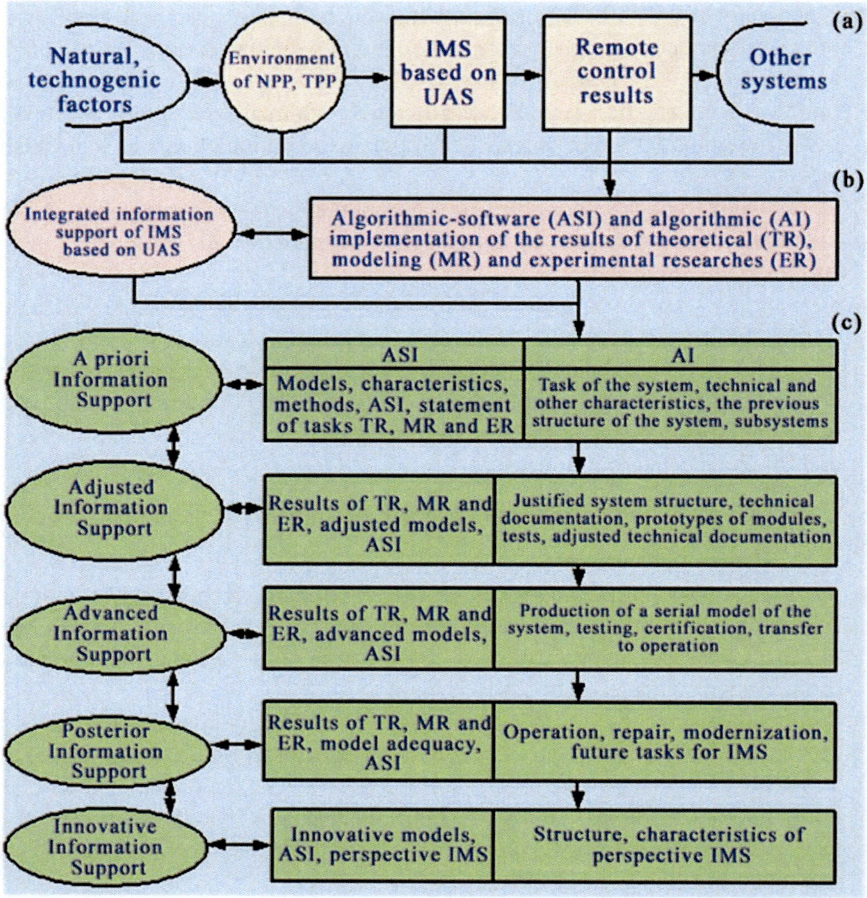

Fig. 7.2 The structure of the stages of creating and using information support during the development and operation of IMS based on the UAS: **a** objects of research; **b** development items; **c** stages of creation and use of information support

fields, they do not fully correspond to the real objects of research, their role is very important and has a fundamental character.

In the process of creating a physical model, physical properties, characteristics of the object of environmental measurements, its elements, modules, their actions and interactions with each other taking into account physical laws are primarily reflected. In this case, the model is written using mathematical formulas with dimensions of a system of units of physical quantities.

The mathematical model formalizes the process of reflecting the properties of the measurement object. Mathematical methods are effective research methods and, depending on the formulation of measurement tasks for one measurement object, it can be used different mathematical models on the one hand, and on the other hand, use the same model to study objects of different physical nature.

Each of the models, physical and mathematical, complement each other and make it possible to conduct the measurement process more efficiently, for example, justify ways to improve the accuracy of measurement results.

7.3 Atmospheric Field Models

7.3.1 General Models

During classifying models of atmospheric pollution fields, two classes of models are distinguished:

- normal operation of energy facilities (Model A);
- abnormal mode of operation of energy facilities caused by accidents, disasters, other natural and man-made causes (Model B).

For emergency conditions, depending on the degree of damage, they are isolated separately:

- B—abnormal mode (failure to perform normal operation);
- B1—emergency condition (accident, conditional minor damage);
- B2—catastrophic state (catastrophe, significant damage).

In substantiating the physical and mathematical models of atmospheric pollution fields, statistical field models are used, the formation of which is due to the action of a significant number of factors in space and time, most of which are stochastic in nature [35–38].

The terminology of random fields used in this chapter is as follows.

The random function $\xi(\omega; \mathbf{r}; t)$ is called a random field, where $\omega \in \Omega$ is a random event in the space of random events Ω, $\mathbf{r} = (x, y, z)$, $\mathbf{r} \in Q$ are the spatial coordinates of the region of the space Q, and $t \in T$ is the current time on the interval T.

It will be further considered Hilbert random fields $\xi(\omega; \mathbf{r}; t)$ for which condition $\mathrm{M}\{\xi^2(\omega; \mathbf{r}; t)\} < \infty$ is satisfied, the physical interpretation of which is that the field has finite power and energy, and $\mathrm{M}\{\cdot\}$ is the expectation operator with a probability measure.

The analysis of a Hilbert random field is carried out in the framework of the correlation or energy theory of random functions, which is limited to the definition of the first two moment functions, namely:

- mathematical expectation

$$a(\mathbf{r}, t) = \mathrm{M}\{\xi(\omega; \mathbf{r}; t)\}; \tag{7.1}$$

- correlation (autocorrelation) function

$$R(\mathbf{r}_1, \mathbf{r}_2, t_1, t_2) = \mathrm{M}\{[\xi(\omega; \mathbf{r}_1; t_1) - a(\mathbf{r}_1, t_1)] \times [\xi(\omega; \mathbf{r}_2; t_2) - a(\mathbf{r}_2, t_2)]\}. \quad (7.2)$$

Together with the function $R(\mathbf{r}_1, \mathbf{r}_2, t_1, t_2)$, in the analysis of random fields, the so-called structure function of the random field $\xi(\omega; \mathbf{r}; t)$ is used, namely:

$$B(\mathbf{r}_1, \mathbf{r}_2, t_1, t_2) = \mathrm{M}\left\{[[\xi(\omega; \mathbf{r}_1; t_1) - a(\mathbf{r}_1, t_1)] - [\xi(\omega; \mathbf{r}_2; t_2) - a(\mathbf{r}_2, t_2)]]^2\right\}^2. \quad (7.3)$$

The function $B(\mathbf{r}_1, \mathbf{r}_2, t_1, t_2)$ characterizes not only the correlation properties of the studied random field, but also the correlation properties of its growth.

The most studied particular cases of a random field $\xi(\omega; \mathbf{r}; t)$:

- stationary in time or just stationary if

$$R(\mathbf{r}_1, \mathbf{r}_2, t_1, t_2) \equiv R(\mathbf{r}_1, \mathbf{r}_2, t_2 - t_1) = R(\mathbf{r}_1, \mathbf{r}_2, \tau) \quad (7.4)$$

or

$$B(\mathbf{r}_1, \mathbf{r}_2, t_1, t_2) \equiv R(\mathbf{r}_1, \mathbf{r}_2, \tau); \quad (7.5)$$

- homogeneous in space or just homogeneous if

$$R(\mathbf{r}_1, \mathbf{r}_2, t_1, t_2) \equiv R(\mathbf{r}_2 - \mathbf{r}_1, t_1, t_2) \quad (7.6)$$

or

$$B(\mathbf{r}_1, \mathbf{r}_2, t_1, t_2) \equiv B(\mathbf{r}_2 - \mathbf{r}_1, t_1, t_2); \quad (7.7)$$

- homogeneous and isotropic in space or homogeneous and isotropic if

$$R(\mathbf{r}_1, \mathbf{r}_2, t_1, t_2) \equiv R(|\mathbf{r}_2 - \mathbf{r}_1|, t_1, t_2) \quad (7.8)$$

or

$$B(\mathbf{r}_1, \mathbf{r}_2, t_1, t_2) \equiv B(|\mathbf{r}_2 - \mathbf{r}_1|, t_1, t_2). \quad (7.9)$$

Moreover, such special cases with limited regions of space and time are called locally homogeneous, locally homogeneous and isotropic, or locally homogeneous and stationary.

Such terminology objectively reflects the natural and objective factors in the formation of atmospheric random fields with corresponding restrictions, namely:

such models of the studied fields can adequately describe physical fields only in limited spatial areas and at limited time intervals.

7.3.2 *Development of Models of Atmospheric Fields of Environmental Pollution*

The development of models of atmospheric pollution fields, as one of the main components of information support, is based on the results of theoretical, simulation (modeling) and experimental studies of environmental fields, taking into account the latest achievements of science and technology. This is the use of existing models of environmental fields and the results of their research, the possibility of using a new measuring instrument—IMS based on unmanned aerial systems (UAS) to determine the spatio-temporal and spectral characteristics of diffusion fields of the environment [39–41].

First, it can be dwelt on the models of the physical fields of the environment during the normal operation of energy facilities—models of A type.

In the research of physical atmospheric fields of the environment, mainly 4 directions of building models of the studied fields are used:

- statistical;
- deterministic;
- combined deterministic-statistical;
- based on the results of experimental (full-scale) studies.

The following is given as examples of the implementation of these areas of research into the physical fields of the environment.

Statistical models of atmospheric meteorological fields of the environment began to be created in the 30s of the twentieth century based on the use of theoretical studies of random functions—the theory of random processes and fields. Due to the lack of real data of measuring the characteristics of meteorological fields in time and space, specific results of studies of statistical models were obtained for a particular case, namely for locally homogeneous and isotropic random fields.

In general, *physical atmospheric pollution fields* can be described by a random field model based on the results of theoretical studies of the theory of random functions [42].

A promising model of the physical atmospheric field of the environment can be used *random field* based on the following physical justifications.

The random field of pollution describes the movement of particles, for our case, the movement of particles of the products of atmospheric pollution in the process of their interaction with the air. The movement of particles is chaotic and is due to thermal, turbulent, wind processes of movement of air masses. This chaotic motion is reminiscent of the Brownian motion of particles, and this fact is confirmed by the use of Wiener field pollution in the model of the atmospheric field as a model of

Brownian motion of particles in space and time. It is known that a Wiener field is a random field with independent increments, bases on Gaussian laws of the distribution of field values in time and space, and is characterized by Markov probabilities of transition from the previous to the next state of the Wiener field.

The random field $\xi(\omega; \mathbf{r}; t)$, $\omega \in \Omega, \mathbf{r} \in Q, t \in T$ is specified in the form of a stochastic differential or integral equation, the corresponding coefficients of which are determined are the transport $a(\mathbf{r}; t)$ and diffusion $b(\mathbf{r}; t)$ coefficients of the particles of their chaotic motion. In the particular stationary case, the increments of the field $\xi(\omega; \mathbf{r}; t)$ in the time interval $t \in [t, t+h]$ can be written as:

$$\xi(\omega; \mathbf{r}; t+h) - \xi(\omega; \mathbf{r}; t) = a(\mathbf{r})h + b(\mathbf{r})[w(\omega; \mathbf{r}; t+h) - w(\omega; \mathbf{r}; t)], \quad (7.10)$$

where the corresponding transport and diffusion coefficients of the particles are the functions $a(\mathbf{r})$ and $b(\mathbf{r})$ of the space arguments, and $w(\omega; \mathbf{r}, t)$ is the Wiener field with independent increments and Gaussian distribution laws that describes the random motion of particles in air [43].

If it talks about the practical use of the model of the physical atmospheric field of the environment—a random field, then it is first necessary to conduct a full-scale measurement experiment using the IMS based on the UAS and evaluate the transfer $a(\mathbf{r}; t)$ and diffusion $b(\mathbf{r}; t)$ coefficients as the corresponding statistical estimates of the mathematical expectation $\mathbf{M}\{\xi(\omega; \mathbf{r}; t)\}$ and variance $\mathbf{D}\{\xi(\omega; \mathbf{r}; t)\} = b^2(\mathbf{r}; t)$.

The deterministic direction of creating models of the physical field of the environment is based on various differential or integral equations that have a clear physical interpretation. This is the equation of continuity, thermal conductivity, gas turbulence and others.

It should be noted that today during creating models of the physical fields of environmental pollution of energy facilities, the deterministic direction of research in its “pure” form is almost not used. This is due to the fact that at constructing deterministic models of the studied fields, the results of real measurements of the field characteristics obtained on the basis of statistical methods for processing the measurement data are used [43, 44].

In most practical cases, the creation of models of the studied processes, fields is based on the use of a combined deterministic-statistical direction of research. This is also justified by the general approach to conducting practical research—it is practically impossible to take into account in the model of processes and action fields of all possible factors, therefore, a certain number of actions factors whose characteristics and parameters are unknown and cannot be estimated, measured in total, are assigned to the random variable $\omega \in \Omega$.

Among the models that are considered, it is worth highlighting models of the abnormal mode of power facilities, primarily in the event of nuclear accidents.

For analyzing and predicting the impact of radioactive emissions from nuclear facilities, both on the environment and on public health, one of the decisive elements is the assessment of the spread of radionuclides in the atmosphere and their deposition on the surface.

This is due to the fact that the conditions of air transport play a decisive role in the formation of fields of radioactive contamination of air and soil in the case of radioactive emissions of NPPs and, as a consequence, the values of dose loads on the population.

Mathematical models of the spread of radioactivity in the atmosphere is an important tool for assessing radioactive contamination of the atmosphere and the surface of the earth during accidental releases of NPPs during the early phase of a radiation accident. The choice of an adequate model of atmospheric transport, taking into account the characteristics of both the source of the emission and the territory of the spread of radioactivity, is crucial in solving this problem.

In the event of accidents and catastrophes at NPPs, there is a lack of time and instrumental resources for carrying out full-scale measurements in the zone of influence of the emission source, as consequences of radioactive pollution of the environment.

Therefore, mathematical models of the distribution of radionuclides in the atmosphere and their deposition to the surface is an important tool that allows to get the first assessment of the radiation situation near the emission source and to give a forecast for its development for the near period.

It is important to note that the use of IMS based on the UAS at the time of accidents, as well as in assessing its consequences, makes it possible to remotely measure the necessary characteristics, parameters of atmospheric fields of radionuclides, while conducting a full-scale measurement experiment is dangerous.

The direction of developing models of atmospheric pollution fields according to the results of experimental studies, that is, according to the results of measuring the real characteristics of the studied fields using IMS based on the UAS, is the most promising. This conclusion is due to the fact that:

- the results of the measurement of characteristics make it possible to confirm the adequacy of a particular model of the studied physical field;
- the dynamics of changes in the composition of atmospheric pollution products requires appropriate adaptation of the use of not only traditional, but also new IMS sensor subsystems to ensure measurement and control processes;
- in general, the solution of problems for today and in the future, remote monitoring of the state, characteristics of the atmospheric fields of the environment requires the need to adapt the hardware-software structure of IMS, which leads to corresponding changes and correction of the hardware and information component of the system.

Thus, it can be justified the general model of the physical atmospheric field of the environment of energy facilities in the form of a vector random field:

$$\Xi_m(\boldsymbol{\omega}; \mathbf{r}; t) = (\xi_1(\omega_1; \mathbf{r}; t), \ldots, \xi_n(\omega_n; \mathbf{r}; t)), \boldsymbol{\omega} = (\omega_1, \ldots, \omega_m), \mathbf{r} \in Q, t \in T, \tag{7.11}$$

where $\left\{\xi_i(\omega_i; \mathbf{r}; t), i = \overline{1, n}\right\}$ is the sequence of components of the vector field $\Xi_m(\omega; \mathbf{r}; t)$, the characteristics of which are measured by modern means, systems and components, including IMS, and each component is formed by one source of pollution.

7.4 Example of Remote Monitoring

7.4.1 Structure of the Multifunctional UAS

An analysis of the composition and purpose of the on-board equipment of modern UASs indicates that their effective use largely depends on the functionality of the automated UAS flight control systems and its payload in the form, as a rule, of multifunctional control and measuring equipment [45–48]. The experience of using the UAS during the liquidation of the accident at the Fukushima 1 NPP in Japan draws attention to the role and place of automatic flight control systems (AFCS) and additional IMS in the performance of tasks that are assigned to these systems for environmental control of NPPs and TPPs, especially in conditions possible technological accidents.

Despite the modern achievements in the field of automation, computerization and the creation of robotic systems, the UAS application remains fully automatic. UAV flights are in most cases controlled and will be operated by humans in the near future.

For performing monitoring tasks, various UAV control methods are used.

Manual operator control (or remote piloting) is carried out from a remote control within the optical surveillance zone or using visual information coming from a front-view video camera. In the case of such control, the operator first of all solves the tasks of piloting: maintaining the desired course, altitude, etc.

Automatic control provides the possibility of a fully autonomous UAV flight along a given trajectory at a given height at a given speed and with stabilization of orientation angles. Automatic control is carried out using on-board software devices.

With semi-automatic control (or remote control), the flight is carried out automatically without human intervention using autopilot and predefined parameters, but the operator can make changes to the route interactively. Thus, the operator has the ability to influence the result of the functioning of the system without being distracted by solving piloting problems.

Manual control may be one of the modes for a UAV, or it may be the only way to control it. UAVs that are devoid of any means of automatic flight control cannot be considered as platforms for fulfilling serious targets. The last two methods are currently the most demanded by the operators of unmanned systems, since they put forward less stringent requirements for personnel training and contribute to the safe and efficient operation of UAV systems.

Fully automatic control may be the optimal solution for the tasks of aerial photography of a given area, if it will be needed to provide photographing at a great distance from the base, that is, out of contact with the ground station. At the same time, since

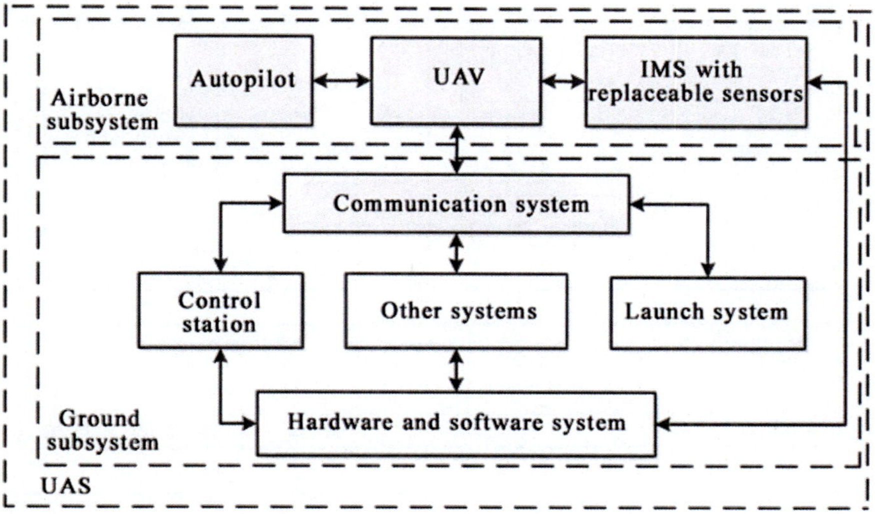

Fig. 7.3 Generalized structure of the UAS

the launch person is responsible for the flight, the ability to influence the flight parameters from the ground station can help to avoid emergency situations.

As a rule, a UAV-based monitoring complex is developed as an integrated UAS, consisting of a number of subsystems (Fig. 7.3), and includes: a control station where the operator's workstation and software applications that provide the operator with control over the complex's work are located; UAV that carries payload equipment of various types; a communication system that provides the transfer of control commands from the control station to the UAV, as well as the transfer of useful information from the UAV to the ground control station in real time; additional equipment designed to support the planned research program (for example, IMS with interchangeable sensors).

UAVs have a high level of automation. They have the ability to "communicate" with the operator of the ground control station and can transmit data to it, for example, optical or thermal images of the area, along with primary information about the UAV flight parameters—altitude, course, speed, incline angle, etc. [49–53].

The generalized structure of the IMS based on the UAS is presented in Fig. 7.4 and includes airborne and ground subsystems of IMS, which include elements and modules that provide the implementation of a research program, accumulation and registration of measurement data, preliminary processing and transmission of information, as well as the formation of data about the location to the possibility of displaying information in 3D or 2D format.

Avionics of UAV is a complex of hardware and software located on its board, that is, on-board control equipment, which provides all flight modes and perform functional tasks. Avionics has a radio channel with a ground control station [54].

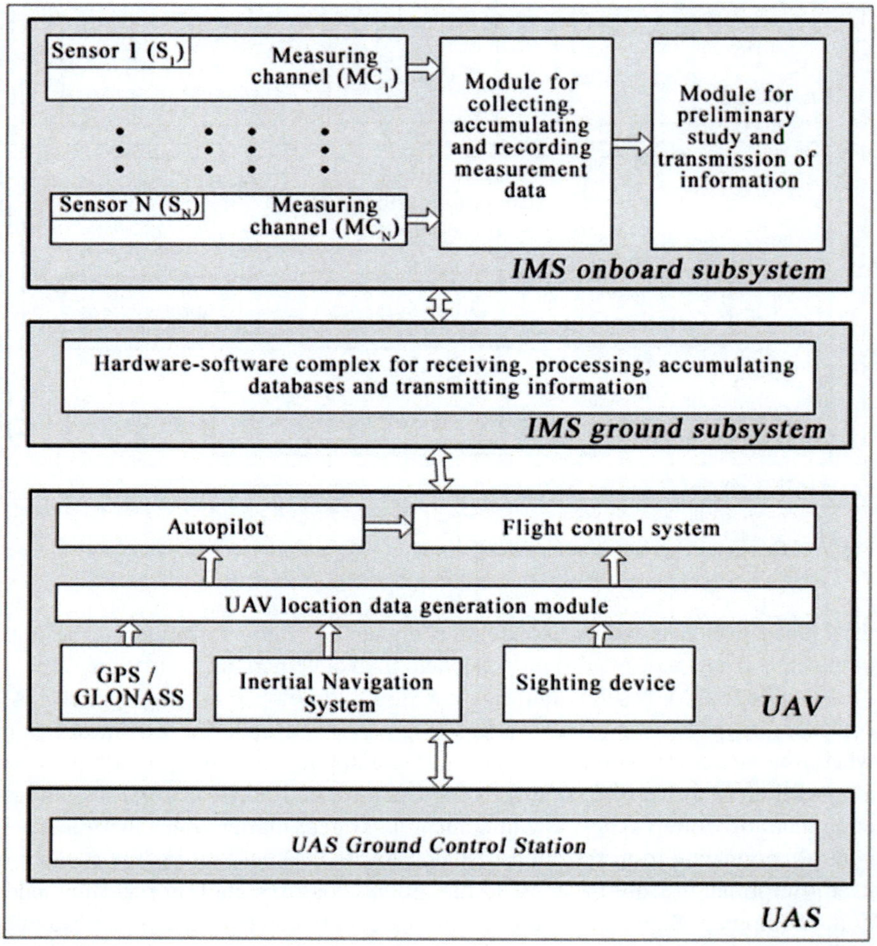

Fig. 7.4 Generalized structure of IMS based on UAS

UAV, as a control object, rudders of control elements, a radio channel block, sensors and a navigator form an automatic process control system (APCS) (Fig. 7.5). Onboard and ground control equipment should provide the following UAV flight modes: take-off and landing in automatic mode (manual take-off and landing with radio control by an operator are also possible); semi-automatic flight with radio control with adjustment of the operator's actions by on-board control equipment (OCE); automatic flight by control points with simultaneous transmission of telemetry to ground control equipment (GCE).

In manual mode, the operator, visually assessing the behavior in the space of the UAV, with the help of the GCE tills the controls (steering wheel, engine controls) driven by steering mechanisms (SM). Semi-automatic mode is possible within the

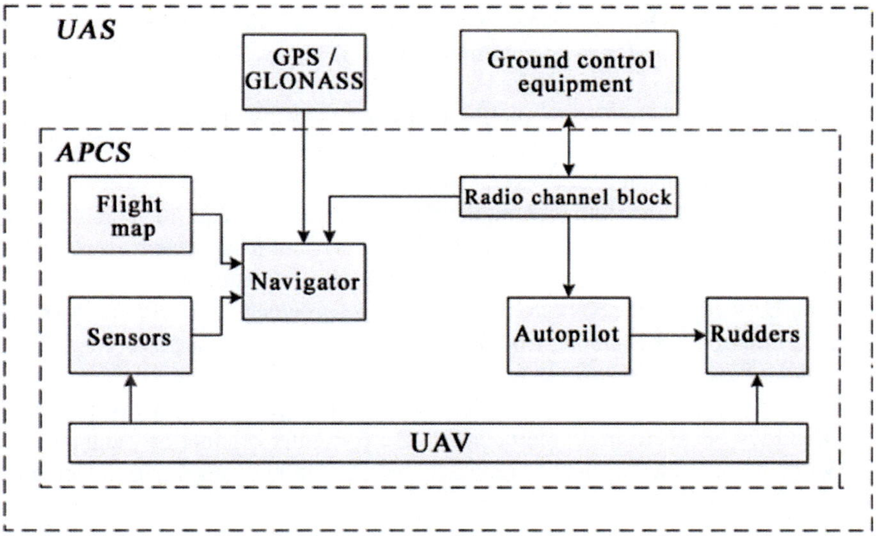

Fig. 7.5 Scheme of the automatic flight control system as part of the UAS

radius of the radio channel, which for small UAVs without the use of special radio antenna means is within 2500 m [47].

The semi-automatic control (piloting) mode in this case is carried out using information about the spatial position of the UAV, which is received via the radio channel and displayed on the virtual instrument panel of the GCU. Operator actions in this control mode are corrected using APCS, which performs the functions of an autopilot, which excludes the possibility of working out potentially dangerous UAV motion parameters.

Automatic mode is the provision of flight at predetermined control points of the route. In this mode, the UAV may not have radio communications with control and communication equipment. It turns on automatically if the UAV leaves the radio channel visibility range. In this flight mode, commands from the control unit are ignored. But it automatically controls the altitude, speed, and course according to the indications of orientation and navigation systems and sensors; it is also possible to control the deviation from a given path. In the automatic flight mode, control is carried out on the basis of the "guidance-stabilization" principle. The navigator module solves the guidance tasks, that is, it generates a guidance command (including the necessary flight direction and the current flight direction, calculated from the signals of orientation systems, navigation and sensors), which is transmitted to the "autopilot".

"Autopilot" module solves the stabilization problem, that is, processing the guidance command and ensuring stability of movement by generating SM control commands by the autopilot algorithm. In case of exceeding the specified thresholds (at angles and angular speeds), a horizontal flight stabilization command is issued, which is provided for by the autopilot operation algorithm as in the case of semi-automatic control. In automatic mode, the navigator module is tasked with periodically checking for radio communications. If it's available, this module sends telemetry data to the ground station.

During a flight in automatic mode, the UAV should fly at low altitudes with an envelope of the terrain. In this case, an error in maintaining the height within three meters should be ensured.

The creation of an intelligent automatic UAV control system, the core of which is OCE (avionics), is possible only if the control system (as a system for assessing the UAV state) has an integrated orientation system and autopilot coefficients for specific UAV flight modes. So, options should be provided for overcoming critical flight conditions, for example, from a critical roll, which may occur as a result of a gust of wind during a maneuver from a turn [55–57].

The main goal of integration of orientation and navigation systems is to increase the accuracy of determining the navigation and angular orientation parameters of UAVs. Not only systems can be combined, but also individual primary information sensors (pressure sensors, magnetoresistors, accelerometers, etc.) that produce the same parameters.

At combining several navigational measuring instruments, 2 complexing schemes, known as a compensation and filtering method, were most widely used.

The functional orientation of the UAV necessitated the following improvements in the construction of avionics using micromechanical elements and systems. On-board controllers (processors) and communication lines are complexed as the core of the system, providing the main (critical) communication lines between the on-board sensors and ground equipment, as well as engine control systems. Multifunctional capabilities, the need for which is determined by the limitations of weight and power, are achieved using a highly integrated design with physical multifunctional components. For example, the wing also serves as an antenna or sensor aperture. The power source is integrated with the fuselage design. A similar degree of constructive synergy has not yet been achieved in the design of traditional aircraft.

To ensure navigation, the algorithm of sliding orientation is used. The principle of its work is that the geographic coordinates of the UAV are calculated not in relation to the initial position, but relative to the previous position, calculated earlier.

To improve the accuracy of determining the parameters of the UAV's angular orientation, it is possible to combine different types of orientation systems. Figure 7.6 shows the general structure of combining the module of a triaxial magnetometer, pyrohorizon and the module of the GPS/GLONASS satellite navigation system.

A triaxial magnetometer gives normalized indications X, Y, Z for each channel. The pyrohorizont determines pitch θ_n and roll γ_n angles. The GPS/GLONASS module receives signals from the satellite navigation system and transmits to the block of calculation of orientation angles the value of the geographical coordinates φ, λ, the path angle ψ and the angles of the magnetic slope D and inclination I.

Based on the pitch and roll angles determined by the pyrometric sensors, the heading angle is calculated in the calculation unit. In this way, course angle values calculated from the indications of pyrometers and magnetic sensors and the course

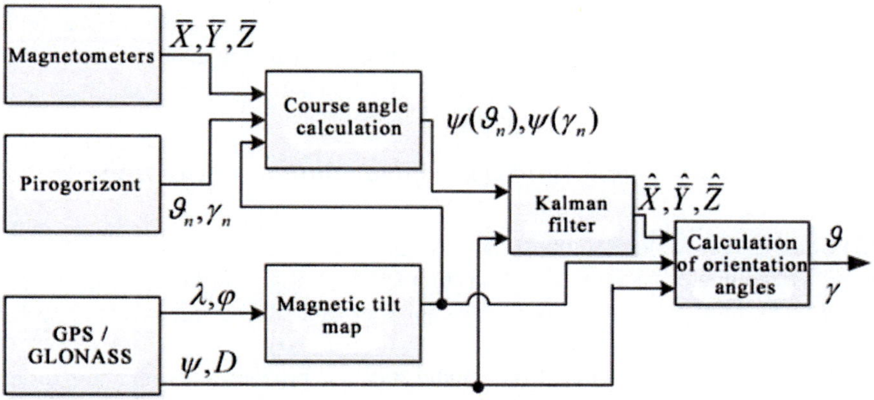

Fig. 7.6 General structure of integration of the magnetometer module, pyrogorizont and GPS/GLONASS module

that is issued by the GPS module are obtained. This allows to implement a Kalman filter to evaluate the systematic errors of magnetic sensors.

Thus, the combination increases the accuracy of determining the pitch and roll angles and eliminates the component errors associated with the presence of residual uncompensated magnetic interference and errors due to the UAV sliding angle.

The use of Kalman optimal filtering methods provides that the parameters of the stochastic description of disturbances and measurement errors are known exactly.

In practice, in case of uncertainty of the estimated parameters, the filter settings are made on an a priori model of perturbations and measurement errors, which leads to additional loss of estimation accuracy. In this case, the covariance matrix is not an estimate of the accuracy of the state vector, that is, the filter incorrectly generates an accuracy characteristic (covariance of the estimation error) simultaneously with the estimation of the state vector.

To improve the accuracy of determining the state vector, along with Kalman filtering, an algorithm for neural network approximation of an arbitrary probability distribution density has been developed. Neural networks can learn from the results of real measurements, and not from the error model, which can significantly reduce the negative impact of model uncertainty.

The implementation of the above functions is ensured by the use of a variable payload of modular construction, which is installed on the internal and / or external suspensions of the UAV.

In essence, a UAV is an aviation platform for transporting payloads, which may include: information collection sensors, including: photo, television, thermal imaging and multispectral cameras; laser scanners and rangefinders; SAR; multifunctional IMS and the like [58, 59].

At choosing the layout option of the payload modules, their technical characteristics, advantages and disadvantages, as well as the conditions for performing flight, control or other tasks according to their purpose and operating conditions (weather, terrain, time of day and time of year and other factors) are taken into account.

7.4.2 Hardware and Software Implementation of the UAS

APCS of UAV is a complex, multi-level integrated system, which includes (Fig. 7.7): on-board computer system; navigation complex; UAV flight control system; communication system; information exchange equipment and other components necessary for UAV functioning.

The on-board computer complex should solve the problems of navigation, orientation and vector gravimetry, the optimal assessment of various parameters and their correction, as well as the tasks of polling sensors and systems, storing arrays of initial information in memory, ensuring the functioning of an indication and monitoring system for peripheral devices, self-monitoring, etc.

The UAV navigation complex is an integrated system that processes navigation information receiving from the inertia system and the receiver of the satellite radio navigation system (SRNS).

The key element of the APCS for ensuring the autonomy of its functioning is the inertial component. This, as a rule, is a platformless inertial navigation system that

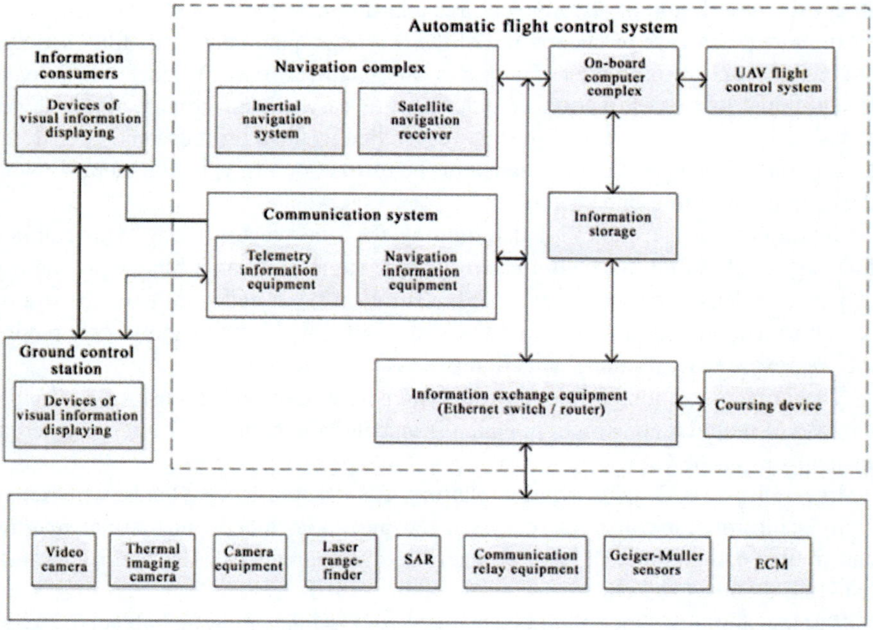

Fig. 7.7 Structure of the flight control system and UAV multifunctional control and measuring equipment and the management of information collection and accumulation modes

performs the function of determining the position of a UAV in space and incorporates inertial sensors, a barometric altimeter and a three-dimensional magnetometer. After comparing the data from the above-mentioned sensors with the data of the APCS receiver, the system produces a complete navigation solution for the coordinates and orientation angles.

The satellite component provides the determination of UAV coordinates and reconnaissance targets based on the signals of the global SRNS (GPS/GLONASS), as well as the determination of the UAV coursing angle.

The communication system is designed to ensure stable communication between the personnel of the ground control station and consumers of information from the UAV.

The main elements of the UAV communication system are follows.

Telemetry information equipment (transmitter and antenna-feeder device), which is designed to transmit real-time telemetry information to a ground control station and distribute it in radio visibility range.

Equipment for command and navigation information (receiver and antenna-feeder device). It's designed to receive UAV flight control commands and its payload. To connect over long distances and to increase the noise immunity of the equipment both on the UAV and on the ground control station, highly directional antenna systems are used.

Information exchange equipment provides switching, transmission and routing of information flows between the components and elements of the on-board and target equipment of the UAS, including information distribution between airborne information collection sensors, a video information transmitter and its storage device. The introduction of the flight mission and pre-launch control of the operation of the main components and systems of the UAV are carried out before its take-off through the external port of this equipment.

The information storage device is designed to accumulate information from the moment of start to the moment of UAV landing. This device can be removable or integrated. The information that is read from it allows to conduct a detailed analysis of the information received by UAV sensors during flight missions.

The coursing device is designed for visual control of the UAV flight control process and provides the necessary coverage area over which it flies.

The information that is received from the coursing device is transmitted to the UAV operator (payload operator) and used to control UAV flight and the operation of the multifunctional instrumentation. The coursing device includes a television camera with a wide-angle lens, which, depending on the tasks, can be replaced or supplemented by a thermal imaging camera, digital camera or SAR.

One of the priority tasks of creating promising UAS should be considered the development of more advanced APCS that will improve flight safety, reduce UAV losses and ensure the effective use of the payload.

A perspective direction in the development of APCS is the creation of "intelligent" avionics, which has software capable of selecting alternative control algorithms for continued flight upon failure of all systems. At the same time, an interactive mode is provided for controlling flight parameters and collecting measurement information,

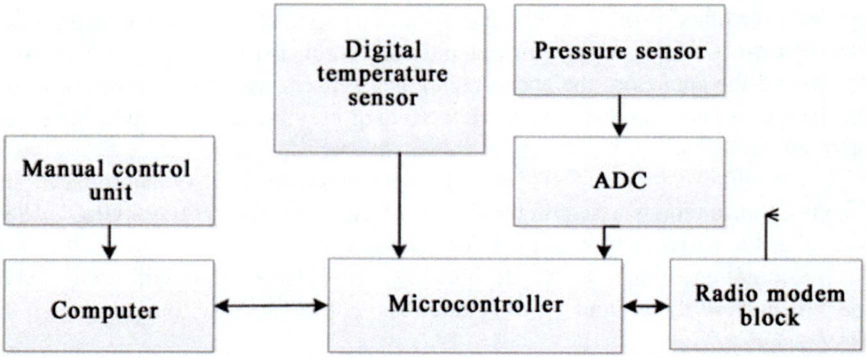

Fig. 7.8 Scheme of ground control equipment UAV

accumulating and transmitting data from various IMS sensors to obtain optical or thermal images of the area, along with primary information about the current state of the UAV—altitude, course, speed, roll, etc.

The main tasks of the UAS ground equipment (Fig. 7.8) can be formulated as follows: creating a flight map with the subsequent transfer to the storage unit of the specified parameters and flight path on board of the UAV; UAV control in manual and semi-automatic modes in the radio system coverage area; monitoring data from UAVs (data from UAV on-board systems, values of operating parameters of the power supply system, UAV coordinates, flight direction, speed); reading flight data [60, 61].

The microcontroller performs the functions of controlling data flows and generating UAV control command packets. During determining the layout options for the payload modules, control tasks, technical specifications of the sensors, their advantages and disadvantages, as well as the conditions for performing flight, instrumentation or other tasks for the intended purpose (weather, terrain, time of day and time of year, and other factors) are taken into account.

Control of atmospheric air is performed in two stages. The program includes monitoring the pollutants concentration in the upper atmosphere and the total contribution of the pollution source in the lower atmosphere.

The flight course is determined by the area of the pollution source, as well as the technical capabilities of the UAV, which determine the maximum UAV flight time.

It is rational to apply such a technique in the case of monitoring the state of atmospheric air in areas where point and area sources of air pollution are located.

The software of the ground control station allows to use any topographic basis as a map. Snapping can be done at two or more points. It is also possible to use digital maps as the topological basis.

The program provides input, automatic control and editing of the flight route. For each waypoint, a height can be set. There is the possibility of setting the landing point, as well as an algorithm for the behavior of UAVs in emergency situations.

At the ground control station, the sampling frequency of the air and the overlap rate of the frames throughout the flight course are also programmed.

If it is necessary to record points of any actions associated with changing the UAV control mode, an event log is kept. Telemetry data is also logged as it arrives.

The operational display panel allows to control all the important parameters for analyzing the functioning and monitoring the health of on-board systems.

To view flight logs (including in real time), analyze the behavior of UAVs and on-board equipment in flight, and promptly adjust parameters, the built-in software is used.

A system has been developed for measuring radiation background (exposure dose rate—EDR) (Fig. 7.9), which allows automatic determination of EDR in given spatial coordinates with simultaneous registration of longitude, latitude and height of the measurement site.

To study the composition of the air medium near the emission sources, it was proposed to install a filter-injection device (FID) on the UAV in the form of an elongated mechanical structure (Fig. 7.10) with an inhomogeneous material and an arbitrary internal structure.

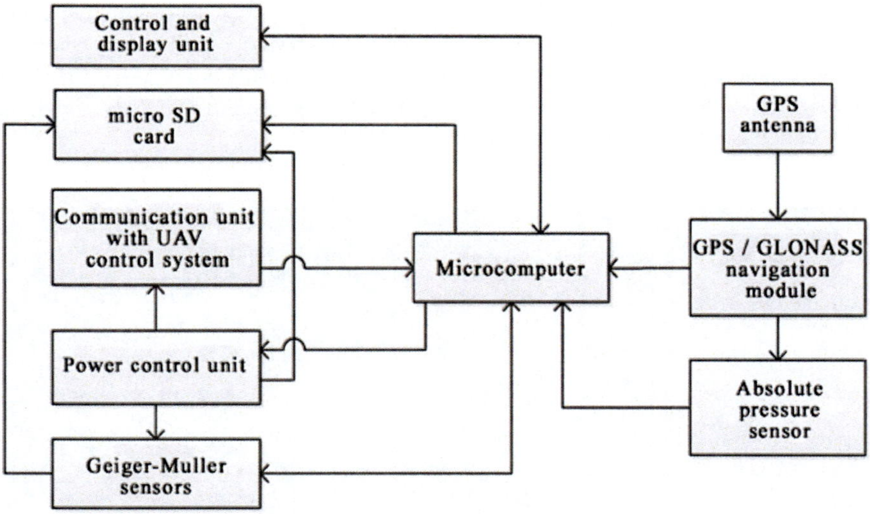

Fig. 7.9 EDR remote control system

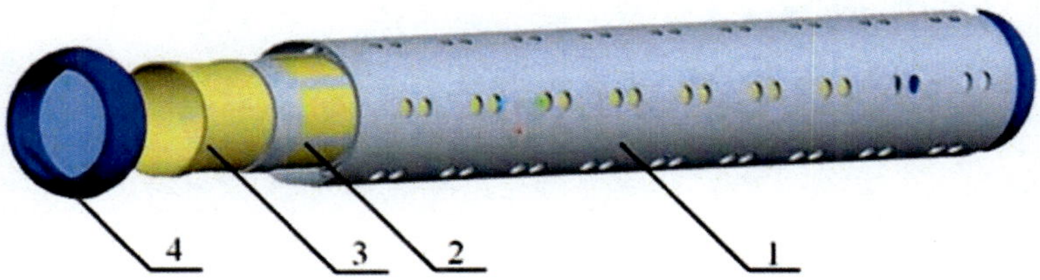

Fig. 7.10 FID design: **a** outer casing, **b** inner mesh casing, **c** filter cloth, **d** inlet flange

7.4.3 Experimental Results

For conducting experimental studies of atmospheric pollution using FID (isokinetic aerosol sampler) and measure EDR using a remote control system based on the UAS, a location 30 km from the Chernobyl NPP in the south-west direction was selected (Fig. 7.11). Experimental studies were carried out in 2016 at an altitude of about 100 m. The system of remote EDR control was activated by the operator remotely from the UAS control panel when the UAV reached the required height.

The flights were performed in the northeast direction of the wind along a path close to the circle (UAV tracks are shown in violet color in Fig. 7.11). The wind speed was 5–7 m/s. Three flights were performed within 1 h each at a speed of 70 km/h at altitudes of 70, 90 and 150 m. The calculations established that about 30 m^3 of atmospheric air passed through each of the three filters. The obtained samples were investigated by gamma spectroscopy (sensitivity 0.5 Bq/m^3) at the Department of Nuclear and Radiation Safety of the Institute for Safety Problems of Nuclear Power Plants of the National Academy of Sciences of Ukraine. The results showed that the activity is less than 10^{-2} Bq/m^3. This was to be expected, since according to the data of the above institute and the Central Geophysical Observatory in the surface layer of the atmosphere in this region, the volume activity of aerosols is 10^{-3}–10^{-5} Bq/m^3.

FID is installed on the UAV (Fig. 7.12) parallel to the longitudinal axis so that the inlet flange is directed towards the flight.

Fig. 7.11 UAV flight track

Fig. 7.12 FID installed on the UAV

Remote radiation safety monitoring systems typically measure the parameters of radioactive contamination by a predetermined measurements number or measurements time. This measurement technique is suitable for analyzing the level of pollution under normal operating conditions of pollution sources, but has a significant disadvantage, since all information about the radiation situation refers to the past time, which is not permissible in emergency situations.

However, in the case of activity of 1 Bq/m^3 or more (which is typical for emergency situations), it is enough to filter 1 m^3 of air, which is achieved in 2 min of a UAV flight.

The productivity of the created FID depends on the air speed, a graph of the corresponding dependence is shown in Fig. 7.13.

Considering UAVs as a tool for monitoring environmental pollution, it is possible to determine the pollutants concentration $C(x, y, z)$, and from its flight data to determine the value of wind speed and direction at the UAV flight altitude relative to the reference coordinate system.

Let show by the example of a UAV flight around a possible source of radionuclide release the possibility of determining their concentration directly at the outlet of the pollution source and establishing a picture of their distribution in the surface layers of the atmosphere. The direction of the wind and its speed are determined on the basis of the flight and navigation parameters of the UAV from the following relationships:

$$K_w = \arcsin\left(\frac{V}{U}\sin(K_\varphi)\right),\ U = \sqrt{V^2 + W^2 - 2VW\cos(K_\varphi)}, \quad (7.12)$$

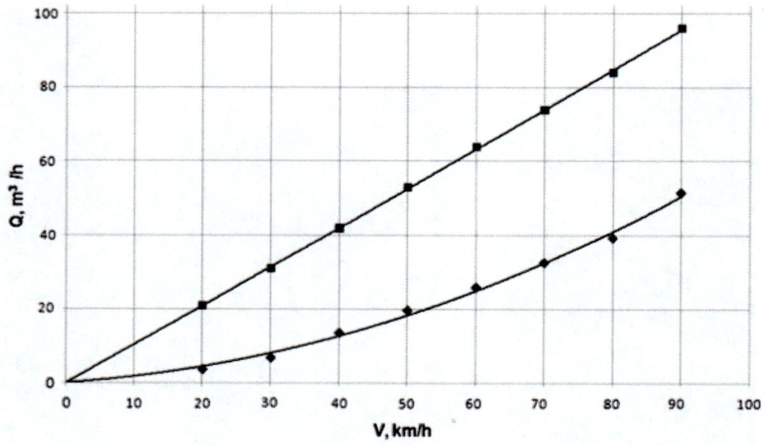

Fig. 7.13 Dependence of FID Q productivity on airspeed V

where K_w is the angle of the wind direction relative to the north direction, V is the airspeed of the UAV, U is the value of the wind speed; W is the UAV speed relative to the earth's surface (V and W values are measured by the UAV's onboard navigation system), K_φ is the drone drift angle, defined as the difference in the coursing angle and UAV path angle (based on data from the UAV's onboard navigation system).

Relation (7.12) allows to solve the direct problem of determining pollutants concentration in the atmosphere—the case when it is possible to determine the pollutants concentration directly at the emissions source.

In the absence of the possibility of direct measurement near the source emission, the "inverse" problem is solved—during concentration measurements are performed at a certain distance from the location of the emissions source. In this case, on the basis of UAV flight data and relationships (7.12), the direction and speed of the wind are determined, and the pollutants concentration in a given area of space is measured with the help of FID. It can be calculated the pollutants concentration at the output of a remote source:

$$c = C(\rho, \varphi, 0)2\pi\rho\cos\varphi\sqrt{k_\varphi k_z}\left[e^{\frac{V_\rho h^2}{4k_z\rho\cos\varphi}}\right]. \tag{7.13}$$

Here, at a known distance from the source, the pollution concentration at the source output is determined along the direction of the wind.

For the case when the location of the emission source does not lie on the line of the wind, it can be obtained the next relation:

$$c = C(\rho, \varphi, z)4\pi\rho\cos\varphi\sqrt{k_\varphi k_z}\times \\ \times\left[\exp\left(-\frac{V_\rho(\rho\sin\varphi)^2}{4k_\varphi\rho\cos\varphi}\right)\left(e^{\left[-\frac{V_\rho(H-h)^2}{4k_z\rho\cos\varphi}\right]}+e^{\left[-\frac{V_\rho(H+h)^2}{4k_z\cos\varphi}\right]}\right)\right]^{-1} \tag{7.14}$$

Expressions (7.13), (7.14) allow to estimate the radioactive aerosols concentration in the zone of the emission source. The initial data in this case are: the value of the local measured aerosol concentration and the coordinates of the sampling relative to the place of aerosol emission.

To solve the direct problem of determining the concentration of radioactive aerosols, provided that the source of emissions, for example, is at a height of 130 m with an intensity of 1000 Bq/m^3 at a wind speed of 5 m/s, a distribution diagram of the emission concentration is obtained as a function of distance from the emission source at an altitude of 70 m. Figure 7.14 shows the propagation curve of a radioactive aerosol in the wind direction at distances up to 10 km from the emission source.

Based on the results of mathematical modeling of the radioactive aerosols concentration, its 3D-dimensional distribution was obtained (Figs. 7.15 and 7.16). The level line diagrams showed the greatest practical interest corresponding to certain constant concentrations of radioactive aerosol (Fig. 7.16).

Thus, a mathematical model, a technique for conducting a full-scale experiment, and the results of mathematical modeling are presented as a rather convenient approach for analyzing the distribution of radioactive aerosols in air.

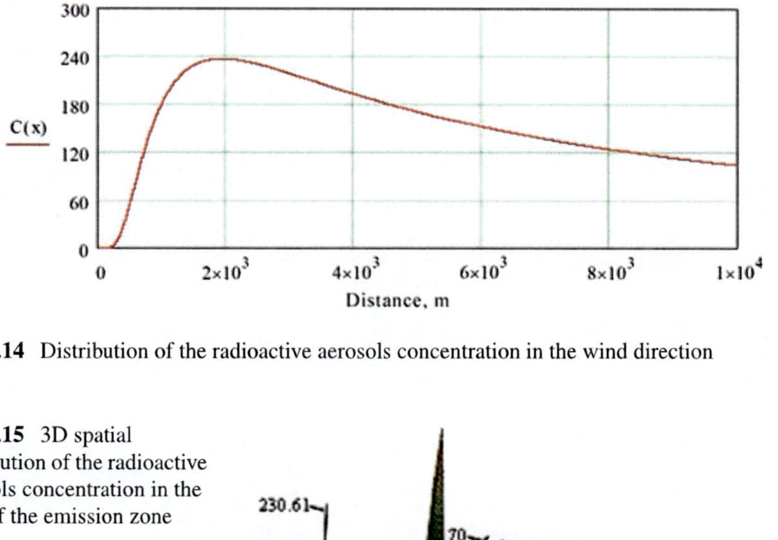

Fig. 7.14 Distribution of the radioactive aerosols concentration in the wind direction

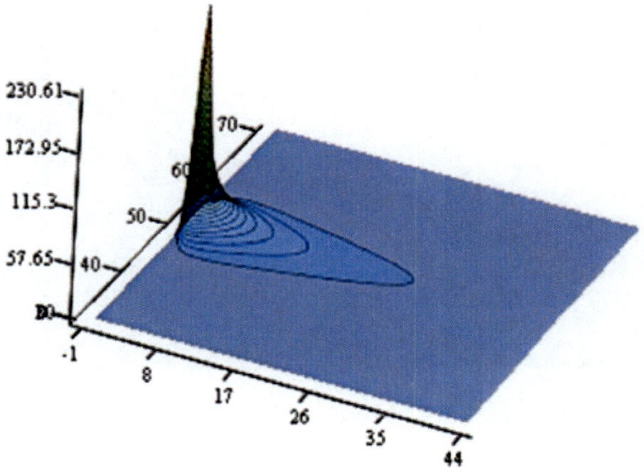

Fig. 7.15 3D spatial distribution of the radioactive aerosols concentration in the area of the emission zone

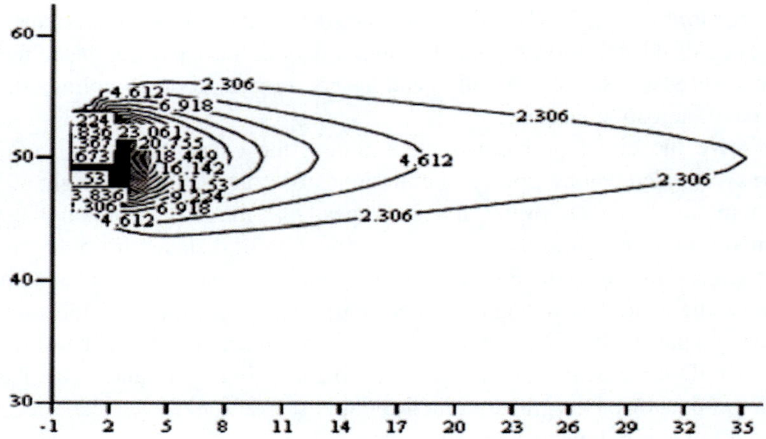

Fig. 7.16 Lines of concentration levels of radioactive aerosols

The same area was chosen for experimental studies and EDR measurement using a computer-based remote control system based on the UAS. Experimental researches were performed at an altitude of about 100 m (Fig. 7.17). The computerized system

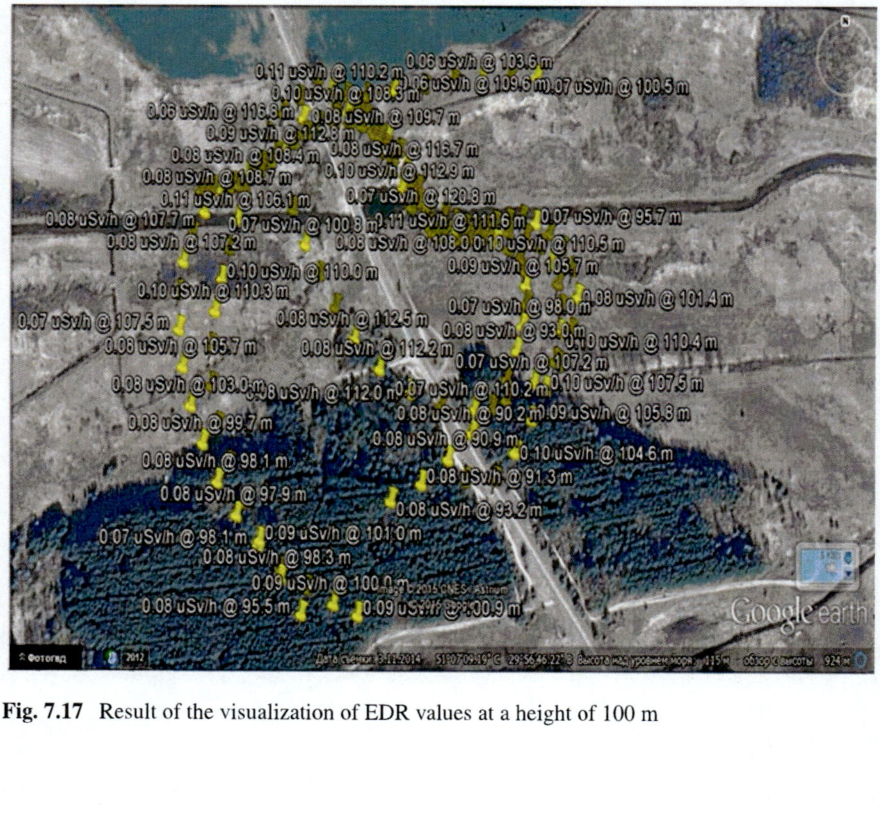

Fig. 7.17 Result of the visualization of EDR values at a height of 100 m

of remote control of EDR was activated by the operator remotely from the UAS control panel when the UAV reached the required height.

The UAV was controlled remotely by the operator due to the fact that it was located at a short distance from the ground control equipment and there was no need for a long flight course.

Measurement data was automatically recorded on a memory card integrated into the EDR monitoring system. After each completed flight, the data from the memory card was transferred to a PC and visualized using the Google Earth program (Fig. 7.17), where it can be seen the UAV flight track during measurements, the EDR values at each measurement point with height fixation. Experimental studies at a height of 100 m recorded a EDR value in the range of 5 … 12 μR/h, which is within the normal radiation background.

Figure 7.18 presents the results of a 3D graphical representation of experimental EDR measurement data.

The above possibilities of graphical interpretation of the results of measurements of EDR values indicate the prospect of using the UAS during monitoring environmental characteristics in hard-to-reach places and in conditions of technogenic hazard, including radiation hazard.

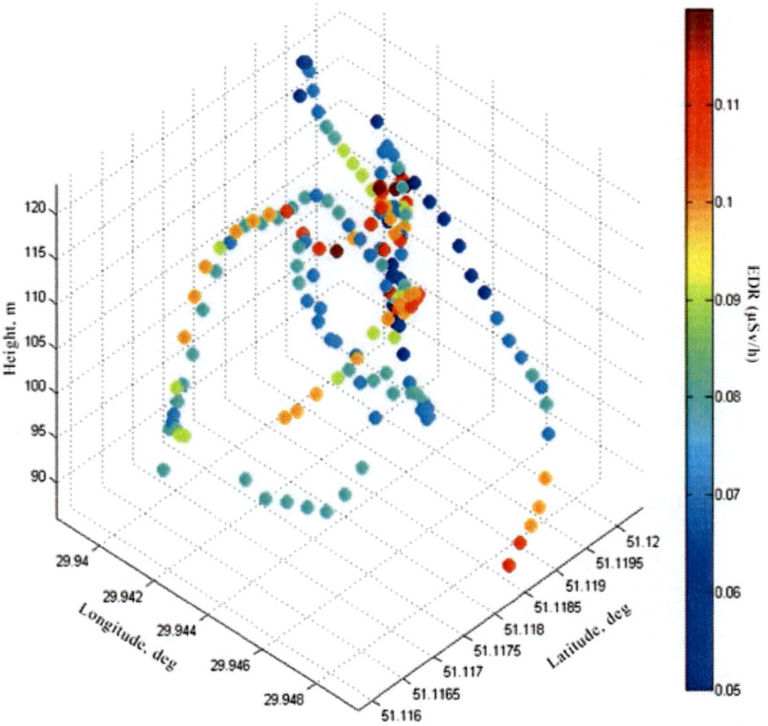

Fig. 7.18 3D graphical representation of EDR measurement results

References

1. Babak, S., Myslovych, M.: Practical application peculiarities of autonomous diagnostic complexes for thermal control of overhead power lines. Tekhnichna elektrodynamika **1**, 73–80 (2016). https://doi.org/10.15407/techned2016.01.073
2. Zaporozhets, A., Eremenko, V., Serhiienko, R., Ivanov, S.: Methods and hardware for diagnosing thermal power equipment based on smart grid technology. In: Shakhovska, N., Medykovskyy, M. (eds.), Advances in Intelligent Systems and Computing III. CSIT 2018. Advances in Intelligent Systems and Computing, vol. 871, pp. 476–489. Springer, Cham (2019). https://doi.org/10.1007/978-3-030-01069-0_34
3. Zaporozhets, A., Eremenko, V., Serhiienko, R., Ivanov, S.: Development of an intelligent system for diagnosing the technical condition of the heat power equipment. In: 2018 IEEE 13th International Scientific and Technical Conference on Computer Sciences and Information Technologies (CSIT), pp. 48–51. Ukraine (2018). https://doi.org/10.1109/stc-csit.2018.8526742
4. Popov, O., Iatsyshyn, A., Kovach, V., Artemchuk, V., Taraduda, D., Sobyna, V., Sokolov, D., Dement, M., Yatsyshyn, T., Matvieieva, I.: Analysis of possible causes of NPP emergencies to minimize risk of their occurrence. Nucl. Radiat. Saf. **1**(81), 75–80 (2019). https://doi.org/10.32918/nrs.2019.1(81).13
5. Popov, O., Iatsyshyn, A., Kovach, V., Artemchuk, V., Taraduda, D., Sobyna, V., Sokolov, D., Dement, M., Hurkovskyi, V., Nikolaev, K., Yatsyshyn, T., Dimitriieva, D.: Physical features of pollutants spread in the air during the emergency at NPPs. Nucl. Radiat. Saf. **4**(84), 88–98 (2019). https://doi.org/10.32918/nrs.2019.4(84).11
6. Zaporozhets, A.O., Khaidurov, V.V.: Mathematical models of inverse problems for finding the main characteristics of air pollution sources. Water Air Soil Pollut. **231**, 563 (2020). https://doi.org/10.1007/s11270-020-04933-z
7. Zaporozhets, A.O.: Correlation analysis between the components of energy balance and pollutant emissions. Water Air Soil Pollut. **232**, 114 (2021). https://doi.org/10.1007/s11270-021-05048-9
8. Villa, T.F., Salimi, F., Morton, K., Morawska, L., Gonzalez, F.: development and validation of a UAV based system for air pollution measurements. Sensors **16**(12), 2202 (2016). https://doi.org/10.3390/s16122202
9. Xiang, H., Tian, L.: Development of a low-cost agricultural remote sensing system based on an autonomous unmanned aerial vehicle (UAV). Biosys. Eng. **108**(2), 174–190 (2011). https://doi.org/10.1016/j.biosystemseng.2010.11.010
10. Kharchenko, V., Sanchenko, A., Kochan, V., Fesenko, H.: Reliability and survivability models of integrated drone-based systems for post emergency monitoring of NPPs. In: 2016 International Conference on Information and Digital Technologies (IDT), pp. 127–132. Poland (2016). https://doi.org/10.1109/dt.2016.7557161
11. Babak, V.P., Kanchenko, V.A., Klyuchnikov, A.A., Krasnov, V.A., Chepur, N.L.: Unmanned aerial complexes as a way of NPP and environment radiation monitoring. In: Problemi Bezpeki Atomnikh Elektrostantsyij yi Chornobilya, vol. 19, pp. 60–69 (2012). https://inis.iaea.org/search/search.aspx?orig_q=RN:44085230
12. Fesenko, H., Kharchenko, V.: Determining the optimum number of single operator controlled unmanned aerial vehicles for NPP monitoring missions: human error issues. In: 2018 International Scientific-Practical Conference Problems of Infocommunications. Science and Technology (PIC S&T), pp. 71–714. Ukraine (2018). https://doi.org/10.1109/infocommst.2018.8632029
13. Kharchenko, V., Fesenko, H., Sanchenko, A., Hiromoto, R.E., Kochan, V.: Reliability issues for a multi-version post-severe NPP accident monitoring system. In: 2017 9th IEEE International Conference on Intelligent Data Acquisition and Advanced Computing Systems: Technology and Applications (IDAACS), pp. 942–946. Romania (2017). https://doi.org/10.1109/idaacs.2017.8095225

14. Zaporozhets A., Kovtun S., Dekusha O.: System for monitoring the technical state of heating networks based on UAVs. In: Shakhovska N., Medykovskyy M.O. (eds.) Advances in Intelligent Systems and Computing IV. CSIT 2019. Advances in Intelligent Systems and Computing, vol 1080. Springer, Cham, pp. 935–950 (2020). https://doi.org/10.1007/978-3-030-33695-0_61
15. Sato, Y., Ozawa, S., Terasaka, Y., KAburagi, M., Tanifuji, Y., Kawabata, K., Nakamura, H., Miyamura, H.N., Izumi, R., Suzuki, T., Toril, T.: Remote radiation imaging system using a compact gamma-ray imager mounted on a multicopter drone. J. Nucl. Sci. Technol. **55**(1), 90–96 (2018). https://doi.org/10.1080/00223131.2017.1383211
16. Martin, P.G., Moore, J., Fardoulis, J.S., Payton, O.D., Scott, T.B.: Radiological assessment on interest areas on the sellafield nuclear site via unmanned aerial vehicle. Remote Sens. **8**(11), 913 (2016). https://doi.org/10.3390/rs8110913
17. Nagatani, K., Kiribayashi, S., Okada, Y., Otake, K., Yoshida, K., Tadokoro, S., Nishimura, T., Yoshida, T., Koyanagi, E., Fukushima, M., Kawatsuma, S.: Emergency response to the nuclear accident at the Fukushima Daiichi Nuclear Power Plants using mobile rescue robots. J. Field Robot. **30**(1), 44–63. https://doi.org/10.1002/rob.21439
18. Cerba, S., Luley, J., Vrban, B., Osusky, F., Necas, V.: Unmanned radiation-monitoring system. IEEE Trans. Nucl. Sci. **67**(4), 636–643 (2020). https://doi.org/10.1109/TNS.2020.2970782
19. Zaporozhets, A.: Hardware and software implementation of modules of the system of the fuel combustion control process. In: Control of Fuel Combustion in Boilers. Studies in Systems, Decision and Control. Springer, Cham, pp. 61–87 (2020). https://doi.org/10.1007/978-3-030-46299-4_3
20. Zaporozhets, A.: Experimental research of a computer system for the control of the fuel combustion process. In: Control of Fuel Combustion in Boilers. Studies in Systems, Decision and Control. Springer, Cham, pp. 89–123 (2020). https://doi.org/10.1007/978-3-030-46299-4_4
21. Zaporozhets, A., Babak, V., Isaienko, V., Babikova, K.: Analysis of the air pollution monitoring system in Ukraine. In: Systems, Decision and Control in Energy I. Studies in Systems, Decision and Control, vol. 298. Springer, Cham (2020). https://doi.org/10.1007/978-3-030-48583-2_6
22. Zaporozhets, A.: Review of quadrocopters for energy and ecological monitoring. In: Systems, Decision and Control in Energy I, Studies in Systems, Decision and Control, vol. 298. Springer, Cham. https://doi.org/10.1007/978-3-030-48583-2_2
23. Bolla, G.M., Casagrande, M., Comazzetto, A., Dal Moro, R., Destro, M., Fantin, E., Colombatti, G., Aboudan, A., Lorenzini, E.C.: ARIA: air pollutants monitoring using UAVs. In: 2018 5th IEEE International Workshop on Metrology for AeroSpace (MetroAeroSpace), pp. 225–229. Italia (2018). https://doi.org/10.1109/metroaerospace.2018.8453584
24. Zaporozhets, A.: Methods and means for the control of the fuel combustion process. In: Control of Fuel Combustion in Boilers. Studies in Systems, Decision and Control. Springer, Cham, pp. 1–33 (2020). https://doi.org/10.1007/978-3-030-46299-4_1
25. Gonzalez, L.F., Montes, G.A., Puig, E., Johnson, S., Mengersen, K., Gaston, K.J.: Unmanned aerial vehicles (UAVs) and artificial intelligence revolutionizing wildlife monitoring and conservation. Sensors **16**(1), 97 (2016). https://doi.org/10.3390/s16010097
26. Manfreda, S., McCabe, M.F., Miller, P.E., Lucas, R., Madrigal, V.P., Mallinis, G., Dor, E.B., Helman, D., Estes, L., Ciraolo, G., Mullerova, J., Tauro, F., De Lima, M.I., De Lima, J.M.P., Maltese, A., Frances, F., Caylor, K., Kohv, M., Perks, M., Ruiz-Perez, G., Su, Z., Vico, G., Toth, B.: On the use of unmanned aerial systems for environmental monitoring. Remote Sens. **10**(4), 641 (2018). https://doi.org/10.3390/rs10040641
27. Zaporozhets, A.: Research of the process of fuel combustion in boilers. In: Control of Fuel Combustion in Boilers. Studies in Systems, Decision and Control. Springer, Cham, pp. 35–50 (2020). https://doi.org/10.1007/978-3-030-46299-4_2
28. Smidl, V., Hofman, R.: Tracking of atmospheric release of pollution using unmanned aerial vehicles. Atmos. Environ. **67**, 425–436 (2013). https://doi.org/10.1016/j.atmosenv.2012.10.054
29. Saha, H.N., Auddy, S., Chatterjee, A., Pal, S., Pandey, S., Singh, R., Singh, R., Sharan, P., Banerjee, S., Ghosh, D., Maity, A.: Pollution control using internet of things (IoT). In: 2017 8th Annual Industrial Automation and Electromechanical Engineering Conference (IEMECON), pp. 65–68. Thailand (2017). https://doi.org/10.1109/iemecon.2017.8079563

30. Babak, S., Myslovych, M., Sysak, R.: Module structure of UAV-based computerized systems for remote environment monitoring of energy facilities. In: 2016 17th International Conference Computational Problems of Electrical Engineering (CPEE), pp. 1–3. Poland (2016). https://doi.org/10.1109/cpee.2016.7738752
31. Maiorov, A.A., Materuhin, A.V.: A conceptual model of an information measurement system based on distributed networks of smart geosensors. Meas. Tech. **61**, 452–458 (2018). https://doi.org/10.1007/s11018-018-1451-0
32. Selezneva, M.S., Neusypin, K.A.: Development of a measurement complex with intelligent component. Meas. Tech. **59**, 916–922 (2016). https://doi.org/10.1007/s11018-016-1067-1
33. Xiao, B., Yin, S.: A new disturbance attenuation control scheme for quadrotor unmanned aerial vehicles. IEEE Trans. Industr. Inf. **13**(6), 2922–2932 (2017). https://doi.org/10.1109/TII.2017.2682900
34. Nolan, P.J., Pinto, J., Gonzalez-Rocha, J., Jensen, A., Vezzi, C.N., Bailey, S.C.C., De Boer, G., Diehi, C., Ill, R.L., Powers, C.W., Foroutan, H., Ross, S.D., Ill, S.G.S.: Coordinated unmanned aircraft system (UAS) and ground-based weather measurements to predict Lagrangian coherent structures (LCSs). Sensors **18**(12), 4448 (2018). https://doi.org/10.3390/s18124448
35. Sajid, M., Yang, Y.J., Kim, G.B., Choi, K.H.: Remote monitoring of environment using multi-sensor wireless node installed on quad-copter drone. In: 2016 IEEE International Symposium on Robotics and Intelligent Sensors (IRIS), pp. 213–216. Japan (2016). https://doi.org/10.1109/iris.2016.8066093
36. Alvear, O., Calafate, C.T., Zema, N.R., Natalizio, E., Hernandez-Orallo, E., Cano, J.-C., Manzoni, P.: A discretized approach to air pollution monitoring using UAV-based sensing. Mobile Netw. Appl. **23**, 1693–1702 (2018). https://doi.org/10.1007/s11036-018-1065-4
37. Conti, G.O., Heibati, B., Kloog, I., Fiore, M., Ferrante, M.: A review of AirQ models and their applications for forecasting the air pollution health outcomes. Environ. Sci. Pollut. Res. **24**, 6426–6445 (2017). https://doi.org/10.1007/s11356-016-8180-1
38. Alvarez-Vazquez, L.J., Garsia-Chan, N., Martinez, A., Vazquez-Mendez, M.E.: Numerical simulation of air pollution due to traffic flow in urban networks. J. Comput. Appl. Math. **326**, 44–61 (2017). https://doi.org/10.1016/j.cam.2017.05.017
39. Babak, S.: Radiation monitoring of environment using unmanned aerial complex. Adv. Sci. J. **12**, 41–44 (2014). https://doi.org/10.15550/ASJ.2014.12.041
40. Yang, Y., Zheng, Z., Bian, K., Jiang, Y., Song, L., Han, Z: Arms: A fine-grained 3D AQI realtime monitoring system by UAV. In: GLOBECOM 2017—2017 IEEE Global Communications Conference, pp. 1–6. Singapore (2017). https://doi.org/10.1109/glocom.2017.8253968
41. Huang, R., Zhai, X., Ivey, C.E., Friberg, M.D., Hu, X., Liu, Y., Di, Q., Schwartz, J., Mulholland, J.A., Russel, A.G.: Air pollutant exposure field modeling using air quality model-data fusion methods and comparison with satellite AOD-derived fields: application over North Carolina, USA. Air Qual. Atmos. Health **11**, 11–22 (2018). https://doi.org/10.1007/s11869-017-0511-y
42. Lu, D., Ge, Y., Wang, L., Zhu, D., Wang, Z., Yuan, X.: Multivariate visualization for atmospheric pollution. J. Visual. **22**, 1093–1105 (2019). https://doi.org/10.1007/s12650-019-00588-z
43. Babak, S., Sherbak, L.: Research of atmospheric fields around industrial facilities. Adv. Sci. J. **6**, 61–64 (2015). https://doi.org/10.15550/ASJ.2015.06.061
44. Jacobson, M.Z.: Review of solutions to global warming, air pollution, and energy security. Energy Environ. Sci. **2**, 148–173 (2009). https://doi.org/10.1039/B809990C
45. Mohammed, F., Idries, A., Mohamed, N., Al-Jaroodi, J., Jawhar, I.: UAVs for smart cities: Opportunities and challenges. In: 2014 International Conference on Unmanned Aircraft Systems (ICUAS), pp. 267–273. USA (2014). https://doi.org/10.1109/icuas.2014.6842265
46. Wivou, J., Udawatta, L., Alshehhi, A., Alzaabi, E., Albeloshi, A., Alfalasi, S.: Air quality monitoring for sustainable systems via drone based technology. In: 2016 IEEE International Conference on Information and Automation for Sustainability (ICIAfS), pp. 1–5. Sri Lanka (2016). https://doi.org/10.1109/iciafs.2016.7946542
47. Gupta, S.G., Ghonge, M., Jawandhiya, P.M.: Review of unmanned aircraft system (UAS). Int. J. Adv. Res. Comput. Eng. Technol. **2**(4) (2013). https://doi.org/10.2139/ssrn.3451039

48. Jun, M., D'Andrea, R.: Path planning for unmanned aerial vehicles in uncertain and adversarial environments. In: Butenko S., Murphey R., Pardalos P.M. (eds.), Cooperative Control: Models, Applications and Algorithms. Cooperative Systems, vol. 1, pp. 95–110. Springer, Boston, MA (2003). https://doi.org/10.1007/978-1-4757-3758-5_6
49. Gu, Q., Michanowicz, D.R., Jia, C.: Developing a modular unmanned aerial vehicle (UAV) platform for air pollution profiling. Sensors **18**(12), 4363 (2018). https://doi.org/10.3390/s18124363
50. Wada, A., Yamashita, T., Maruyama, M., Arai, T., Adachi, H., Tsuji, H.: A surveillance system using small unmanned aerial vehicle (UAV) related technologies. NEC Tech. J. **8**(1), 68–72 (2015). https://pdfs.semanticscholar.org/c8a8/3edc2d261b012ba0f2b36895a42a5315f438.pdf
51. Colomina, I., Molina, P.: Unmanned aerial systems for photogrammetry and remote sensing: a review. ISPRS J. Photogram. Remote Sens. **92**, 79–97 (2014). https://doi.org/10.1016/j.isprsjprs.2014.02.013
52. Necsulescu, D., Jiang, Y.-W., Kim, B.: Neural network based feedback linearization control of an unmanned aerial vehicle. Int. J. Autom. Comput. **4**, 71–79 (2007). https://doi.org/10.1007/s11633-007-0071-y
53. Dalamagkidis, K.: Classification of UAVs. In: Valavanis, K., Vachtsevanos, G. (eds.), Handbook of Unmanned Aerial Vehicles, pp. 83–91. Springer, Dordrecht (2015). https://doi.org/10.1007/978-90-481-9707-1_94
54. Babak, S., Babak, V., Zaporozhets, A., Sverdlova, S.: Method of statistical spline functions for solving problems of data approximation and prediction of objects state. In: CEUR Workshop Proceedings, vol. 2353, pp. 810–821 (2019). http://ceur-ws.org/Vol-2353/paper64.pdf
55. Menouar, H., Guvenc, I., Akkaya, K., Uluagac, A.S., Kadri, A., Tuncer, A.: UAV-enabled intelligent transportation systems for the smart city: applications and challenges. IEEE Commun. Mag. **55**(3), 22–28 (2017). https://doi.org/10.1109/MCOM.2017.1600238CM
56. Emel'yanov, S., Makarov, D., Panov, A.I., Yakovlev, K.: Multilayer cognitive architecture for UAV control. Cognit. Syst. Res. **39**, 58–72 (2016). https://doi.org/10.1016/j.cogsys.2015.12.008
57. Huang, R.: Maritime intelligent real-time control system based on UAV. In: 2018 International Conference on Robots & Intelligent System (ICRIS), pp. 10–12. China (2018). https://doi.org/10.1109/icris.2018.00011
58. Gonzalez-deSantos, L.M., Martinez-Sanchez, J., Gonzalez-Jorge, H., Navarro-Medina, F., Arias, P.: UAV payload with collision mitigation for contact inspection. Autom. Construct. **115**, 103200 (2020). https://doi.org/10.1016/j.autcon.2020.103200
59. Hashemi, D., Heidari, H.: Trajectory planning of quadrotor UAV with maximum payload and minimum oscillation of suspended load using optimal control. J. Intell. Rob. Syst. (2020). https://doi.org/10.1007/s10846-020-01166-4
60. James, M.R., Robson, S., d'Oleire-Oltmanns, S., Niethammer, U.: Ground control quality, quantity and bundle adjustment. Geomorphology **280**, 51–66 (2017). https://doi.org/10.1016/j.geomorph.2016.11.021
61. Martinez-Carricondo, P., Aguera-Vega, F., Carnajal-Ramirez, F., Mesas-Carrascosa, F.-J., Garcia-Ferrer, A., Perez-Porras, F.-J.: Assessment of UAV-photogrammetric mapping accuracy based on variation of ground control points. Int. J. Appl. Earth Observ. Geoinform., 1–10 (2018). https://doi.org/10.1016/j.jag.2018.05.015

Chapter 8
Models and Measures for Atmospheric Pollution Monitoring

Abstract The features of monitoring atmospheric air pollution are considered and a promising direction for the development of an extensive network of monitoring systems is proposed. The results of the analysis of the composition and type of pollutants during the operation of thermal power plants (TPPs) and technologically hazardous enterprises of various industries according to statistics of retrospective information are presented. The features of the formation and distribution of pollutants in the air are described, methods for calculating their concentration are presented. A method for localizing the source of pollution according to monitoring data in the polar coordinate system is proposed. The methods of statistical processing of monitoring data within the framework of the correlation theory are substantiated. The structure of a spatially branched air pollution monitoring network using modern information technologies is described. The measuring modules of the network are developed and the results of their experimental studies are presented. A method for assessing the quality of air pollution based on the results of monitoring the statistical characteristics of local atmospheric pollution fields is proposed.

Keywords Pollutants · Methods for calculating the pollutants concentration · Processing monitoring data · Localization of the pollution source · Air pollution monitoring network

8.1 Main Pollutants in the Products of Fuel Combustion and Their Distribution in the Atmosphere

Integral estimates of the composition and type of the main pollutants (pollutants) of atmospheric air are considering, the statistical data of which are accumulated over a significant observation interval in the form of retrospective information.

An increase in the concentration of various pollutants in the air negatively affects the whole complex of wildlife [1–5]. Long-term impact of pollutants in the atmosphere leads to a deterioration in the health of people and animals, and a decrease in crop yields. Atmospheric pollution can significantly accelerate corrosion processes in building structures and infrastructure elements that contact with air [6–9].

V. P. Babak et al., *Models and Measures in Measurements and Monitoring*,
Studies in Systems, Decision and Control 360,
https://doi.org/10.1007/978-3-030-70783-5_8

Table 8.1 The ratio of emissions of pollutants into the atmosphere between industries

Objects and industries	Share of pollutant emissions into the atmosphere (%)
TPPs and boiler rooms	27
Ferrous metallurgy	17
Oil production and petrochemistry	16
Automobile transport	12
Non-ferrous metallurgy	10
Construction industry	5
Coal industry	2.5
Chemical industry	1.5
Other industries	9
Total	100

Not only pollutants get into the atmosphere from thermal power plants (TPPs), but also from other industrial and energy enterprises, transport, and utilities. Table 8.1 shows the distribution of pollutant emissions into the atmosphere from objects of various industries.

The real situation of the technical equipment of energy facilities in Ukraine, which basically worked out their technical resources, requires the creation of new monitoring systems for remote environmental monitoring. This requires the use of modern achievements of science and technology, the latest information and measurement technologies, and their effectiveness depends on the development and practical use of hardware-software measuring complexes and systems [10–13].

The type of fuel affects on the composition of pollutants that are formed after its combustion. TPPs use solid, liquid and gaseous fuels. The main pollutants contained in the exhaust gases are: sulfur oxides (SO_2 and SO_3), nitrogen oxides (NO and NO_2), carbon monoxide (CO), vanadium compounds (mainly vanadium pentaxide V_2O_5). Ash also refers to pollutants [14–19].

Coal (brown, stone, anthracite), oil shale and peat belong to solid fuels used in the energy sector. The schematic composition of solid fuels is shown in Fig. 8.1. As can be seen from Fig. 8.1, the organic part of the fuel consists of carbon C, hydrogen H, oxygen O, organic sulfur S. The mineral part of the fuel consists of water H_2O and ash. The main part of the mineral component of the fuel transforms into the fly ash during combustion.

Fuel oil, shale oil, diesel and boiler fuel belong to the liquid fuels used in the energy sector. Pyrite sulfur is absent in liquid fuel. Fuel oil ash contains V_2O_5, Ni_2O_3, Al_2O_3, Fe_2O_3, SiO_2, MgO and oxides. The ash content of fuel oil does not exceed 0.3%. The solids content during complete combustion in the exhaust gas is about 0.1 g/m^3.

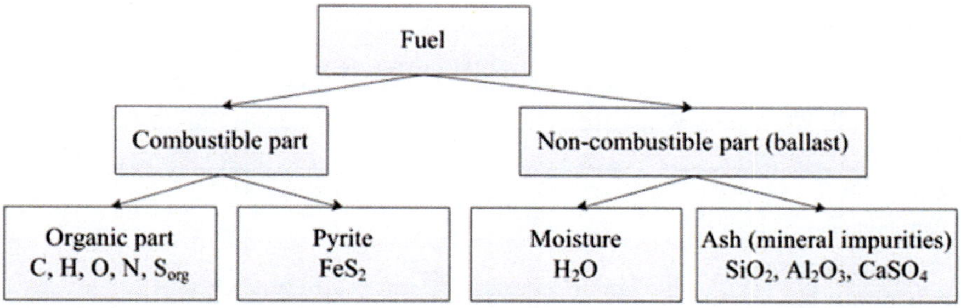

Fig. 8.1 Solid fuel composition

Sulfur in fuel oil is found mainly in the form of organic compounds, "pure" sulfur and hydrogen sulfide. Its content depends on the sulfur content of the oil from which it is made.

Diesel fuel is usually divided by sulfur content into 2 groups: 1—up to 0.2% and 2—up to 0.5%. In low-sulfur boiler-furnace fuel, the sulfur content does not exceed 0.5%, in medium-sulfur—up to 1.1%; in shale oil—not more than 1%.

Gaseous fuel is the "cleanest" organic fuel, and only nitrogen oxides and carbon monoxide CO can be formed during its combustion (in the case of incomplete combustion). Detailed information on the mechanisms of products formation of chemical underburning during the combustion of natural gas is given in [11, 20, 21].

As can be seen from the above data, in the pollution of atmospheric air, the main role is played by the objects of the heat power, metallurgical industry and transport. Their total contribution to environmental pollution is about 2/3 of total industrial emissions.

The amount of permissible concentration of pollutants for humans in the environment is determined on the basis of 2 maximum permissible concentration (MPC) indicators: maximum one-time ($MPC_{m.o.}$) and daily average ($MPC_{d.a.}$) [22]. MPC of pollutants produced by objects of the energy industry are presented in Table 8.2.

The materials presented reflect only the characteristic features of the main pollutants, which form the environmental characteristics of atmospheric pollution fields.

Now let briefly consider the spread of pollutants in the air. Turbulence of atmospheric air is a complex physical phenomenon, the dynamics of which rapidly change in time and space. This phenomenon must be taken into account in studies of the characteristics of local atmospheric pollution fields. Based on the retrospective research information, the main components of atmospheric air turbulence are distinguished. The latter affects the dispersion of pollutants in the atmosphere due to the action of molecular and turbulent diffusion. Characteristic of this process is that the effect of molecular diffusion is relatively small compared with turbulent diffusion.

Turbulent diffusion has 2 components—dynamic and thermal. Thermal diffusion is associated with a vertical temperature gradient of air and provides convective turbulence.

Table 8.2 MPC values (mg/m^3) for some pollutants at the level of human respiration

Pollutant	Chemical formula	$MPC_{m.o.}$	$MPC_{d.a.}$
Nitrogen dioxide	NO_2	0.085	0.04
Nitrogen monoxide	NO	0.6	0.06
Sulfur dioxide	SO_2	0.5	0.05
Benzapyrene	$C_{20}H_{16}$	–	10^{-6}
Vanadium pentaxide	V_2O_5	–	0.002
Carbon monoxide	CO		3
Ammonia	NH_3	0.2	0.04
Hydrogen sulfide	H_2S	0.008	–
Soot	–	0.15	0.05
Inorganic dust containing silicon dioxide (%)			
>70	–	0.15	0.05
20–70	–	0.3	0.1
<20	–	0.5	0.15
TPP coal ash	–	0.05	0.02

Dynamic diffusion, which provides mechanical turbulence, occurs as a result of the movement of air masses under the wind influence. It increases in the lower layers of the troposphere under the influence of macro-irregularities of the relief and is not related to the temperature gradient [23–25].

The main influence on atmospheric turbulence is performed by 2 processes:

- convective turbulence associated with natural convective flows resulting from heating of the atmosphere;
- mechanical turbulence arising from the friction of moving air flows and uneven ground, and is the result of a wind shift.

The following factors have a significant impact on the process of dispersion in the atmosphere of emissions from exhaust pipes and ventilation devices:

- state of the atmosphere;
- physical and chemical properties of substances that produce emissions (density, gas temperature, dispersed dust composition, etc.);
- height and diameter of the emission source;
- location of emission sources;
- terrain.

As an example, a schematic distribution of the pollutants concentration in the atmosphere under a high point source of pollution shown in Fig. 8.2.

High concentrations of pollutants inside the flare transfer zone caused by fugitive emissions.

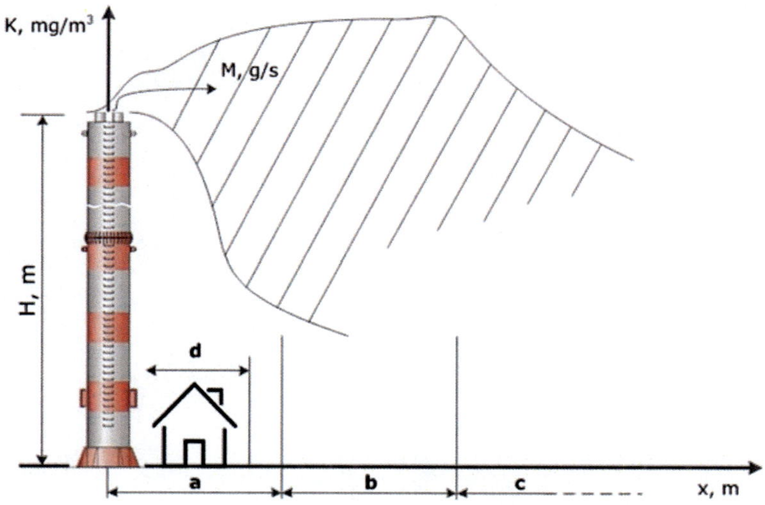

Fig. 8.2 Illustration of the change in the distribution of pollutants in the atmosphere under the torch of a point source: **a** torch tipping zone; **b** smoke zone; **c** zone to reduce pollution; **d** fugitive emission zone

The smoke zone is the most dangerous and unacceptable in the residential area. Its dimensions vary depending on weather conditions and are in the range of 10–50 pipe heights. The maximum concentration of pollutants in the surface layer of the atmosphere is directly proportional to the mass emission of the pollutant M and inversely proportional to the square of the height of the source of emission H^2.

An increase in temperature and velocity of the ejection jet leads to an increase in the height of the ejection plume and a decrease in the surface concentration of pollutants (Fig. 8.3).

At a low wind speed (up to 2 m/s), smoke from the pipe first spreads vertically upwards and only at a certain level begins to move in the horizontal plane. As noted above, this is due to the increased temperature of the exhaust gases compared to atmospheric air and the initial velocity of the jet from the emission source [26].

This causes the effect of raising the ejection source above the pipe by a certain value Δh. Therefore, in the calculations, instead of the actual height of the emission source, the characteristic effective height is used:

$$H = h + \Delta h.$$

The value of Δh is determined by an approximate formula obtained from experimental data:

$$\Delta h = 3.75 \cdot w_0 \cdot R_0 / u + 1.6 \cdot g \cdot V_1 \cdot \Delta T / \left(T_a \cdot u^3\right),$$

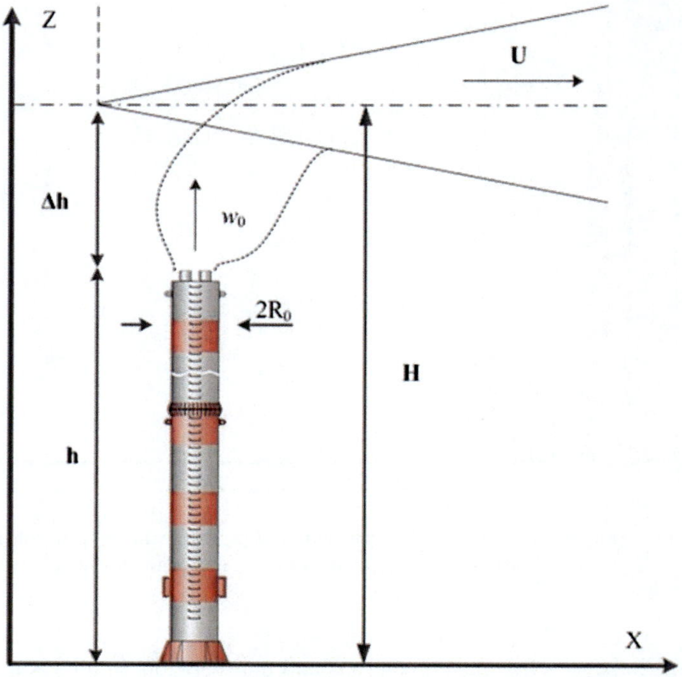

Fig. 8.3 Illustration of pollutant emission sources with effective height H

where w_0 is the gas velocity at the source end (m/s); R_0 is the radius of the ejection jet, (m); u is the wind speed, (m/s); V_1 is the volumetric jet rate of exhaust gases, (m^3/s); ΔT is the temperature difference between the ejection jet and the ambient air, (K); T_a is the ambient air temperature, (K).

The dispersion in the atmosphere of gaseous pollutants and particulate matter (with a diameter of up to 10 μm), which have a relatively low deposition rate, obey the same laws. For particulate matter with a diameter of more than 10 μm, these laws do not apply, because their deposition rate under the influence of gravity increases.

Weather conditions significantly affect the transport and dispersion of pollutants in the atmosphere. The following factors have the greatest influence: temperature, wind direction and strength, level of solar radiation, precipitation, fog.

The main factors providing mixing of air in the atmosphere are the difference in temperature of the jet and ambient air, and mechanical turbulence.

Wind can have various effects on the distribution of pollutants, the level of which depends on the type of emission source and emission characteristics. If the gas temperature significantly exceeds the ambient temperature, this becomes the reason for the formation of a field of vertical velocities, contributing to the rise of the torch and the movement of pollutants up. In the presence of a weak wind, this effect leads to a decrease in the pollutants concentration in the surface layer [27, 28].

The pollutants concentration on the ground also decreases in the case of strong winds, however, this leads to the rapid movement of pollutants in the horizontal plane. The highest concentrations of pollutants in the surface layer are formed for a certain wind speed, which is called "dangerous". The highest concentrations of pollutants in atmospheric air are observed at a certain height H', where the vertical velocity of gas movement w approaches 0.

An increased pollutants concentration in the surface layer is observed for low or cold emissions sources and in the case of a low wind speed (0–1 m/s). Thus, at analyzing air pollution in cities, both the repeatability of low wind speeds and the frequency of "dangerous" wind speeds are taken into account.

The wind direction has a significant effect on air pollution in urban agglomerations. Pollutant concentration levels increase significantly when air masses move from industrial sites. This issue requires additional research and will be considered below.

Inversion stratification (temperature inversion) is called the [29] phenomenon of an increase in air temperature with height. If the inversion begins directly at the surface of the earth, then it is called "surface", and it is "raised" in the case of a certain height. Inversions greatly impede the movement of vertical air masses and the dispersion of pollutants in the atmosphere. Temperature inversions in the lower troposphere result from:

- cooling of the earth's surface due to radiation;
- cooling the surface layer due to the cost of heat for water evaporation or melting snow;
- runoff of cold air to places with low relief;
- action of the sea breeze on the coast of the seas in the warm season.

The local mass of gas during moving in the atmosphere vertically upwards expands, and during moving down—it is compressed in accordance with the distribution of pressure in the atmosphere. If the vertical temperature gradient is 1 °C for every 100 m, then the change in volume proceeds adiabatically:

$$\Delta\Gamma = \left(dT/dz\right)_{ad} = -0.01\,^{\circ}\mathrm{C/m}.$$

Under various weather conditions, the temperature gradient can be either greater or less than the adiabatic value.

The possibility of thermal mixing can be determined by comparing the real temperature gradient in the environment with an adiabatic vertical temperature gradient. Comparison of the $\Delta\Gamma$ value determined in this way with the actual vertical temperature gradient in (°C), which is used to determine the stability of the atmosphere, makes it possible to distinguish the 4 most important categories of atmospheric stability (Fig. 8.4).

In the case shown in Fig. 8.4a, the actual vertical gradient of the ambient temperature $\left(dT/dz\right)_{am}$ is less than $\Delta\Gamma$. In this case, the atmosphere is in an ultra-adiabatic state, that is, its state is unstable.

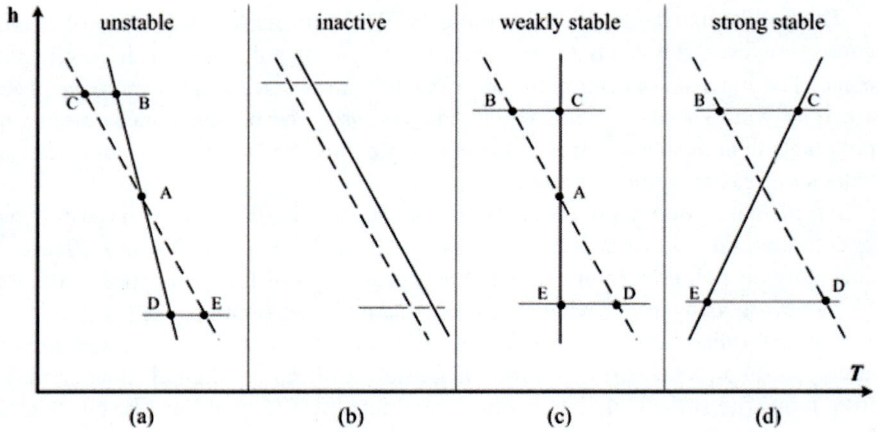

Fig. 8.4 Graphs of a vertical temperature gradient and atmospheric stability: solid line—gradient of the ambient temperature, $(dT/dz)_{am}$; dashed line—dry adiabatic vertical temperature gradient, $\Delta\Gamma$

For explaining this process, it will be considered a small volume of air at point A. If it moves upward with a sufficiently high speed in the atmosphere under the influence of turbulent diffusion, the final position of this volume of air is represented by point B on a direct dry-adiabatic gradient. This position corresponds to a temperature whose value is higher than the actual ambient temperature (point C), and the density is lower compared to the density of the ambient air.

When the air volume moves from point A downward, the elementary air volume adiabatically compresses, as a result of which the air density at point D will exceed the air density value at the ambient air temperature at point E, therefore, the considered air volume will continue to move down. Thus, any movement of the air volume in this case in the vertical plane will be accompanied by a tendency to increase the effect of displacement, so the atmosphere in this case is in an unstable state.

In the case that is shown in Fig. 8.4b ($\Delta\Gamma = (dT/dz)_{am}$), the state of the atmosphere is characterized as inactive, since when a given volume of air moves in a vertical plane, its physical characteristics will not differ from the parameters of the ambient air. Thus, the driving force of vertical displacements will be absent.

If the value of the vertical gradient of the ambient air temperature is greater than the dry adiabatic one (Fig. 8.4c, $\Delta\Gamma < (dT/dz)_{am}$), then the atmosphere is in an adiabatic state. The atmosphere in this case is weakly stable, so any volume of air randomly moved in the vertical direction will return to its original position.

If the air temperature increases with height ($\Delta\Gamma \leq\leq (dT/dz)_{am}$), then the atmospheric conditions are inverse (Fig. 8.4d). The state of the atmosphere is strong. The inverse characteristics of the atmosphere significantly reduce the vertical movement of pollutants and thus contribute to an increase in their concentration in the surface air layer.

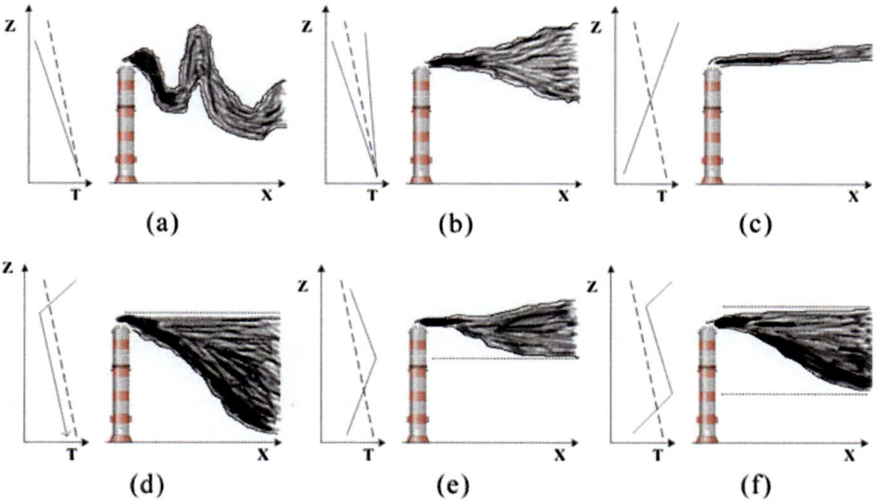

Fig. 8.5 Illustration of the temperature profile and the shape of the source gas jet in the *XZ* plane: dashed line—adiabatic vertical temperature gradient; solid line—ambient temperature gradient

Thus, the degree of atmospheric stability determines the ability of the atmosphere to disperse pollutants. The stability of the atmosphere is determined by the absence of significant vertical movements and mixing.

Variants of the propagation of smoke jet from a single source for various gradients of ambient air temperature are shown in Fig. 8.5 and Table 8.3.

Fog. Fog can affect on the pollutants content in the air as follows. Drops of fog have absorption properties, and are able to absorb contaminants, not only in the surface layer, but also in the upper atmosphere. This leads to the fact that the pollutants concentration increases significantly in the layer of fog and decreases outside it. Dissolution of sulfur dioxide (SO_2) in the droplets of fog leads to the formation of sulfuric acid (H_2SO_4).

Precipitation. Precipitation purifies the air of pollutants. After prolonged intense precipitation, high pollutants concentrations in the air are not observed.

Solar radiation. The influence of solar radiation initiates photochemical reactions in the atmosphere, causing the formation of secondary products. These products often have more toxic properties than substances coming from emission sources. An example is the formation of sulfate aerosols after the oxidation of sulfur dioxide [30, 31].

Table 8.3 Characteristic of the temperature profile and the jet shape according to Fig. 8.6

No.	Characteristic	Description
1(a)	Wave-like jet, strong instability	Puffs of smoke arise, pollution is intensively dispersed in significant volumes. High concentrations of pollutants may form in parts of the surface layer. This position of the atmosphere is typical for clear days with heating the earth's surface, with light winds
2(b)	Cone-shaped jet, stability inactive	Unfavorable conditions for pollutants dispersion. A significant part of pollutants is transported long distances in the direction of the wind before it reaches the earth's surface. Such conditions are typical for cloudy weather, moderate winds. Cloudiness prevents the influx/outflow of solar radiation
3(c)	Fan-shaped jet, surface inversion, strong stability	The source gas jet extends horizontally. The cross section of the jet approaches an elliptical shape, pollution spreads over considerable distances. The pollutants concentration in the surface layer is low. This situation is characteristic of calm, clear nights, when the earth's surface cools
4(d)	Smoky jet, inversion above the pipe throat	Such an effect leads to smoke jet, and emissions spread to the earth's surface. The pollutants concentration in the surface layer can significantly exceed the calculated (in tens times). This situation is typical for morning after night with persistent inversion
5(e)	Raised jet, inversion below the pipe throat	The inversion layer acts as a barrier to prevent the jet from dropping to the ground. Pollutants disperse in the direction of the wind and significant surface concentrations are not formed. This situation is typical in the afternoon - early evening in clear weather
6(f)	Limited jet, inversion below and above the pipe throat	The dispersion of pollutants in this case occurs between two stable layers of the atmosphere. In the case of calm, such an inversion can play the role of a "ceiling", which prevents the rise of emissions. The pollutants concentration in the surface layer increases significantly

8.2 Monitoring and Forecasting of Air Pollution

The development of forecasting methods is based on the results of theoretical and experimental monitoring (studying patterns) of the pollutants' propagation emitted by pollution sources.

The study of the pollutants' propagation in the theory of atmospheric diffusion is carried out on the basis of its mathematical description by the equation of turbulent diffusion. It allows to study the pollutants' propagation from sources of various types with various environmental characteristics [32–36]. In general terms, the problem of forecasting air pollution is mathematically determined as the solution, under certain initial and boundary conditions, of the differential equation [37, 38]:

$$\frac{\partial C}{\partial t} + \sum_{i=1}^{3} U_i \frac{\partial C}{\partial X_i} = \sum_{i=1}^{3} \frac{\partial}{\partial X_i} K_i \frac{\partial C}{\partial X_i} - aC, \tag{8.1}$$

where t—time; X_i—coordinates; U_i—components of the average speed of movement of the pollutant; K_i—the components of the exchange coefficient; $i = 1, 2, 3$—directions of the coordinate axes; a—coefficient determining the change in concentration due to the conversion of pollutant; C—concentration of pollutant.

Using the Cartesian coordinate system, it can be determined the axes X_1, X_2 and X_3 through x, y and z, respectively, the speed $U_1 = u$; $U_2 = v$; $U_3 = w$; exchange coefficients $K_1 = kx$; $K_2 = ky$; $K_3 = kz$.

Equation (8.1) takes the following form

$$\frac{\partial C}{\partial t} + u\frac{\partial C}{\partial x} + v\frac{\partial C}{\partial y} + w\frac{\partial C}{\partial z} = \frac{\partial}{\partial x} k_x \frac{\partial C}{\partial x} + \frac{\partial}{\partial y} k_y \frac{\partial C}{\partial y} + \frac{\partial}{\partial z} k_z \frac{\partial C}{\partial z} - aC. \tag{8.2}$$

During solving practical problems, the form of Eq. (8.2) is simplified. If the x axis is oriented in the direction of the average wind speed, then $v = 0$.

Vertical movements in the atmosphere above a homogeneous horizontal surface are small and $w = 0$ can be taken into account if the pollutant is light and does not have its own speed of movement.

If heavy pollutants are considered, which gradually settles in the atmosphere under the influence of gravitational forces, then w is the deposition rate which entering to Eq. (8.2) with a sign (−).

In the presence of wind, the term kx can be neglected, taking into account diffusion along the x axis, since in this direction the diffuse flow of pollutant is much less than convective.

The change in atmospheric concentrations over time usually acquires a quasistationary character, and therefore, it can be taken $\partial C / \partial t = 0$.

Thus, Eq. (8.2) can be reduced to

$$u\frac{\partial C}{\partial x} - w\frac{\partial C}{\partial z} = \frac{\partial}{\partial y} k_y \frac{\partial C}{\partial y} + \frac{\partial}{\partial z} k_z \frac{\partial C}{\partial z} - aC. \tag{8.3}$$

In case of light pollutant,

$$u\frac{\partial C}{\partial x} = \frac{\partial}{\partial y} k_y \frac{\partial C}{\partial y} + \frac{\partial}{\partial z} k_z \frac{\partial C}{\partial z} - aC. \tag{8.4}$$

In cases where the concentration of the pollutant is maintained

$$u\frac{\partial C}{\partial x} = \frac{\partial}{\partial y}k_y\frac{\partial C}{\partial y} + \frac{\partial}{\partial z}k_z\frac{\partial C}{\partial z}. \tag{8.5}$$

During forecasting air pollution, the main interest is to determine the expected concentrations in the surface layer $h = 1.2$–2 m.

As researches have shown, in the surface air layer to the level $z = h$, the exchange coefficient increases in proportion to the height; speed is a logarithmic function of height.

If $z = 0$ (at the surface of the earth), it can be approximately taken the limiting value $k_z = k_0$—molecular diffusion coefficient for air.

An analytical solution of the diffusion equation can be written for the case when u and k_z are given by power functions of $z(u = u_1 \cdot z^n;\ k_z = k_1 \cdot z)$ for a light pollutant, which preserves the concentration ($w = a = 0$).

The ground concentration of a pollutant for $z = 0$ is defined as:

$$C(x, y) = \frac{M}{2(1-n)k_1\sqrt{\pi k_0 x^3}} e^{\frac{u_1 H^{1+n}}{(1+n)^2 k_1 x} - \frac{y^2}{4k_0 x}}, \tag{8.6}$$

where M—pollutant emission from the source per unit time; H—height of the ejection source.

A characteristic feature of the distribution of ground concentration C along the X axis is the presence of a maximum C_m at a distance X_m from the source.

Visualization of the distribution of the concentration of pollutants from a single point source is shown in Fig. 8.6.

Calculations show that with the same emission parameters, the maximum surface concentration of pollutants from a higher source is less and is observed at a greater distance from the source [39–41].

Consider examples of calculating the concentrations of pollutants in the air from energy facilities in Ukraine [42].

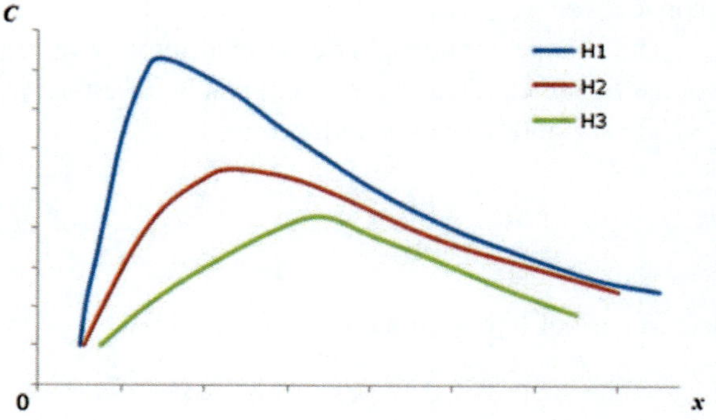

Fig. 8.6 Graph of the theoretical distribution of the pollutants' concentration C in air at a distance x from a point source of different heights H: H1 < H2 < H3

The main document regulating the assessment of dispersion and determination of surface concentrations of pollutants from emission sources in Ukraine is "Methodology for calculating the concentrations of atmospheric pollutants contained in emissions of enterprises OND-86". This methodology for calculating local atmospheric pollution by emissions from industrial enterprises is used not only in Ukraine, but in most of the Commonwealth of Independent States. This technique contains a number of analytical expressions obtained by approximating the turbulent diffusion equation. OND-86 allows to calculate the maximum possible distribution of emission concentrations under conditions of an unstable atmosphere and averaging values over a 20–30 min interval, but does not take into account factors such as atmospheric stability class and type of earth's surface. This technique has been applied to calculate concentrations of pollutants at a distance of up to 100 km from the emission source.

The method allows to calculate the dispersion of pollutants emitted into the atmosphere by single point, linear sources and a group of sources, taking into account the influence of the terrain, to determine the maximum concentration of pollutants in a two-meter layer above the earth's surface, as well as the vertical distribution of concentrations [43–45].

One of the main disadvantages of this methodology is the method of solving the inverse problem to determine the characteristics of the emission source [38]. It directly provides for the presence of such source (i.e., the known location of the energy facility) that produces emissions, and reduces to calculating the emission power M and the pipe height H from the value of the maximum surface concentration C_m.

For substances with the sum of the harmful effects, the dimensionless total concentration q (in fractions of the MPC) and the total concentration C, reduced to substance C_1, are calculated using the emission's power values M_q and M for each source, respectively:

$$\begin{aligned} q &= C_1/MPC_1 + C_2/MPC_2 + \cdots + C_n/MPC_n; \\ M_q &= M_1/MPC_1 + M_2/MPC_2 + \cdots + M_n/MPC_n; \\ C &= C_1 + C_2\frac{MPC_1}{MPC_2} + \cdots + C_n\frac{MPC_1}{MPC_n}; \\ M &= M_1 + M_2\frac{MPC_1}{MPC_2} + \cdots + M_n\frac{MPC_1}{MPC_n}, \end{aligned} \tag{8.7}$$

where C_1, MPC_1—concentration and MPC of the substance to which reduction occurs; C_2, ..., C_n and MPC_2, ..., MPC_n—concentrations and MPC of other substances included in the group are considered.

One of the main attributes of an energy facility is the presence of emission pipes in its territory that directly carry out emissions. It is advisable to consider the further calculation of the pollutants concentrations from energy facilities as from point sources with a round or rectangular hole [46].

Let consider the calculation of atmospheric pollution by emissions of a single source with a round hole. The maximum value of the surface concentration of the pollutant C_m when the gas-air mixture is emitted from one point source with a round hole is achieved under adverse weather conditions at a distance of X_m and is determined by the formula:

$$C_m = \frac{A \cdot M \cdot F \cdot m \cdot n \cdot \eta}{H^2 \cdot \sqrt[3]{V_1} \cdot \Delta T}, \quad (8.8)$$

$$C_m = \frac{A \cdot M \cdot F \cdot n \cdot \eta}{H^{4/3}} \times \frac{D}{8V_1}, \quad (8.9)$$

where A—coefficient depending on the temperature stratification of the atmosphere (within 140 ... 250); M—mass of a pollutant emitted into the atmosphere per unit time; F—coefficient taking into account the rate of sediment of pollutants; m, n—coefficients that take into account the conditions for the exit of the gas-air mixture from the hole of the exhaust source; H—height of the emission source above ground level (for terrestrial sources, $H = 2$ m is assumed); η—coefficient taking into account the influence of the terrain (in the case of equal or slightly rough terrain with a height difference of not more than 50 m/km, $\eta = 1$ is assumed); V_1—volumetric flow rates of the gas-air mixture; ΔT—difference between the temperature of the gas-air mixture that is emitted and the ambient temperature.

Formula (8.9) is intended for cold emissions, i.e., at $\Delta T \approx 0$.

The maximum concentration of a pollutant in the surface layer of the atmosphere is directly proportional to the mass flow rate of the pollutant M and inversely proportional to the square of the source height—H^2.

Coefficient A under adverse weather conditions, providing maximum concentrations of pollutants in the surface layer of the atmosphere, has the following values in Ukraine:

- for sources located north of 52° latitude $A = 160$;
- for sources within 50°... 52° latitude $A = 180$;
- for sources located south of 50° latitude $A = 200$.

At determining ΔT, the temperature of the ambient air is taken equal to the average outdoor temperature of the hottest month at 13:00 local time.

The value of the F coefficient:

- for gaseous pollutants and fine aerosols, the sediment rate of which is close to zero, $F = 1$;
- for other fine-grained aerosols with a degree of gas purification in dust traps $\eta \geq 0.9$—$F = 2$; at $0.75 \leq \eta < 0.9$—$F = 2.5$; at $\eta < 0.75$—$F = 3$;
- in the presence of data on the dispersed composition of dust depending on the ratio V_g/U_m: $F = 1$ at $V_g/U_m \leq 0.015$; $F = 1.5$ at $0.015 < V_g/U_m \leq 0.03$ (V_g is the particle velocity of such diameter d_g that the mass of all particles of a diameter greater than d_g is 5% of the total mass of the particles; U_m is the dangerous wind speed).

To obtain the values of the coefficients m and n, the following intermediate coefficients are determined:

$$f = \frac{1000 \cdot w_0^2 \cdot D}{H^2 \cdot \Delta T}; \tag{8.10}$$

$$V_m = 0.65 \times \sqrt[3]{V_1 \cdot \Delta T / H}; \tag{8.11}$$

$$V_m^* = 1.3 \cdot w_0 \cdot D / H; \tag{8.12}$$

$$f_e = 800 (V_m^*)^3, \tag{8.13}$$

where D—diameter of the hole of the source pipe; w_0—speed of the gas-air mixture from the exhaust pipe.

The coefficient m is determined by the formula

$$\begin{array}{ll} m = 1/\left(0.67 + 0.1\sqrt{f} + 0.34\sqrt[3]{f}\right) & at\ f < 100; \\ m = 1.47/\sqrt[3]{f} & at\ f \geq 100. \end{array} \tag{8.14}$$

If the condition $f_e < f < 100$ is fulfilled, during calculating $m, f = f_e$ is taken.

If $f \leq 100$, the n value is determined depending on V_m:

$$\begin{array}{ll} n = 1 & at\ V_m \geq 2, \\ n = 0.532 V_m^2 - 2.13 V_m + 3.13 & at\ 0.5 \leq V_m < 2, \\ n = 4.4 V_m & at\ V_m < 0.5. \end{array} \tag{8.15}$$

For determining the coefficients m and n, it can be used the graph shown in Fig. 8.7.

If $f \geq 100$ or $\Delta T \approx 0$ and $V_m^1 \geq 0.5$ (cold emissions), during calculating C_m, formula (8.3) is used instead of formula (8.2), where $K = D/8V_1 = 1/7.1\sqrt{W_0 V_1}$ and n are determined by formula (8.9) with the substitution of V_m^1 instead of V_m.

Similarly, for $f < 100$ and $V_m < 0.5$ or $f < 100$ and V_m^1 (cases of extremely small dangerous wind speeds), C_m is calculated instead of formula (8.8) by the formula

$$C_m = A \cdot M \cdot F \cdot m^1 \cdot \eta / H^{7/3}, \tag{8.16}$$

where $m^1 = 2.86$ m at $f < 100$, $V_m < 0.5$; $m^1 = 0.9$ at $f \geq 100$, $V_m^1 < 0.5$.

The distance x_m at which the surface concentration C under adverse weather conditions reaches a maximum value of C_m is determined by the formula:

$$x_m = (5 - H) d H / 4, \tag{8.17}$$

where d is the dimensionless coefficient calculated by the formula (8.18), (8.19).

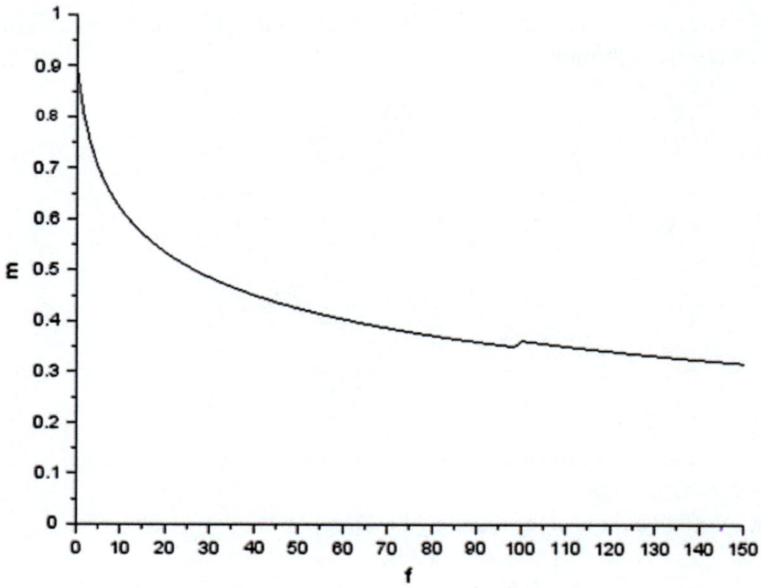

Fig. 8.7 Dependence of the coefficients m on f

At $f < 100$

$$\begin{aligned} d &= 2.48\left(1+0.28\sqrt[3]{f_e}\right) \quad at\ V_m \le 0.5;\ at\ V_m \le 0.5;\\ d &= 4.95V_m\left(1+0.28\sqrt[3]{f_e}\right)\ at\ 0.5 < V_m \le 2;\ at\ 0.5 < V_m \le 2;\\ d &= 7\sqrt{V_m}\left(1+0.28\sqrt[3]{f_e}\right) \quad at\ V_m \ge 2. \end{aligned} \tag{8.18}$$

At $f > 100$ or $\Delta T \approx 0$, the value of d is found by the formulas:

$$\begin{aligned} d &= 5.7d = 5.7\ at\ V_m^1 \le 0.5;\ at\ V_m^1 \le 0.5;\\ d &= 11.4V_m^1 \qquad at\ 0.5 < V_m^1 \le 2;\\ d &= 16\sqrt{V_m^1} \qquad at\ V_m^1 \ge 2. \end{aligned} \tag{8.19}$$

The value of the dangerous wind speed U_m, at which the highest value of the surface concentration of pollutants C_m is reached, is calculated according to formulas (8.20) and (8.21).

At $f < 100$

$$\begin{aligned} U_m &= 0.5V_m && at\ V_m \le 0.5;\\ U_m &= V_m U_m = V_m && at\ 0.5 < V_m \le 2;\\ U_m &= V_m\left(1+0.12\sqrt{f}\right)U_m = V_m\left(1+0.12\sqrt{f}\right) && at\ V_m \ge 2. \end{aligned} \tag{8.20}$$

At $f \ge 100$ or $\Delta T \approx 0$

$$\begin{aligned} &U_m = 0.5 && at\ V_m^1 \le 0.5; \\ &U_m = V_m^1 && at\ 0.5 < V_m^1 \le 2; \\ &U_m = 2.2V_m^1 U_m = 2.2V_m^1 && at\ V_m^1 \ge 2. \end{aligned} \tag{8.21}$$

At a dangerous wind speed U_m, the surface concentration of pollutants C in the atmosphere along the axis of the emission plume at a distance X from the source is determined by the formula:

$$C_x = S_1 \cdot C_m, \tag{8.22}$$

where S_1 is a dimensionless coefficient determined by the formula:

$$\begin{aligned} &S_1 = 3a^4 - 8a^3 + 6a^2 && at\ a \le 1; \\ &S_1 = 1.13/\left(0.13a^2 + 1\right) && at\ 1 < a \le 8; \\ &S_1 = a/\left(3.58a^2 - 35.2a + 120\right) && at\ F \le 1.5,\ a > 8; \\ &S_1 = 1/\left(0.1a^2 + 2.47a - 17.8\right) && at\ F > 1.5,\ a > 8, \end{aligned} \tag{8.23}$$

where $a = X/X_m$. For low and terrestrial sources at 2 m $\le H \le$ 10 m and X/X_m, the value of S_1 will be replaced by $S_{1H} = 0.125 \cdot (10\text{-}H) + 0.125 \cdot (H\text{-}2) \cdot S_1$.

The value of the surface concentration of pollutants in the atmosphere of C_y at a distance y along the perpendicular to the axis of the emission plume is determined by the formula:

$$C_y = S_2 \cdot C_m, \tag{8.24}$$

where S_2 is the dimensionless coefficient (depends on the wind speed U and the ratio y/x), determined by the value of the argument t_y by the formula:

$$\begin{aligned} &S_2 = 1/\left(1 + 5t_y + 12.8t_y^2 + 17t_y^3 + 45.1t_y^4\right)^2; \\ &t_y = U \cdot y^2/x^2 t_y = U \cdot y^2/x^2\ at\ U \le 5; \\ &t_y = 5y^2/x^2 \qquad at\ U > 5. \end{aligned} \tag{8.25}$$

Let turn to the calculation of atmospheric pollution by emissions' source with a rectangular hole. Calculations of air pollution during emissions of a gas-air mixture from a source with a rectangular hole are carried out according to the above formulas at an average speed of ω_0 and values of $D = D_e$ and $V_1 = V_{1e}$.

The average air-gas mixture velocity rate ω_0 is determined by the formula

$$\omega_0 = V_1/(L \cdot b), \tag{8.26}$$

where L—length of the hole; b—width of the hole.

The effective hole diameter D_e is determined by the formula:

$$D_e = 2 \cdot L \cdot b \big/ (L + b). \tag{8.27}$$

The effective volume of the air-gas mixture V_{1e} outing to the atmosphere is determined by the formula:

$$V_{1e} = \pi \cdot D_e^2 \cdot \omega_0 \big/ 4. \tag{8.28}$$

The radius of the zone of influence for each source is calculated as the largest of the two distances X_1 and X_2, where $X_1 = 10.\ X_{\mathrm{m}}$, and the value of X_2 is defined as the distance from the source, starting from which C $\leq$ 0.05 MPC.

The value of X_2 can be found graphically. On the *OY* axis, it can be found the point 0.05 MPC/C_m, a line is drawn through it parallel to the *OX* axis until the intersection with the graph of the function $S_1 = f(X/X_m)$ to the maximum. The perpendicular to the axis *OX* is lowered from the intersection point. The resulting value of X/X_m is multiplied by X_m and get the desired value of X_2.

Solving inverse problems involves determining the emission power *M* [g/s] and pipe height *H* corresponding to a given level of maximum surface concentration C_m (μg/m^3) with other fixed emission parameters.

The emission power *M*, which corresponds to the maximum concentration of C_m, is determined by the formula:

$$M = C_M H^2 \sqrt[3]{V_1 \Delta T} \Big/ AFmn\eta.$$

At $f \geq 100$ or $\Delta T \approx 0$

$$M = 8 C_m H^{4/3} V_1 \Big/ AFn\eta D.$$

The value of the minimum height for the case of hot emissions ($\Delta T > 0$) is determined by the equation

$$H = \left(AMFD\eta \big/ 8V_1 C_M\right)^{0.75}.$$

If the $H \leq w_0 \sqrt{10D \big/ \Delta T}$ value is found, then the decision is final. Otherwise, the previous minimum height value is determined by the formula

$$H = \sqrt{AMF\eta \Big/ C_m \sqrt[3]{V_1 \Delta T}}.$$

According to the found value of *H*, the coefficients f, V_m, V_m^i, f_e are determined by formulas (8.10–8.13) and approximate values of the coefficients $m = m_1$ and $n = n_1$ are obtained. If the multiply is $m_1 \cdot n_1 \neq 1$, then the quantities m_1 and n_1 determine *H* in the second approximation (H_2):

$$H_2 = H_1\sqrt{m_1 n_1}.$$

In the general case, the value of H at the $(i + 1)$ iteration step is determined by the formula

$$H_{i+1} = H_i\sqrt{m_1 n_1 / m_{i-1} n_{i-1}},$$

where m_1 and n_1 correspond to H_i; m_{i-1} and n_{i-1} – H_{i-1}.

The presented iterative algorithm for updating the values of H is performed until the found values of H_i and H_{i-1} differ from each other by no more than 1 m.

8.3 Methods of Analysis and Data Processing of Air Pollution

Convenient mathematical methods for the analysis, research, and construction of pollution data processing models are the methods [47–50] of theory of spectral analysis, Fourier transform, Z-transform and Laplace transform, splines, wavelet transforms and others [42, 51–53].

At presenting a continuous function of the values of the concentrations of pollutants $x(t)$, as the implementation of a random process, in the form of a time series $x(k)$, the sampling process of a continuous function can be represented as:

$$x(kT) = x(t)\sum_{k=0}^{\infty}\delta(t - kT) = \sum_{k=0}^{\infty}x(kT)\delta(t - kT). \quad (8.29)$$

Applying the Fourier transform

$$f(\omega) = \int_0^{\infty} x(t)e^{-i\omega t}dt, \quad (8.30)$$

and after substituting the value of $x(kT)$ into formula (8.30), it can be obtained the spectral function of the time series

$$F_T(\omega) = \sum_{k=0}^{\infty}x(kT)\int_0^{\infty} e^{-i\omega t}\delta(t - kT)dt = \sum_{k=0}^{\infty} f(kT)e^{-i\omega kT}. \quad (8.31)$$

Substituting the complex variable $s = \sigma + i\omega$ instead of the complex frequency $i\omega$, it can be obtained a series according to the Laplace transform

$$F_T(s) = \sum_{k=0}^{\infty} s(kT)e^{-skT}. \tag{8.32}$$

Series (8.32) can be simplified by applying Z-transformations. To do this, it needs to introduce a new variable Z, associated with s by expression $Z = e^{sT}$. After substituting the value of Z in the formula (8.32), it can be obtained the time series in the Z-plane:

$$F_T(z) = \sum_{k=0}^{\infty} s(kT)z^{-k}. \tag{8.33}$$

Z-transformation is used to synthesize transfer functions of filtering and transfer functions of time series processing algorithms:

$$k(z) = \sum_{k=0}^{\infty} g(kT)z^{-k}, \tag{8.34}$$

where $g(kT)$ is the impulse response of the four-terminal network, which implements the signal processing algorithm.

Obtaining the initial image of the time series can be obtained by applying the inverse Z-transformation:

$$g(kT) = \frac{1}{2\pi j} \oint_{|z|=1} k(z)z^{(k-1)}dz. \tag{8.35}$$

Applying the Fourier transform on a finite sample of N in (8.32), it can be obtained the following equation

$$S\left(e^{j\omega}\right) = \sum_{k=0}^{N-1} x(k)e^{-j\omega kT}. \tag{8.36}$$

In this case, the calculation of the values of the discrete Fourier transform at L-points, evenly placed on the unit circle of the Z-plane when ωl changes in the range of values from 0 to $2\pi F_d$, is carried out according to the expression:

$$S\left(e^{j\frac{2\pi F_d}{L}l}\right) = \sum_{l=0}^{L-1}\sum_{k=0}^{N-1} x(k)e^{-j2\pi lkT F_d/L}, \; at \; L < N. \tag{8.37}$$

In the case $L = N$, Eq. (8.37) takes the form

$$S\left(e^{j\frac{2\pi F_d}{N}l}\right) = \sum_{l=0}^{N-1}\sum_{k=0}^{N-1} x(k)e^{-j2\pi lkTF_d/L}, \, at\ L < N. \tag{8.38}$$

Substituting the value $W_N = e^{-j\frac{2\pi}{N}}$ in the above equation of the time series according to the discrete Fourier transform, it can be obtained the equation

$$S(l) = \sum_{k=0}^{N-1} x(k)w_N^{lk}, \ if\ L = N, \ then\ 0 \le l \le N-1, \tag{8.39}$$

where $S(l) = S(j \cdot \Delta\omega \cdot l) = S(j \cdot \omega_l)$.

The inverse Fourier transform is performed by the formula

$$x(k) = \sum_{l=0}^{N-1} S(l)W_N^{-lk}/N. \tag{8.40}$$

As an example, it can be considered the wavelet transform, which is one of the methods of signal analysis and processing, which includes a discrete transform of the continuous function of changes in the pollutants concentrations. According to the wavelet transform methods, any function can be expanded at a given level $i = n$ in a series of the form

$$s(k) = \sum_k C_{n,k}\varphi_{n,k} + \sum_k\sum_{i=1}^{n} d_{i,k}\psi_{i,k}, \tag{8.41}$$

where $\varphi_{n,k}$ weighted (scaling) and $\psi_{i,k}$ wavelet functions respectively have the form

$$\varphi(t) = \sqrt{2}\sum_l h_l\varphi(2t-k),$$
$$\psi(t) = \sqrt{2}\sum_l g_l\varphi(2t-k), \tag{8.42}$$

where $l = 0,1,\ldots, l_0 = 2\ m\text{-}1$, m – wavelet's order.

The coefficients of the scaling function (scaling vector or scaling filter) h_l may represent the transient characteristics of the processing filters.

The values of h_l and g_l for orthonormal bases are determined by the expressions

$$h_l = \sqrt{2}\int \varphi(x)\varphi(2t-k)dt, \quad g_l = (-1)^l h_{2n-l-1}. \tag{8.43}$$

Such a representation with the help of wavelets of each signal component can be formed both in the time and frequency domains. Figure 8.8 schematically shows the "tree" of coefficients; it is formed during a fast wavelet transform.

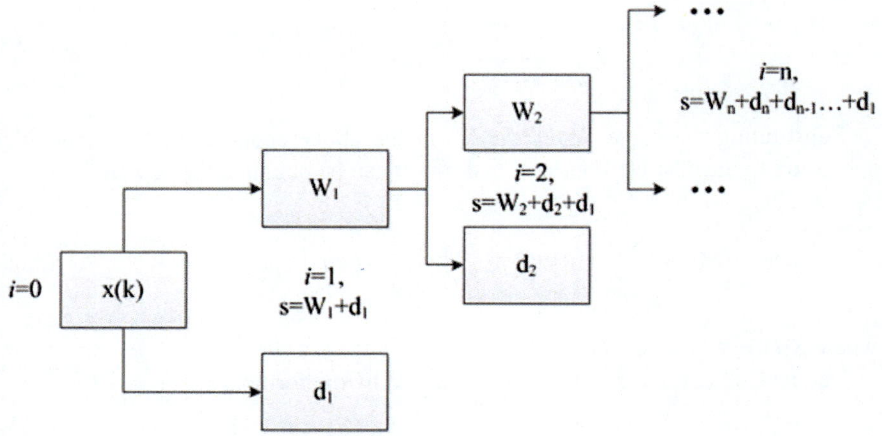

Fig. 8.8 "Tree" of wavelet transform coefficients

The scaling and displacement of the scaling function $\varphi_{i,k}$ and the initial wavelet $\psi_{i,k}$ has the form

$$\varphi_{i,k} = 2^{i/2}\varphi\left(2^i t - k\right), \quad \psi_{i,k} = 2^{i/2}\psi\left(2^i t - k\right). \tag{8.44}$$

In this case, approximating and detailing coefficients are calculated by the formula

$$W_{i,k} = \sum_{i=1}^{n} W_{i-1}\varphi_{i,k}\left(2^i t - k\right)/p, \quad d_{i,k} = \sum_{i=1}^{n} W_{i-1}\psi_{i,k}\left(2^i t - k\right)/p, \tag{8.45}$$

where $p = 2^{i/2}$—orthonormal coefficient, which provides the unit norm of the scaling function. For $i = 1$, it can be written the following equation

$$d_{1,k}(t) = W_{0,k}\psi_{1,k}(2t - k)/\,p, \quad W_{0,k} = x(k)\varphi_{1,k}(t - k) \approx x(k), \tag{8.46}$$

where $x(k) = u(k) + n(k)$, $n(k)$—the noise component of the fluctuation of the samples of the function of changes in the concentrations of emissions, $W_{0,k}$—local average value of the signal $x(k)$ weighted with a scaling function $\varphi_{1,k}$.

Accordingly, when the signal $x(k)$ is decomposed to the n-th level and then restored in the form of (8.41), approximating coefficients of the i-th level are obtained by local averaging of the approximating coefficients of the level $(i - 1)$ with the i-th scaling function on each level of decomposition in the form

$$W_{n,k} = \sum_k W_{n-1}\varphi_{n,k}\left(2^n t - k\right)/\,p, \; W_{n-1,k} = \sum_k W_{n-2}\varphi_{n-1}\left(2^{n-1} t - k\right)/\,p, \ldots,$$
$$W_{i,k} = \sum_k W_{i-1}\varphi_{i,k}\left(2^i t - k\right)/\,p, \ldots, \; W_{0,k} = x(k)\varphi_{1,k}(2t - k)/\,p. \tag{8.47}$$

The above transformations are examples of using a powerful apparatus for discrete processing of time series, the use of which makes it possible to analyze and obtain the results of evaluating various characteristics of monitoring air pollution.

8.4 Localization Models of Air Pollution Source

It was noted in [54–57] that certain characteristics of the wind are important both for assessing wind resources, as well as for the design and operation of wind power plants. These characteristics include the distribution function of wind speed and the probability of direction. The probability density functions (PDFs) of wind speed are fairly common [54].

At assessing the characteristics of wind in a certain area with a difficult topography or with several prevailing wind directions, it can be used the general two-dimensional PDF, that is, use a continuous model of wind rose, which will analyze the change in energy characteristics in terms of speed and direction. Information on these characteristics will make it possible to more accurately localize the source of atmospheric air pollution.

Below it will be considering the isotropic and anisotropic Gaussian models used to construct the density function of the wind direction distribution.

Isotropic Gaussian model of wind characteristics. This model is based on the following hypotheses: (a) the presence of a preferred wind direction; (b) the components of the wind speed for the prevailing wind direction (longitudinal component, v_y) and the direction perpendicular to it (transverse component, v_x) are random variables, and described by the normal distribution; (c) the longitudinal and transverse components of the wind speed are independent values; (d) the variances of the longitudinal σ_y^2 and transverse σ_x^2 components are equal; (e) the average value of the longitudinal component μ_y is nonzero, and the average value of the transverse component is $\mu_x = 0$.

According to this model, the longitudinal and transverse components are described by PDF of the form:

$$f_y(v_y) = \frac{1}{\sqrt{2\pi}\sigma_y} e^{\frac{(v_y-\mu_y)^2}{-2\sigma_y^2}}, \tag{8.48}$$

$$f_x(v_x) = \frac{1}{\sqrt{2\pi}\sigma_y} e^{\frac{(v_x)^2}{-2\sigma_y}}. \tag{8.49}$$

In accordance with the above hypothesis (c), their common two-dimensional PDF is defined as the multiply of functions (8.48) and (8.49).

Passing to the polar coordinate system defined by Eqs. (8.50), the general two-dimensional PDF (8.51) becomes dependent on the speed and angle:

$$v_x = v \sin\theta'; \quad v_y = v \cos\theta'; \tag{8.50}$$

$$f\left(v,\theta'\right)=\frac{v}{2\pi\sigma_y^2}e^{\frac{\mu_y^2}{-2\sigma_y^2}}e^{\frac{v^2-2\mu_y v\cos\theta'}{-2\sigma_y^2}}. \quad (8.51)$$

After integrating Eq. (8.51) over v, it can be obtained the PDF of the angle θ':

$$f\left(\theta'\right)=\tfrac{1}{2\pi}e^{(-\lambda^2/2)}\left[1-\sqrt{\pi}\xi\phi(\xi)e^{\xi^2}\right];\quad 0<\theta'<2\pi, \quad (8.52)$$

where

$$\lambda=\mu_y/\sigma_y;\ \xi=-\lambda\cos\theta'\Big/\sqrt{2};\ \phi(\xi)=2\int_{\xi}^{\infty}e^{-t^2}\mathrm{d}t/\sqrt{\pi}. \quad (8.53)$$

After integration Eq. (8.51) over θ', it can be obtained the PDF of the velocity v:

$$f(v)=I_0\left(v^2\mu_y/\sigma_y^4\right)e^{\left(-v^2-\mu_y^2\right)/2\sigma_y^2};\quad 0<v<\infty, \quad (8.54)$$

where I_0 is the modified Bessel function of the first kind and zero order.

Anisotropic Gaussian model of wind characteristics. This model uses the same hypotheses as the isotropic model, except for hypothesis (d). The anisotropic model assumes that the variances of σ_y^2 and σ_x^2 need not be equal.

According to this model, the longitudinal and transverse components are described by the PDFs defined by Eqs. (8.48) and (8.55):

$$f_x(v_x)=\frac{1}{\sqrt{2\pi}\sigma_y}e^{\frac{(v_x)^2}{-2\sigma_y}}. \quad (8.55)$$

In accordance with the above hypothesis (c), their general PDF is defined as the multiply of functions (8.48) and (8.55).

Passing to the polar coordinate system defined by Eqs. (8.50), the joint PDF (8.51) becomes dependent on the speed and angle, forms the wind speed with the axis of the prevailing wind direction

$$f\left(v,\theta'\right)=ve^{\left[-\frac{(v\cos\theta'-\mu_y)}{2\sigma_y^2}-\frac{v^2\sin\theta'}{2\sigma_x^2}\right]}\Big/2\pi\sigma_y\sigma_x. \quad (8.56)$$

After integrating Eq. (8.56) over v, it can be obtained the PDF of the angle θ':

$$f\left(\theta'\right)=\tfrac{1}{2\pi}\exp\left(-\tfrac{\lambda^2}{2}\right)\left[\tfrac{\gamma}{\gamma^2\sin^2\theta'+\cos^2\theta'}\right]\times\left[1-\sqrt{\pi}\eta\exp\left(\eta^2\right)\phi(\eta)\right];\quad 0<\theta'<2\pi, \quad (8.57)$$

where

$$\gamma = \frac{\sigma_y}{\sigma_x};\ \eta = -\frac{\lambda \cos\theta'}{\sqrt{2\cos^2\theta' + 2\gamma \sin^2\theta'}}. \quad (8.58)$$

After integration Eq. (8.56) over θ', it can be obtained the velocity PDF of the v:

$$f(v) = I_0 \frac{v^2 \mu_y}{\sigma_y^3 \sigma_x} \exp\left(-\frac{\lambda^2}{2}\right)$$
$$\times \int_0^{2\pi} \exp\left[-\frac{v^2}{2\sigma_y^2}\left(\cos^2\theta' + \gamma^2 \sin^2\theta'\right)\right] d\theta';\ 0 < v < \infty. \quad (8.59)$$

According to the isotropic and anisotropic Gaussian models, the PDFs of the velocity and wind direction make it possible to determine the preferred direction of propagation of pollutants. Knowing this direction, it is possible to localize the zone where the source of atmospheric pollution is located and thereby obtain the result of solving the inverse problem [38]. The presented results supplement the indicated method using a statistical approach for processing monitoring data.

In practice, a wind rose is used to determine the direction of wind propagation, which is a statistical estimate (histogram) averaged over the corresponding time interval (hour, day, week) of wind directions in the polar coordinate system. In accordance with the needs of such a wind rose can be 4-, 8-, 16-, 32-directional (the number of sectors of the circle division). Figure 8.9 schematically shows an image of the corresponding lobe diagrams.

As can be seen from Fig. 8.9, the more directions a wind direction measuring device distinguishes, the smaller the difference in angles Δ between these directions. The value of Δ allows to localize the source of air pollution in the wind direction determined using a wind vane. Tables 8.4, 8.5 and 8.6 show the corresponding directions, the range of degrees, and the difference in angles between the nearest rays of a 4-, 8-, and 16-directional wind rose.

Knowing the direction of wind propagation from the corresponding wind rose, it can be determined the area of the sector where the source of pollution is located, according to the formula:

$$S = \pi R^2 \Delta / 360.$$

The corresponding graph of the dependence of the sector area with using different types of wind rose is shown in Fig. 8.10. As can be seen from Fig. 8.10, the use of 32-directional wind rose compared to 4-directional increases the accuracy of determining the source of pollution by 8 times.

Figure 8.11 shows the concentration values of sulfur dioxide in the air obtained from a measuring station located in Kyiv (near the metro station Chernihivska). The concentration of sulfur dioxide was measured twice a day (07:00 and 19:00) from

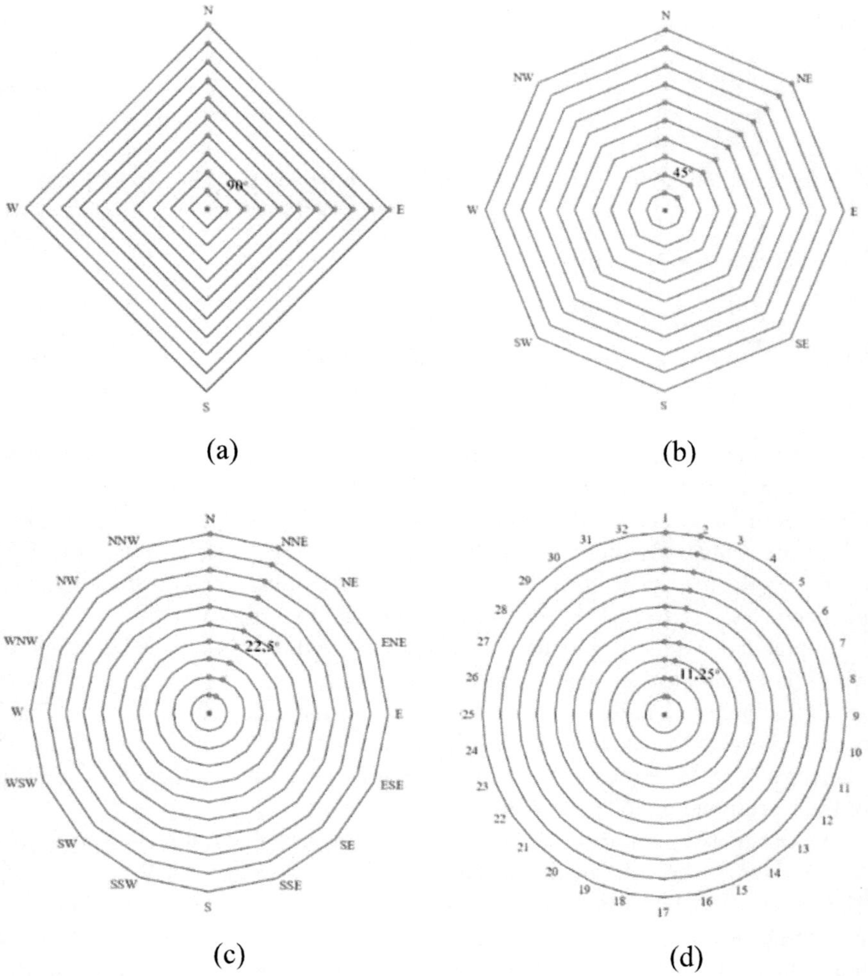

Fig. 8.9 4-, 8-, 16-, 32-directional wind roses

Table 8.4 Characteristic of 4 directional wind rose

No.	Direction	Range of values, °	Δ, °
1	North (N)	315–45	90
2	East (E)	45–135	
3	South (S)	135–225	
4	West (W)	225–315	

Table 8.5 Characteristic of 8 directional wind rose

No.	Direction	Range of values, °	Δ, °
1	North (N)	337.5–22.5	45
2	Northeast (NE)	22.5–67.5	
3	East (E)	67.5–112.5	
4	Southeast (SE)	112.5–157.5	
5	South (S)	157.5–202.5	
6	Southwest (SW)	202.5–247.5	
7	West (W)	247.5–292.5	
8	Morthwest (NW)s	292.5–337.5	

Table 8.6 Characteristic of 16 directional wind rose

No.	Direction	Range of values, °	Δ, °
1	N	348.75–11.25	22.5
2	NNE	11.25–33.75	
3	NE	33.75–56.25	
4	ENE	56.25–78.75	
5	E	78.75–101.25	
6	ESE	101.25–123.75	
7	SE	123.75–146.25	
8	SSE	146.25–168.75	
9	S	168.75–191.25	
10	SSW	191.25–213.75	
11	SW	213.75–236.75	
12	WSW	236.75–258.75	
13	WSW	258.75–281.25	
14	WNW	281.25–303.25	
15	NW	303.25–326.25	
16	NNW	326.25–348.75	

September 2018 to April 2019. However, some data are missing, due to technical problems with the site of the Geophysical Observatory (cgo-sreznevskyi.kyiv.ua).

Figure 8.12 shows a graph of the distribution of sulfur dioxide concentrations according to a 16-directional wind rose. The concentration value of sulfur dioxide in accordance with each direction of the wind rose is the total value of the concentration observed during the calendar month.

Wind direction data taken from wunderground.com database according to UKKK location.

Figure 8.13 shows the concentration values of nitrogen dioxide in the air obtained from a measuring station located in Kyiv (near the metro station Chernihivska). The concentration of nitrogen dioxide was measured twice a day (07:00 and 19:00) from

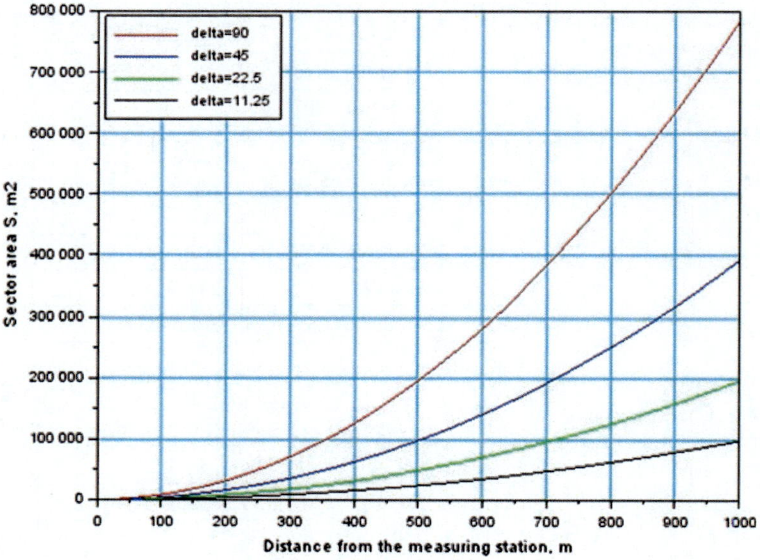

Fig. 8.10 Graph of change in the area of the localization sector of the pollution source for various types of wind direction fixation (4, 8, 16, 32-directional wind roses)

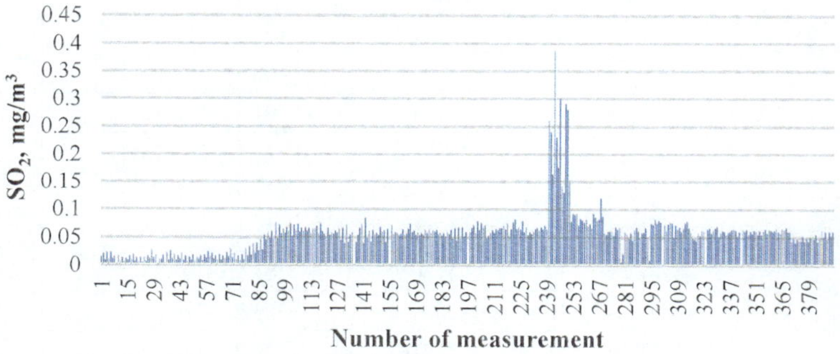

Fig. 8.11 Graph of the values of direct measurements of the sulfur dioxide concentration in Kyiv (metro station Chernihivska) from September 2018 to April 2019

September 2018 to April 2019. However, some data are missing, due to technical problems with the site of the Geophysical Observatory (cgo-sreznevskyi.kyiv.ua).

Figure 8.14 shows a graph of the distribution of nitrogen dioxide concentrations according to a 16-directional wind rose. The concentration value of nitrogen dioxide in accordance with each direction of the wind rose is the total value of the concentration observed during the calendar month.

By analyzing the above graphs, it can be argued about the influence of various pollution sources in different months of the considered time period. It should be noted about a significant increase in the concentration of sulfur dioxide in the period from

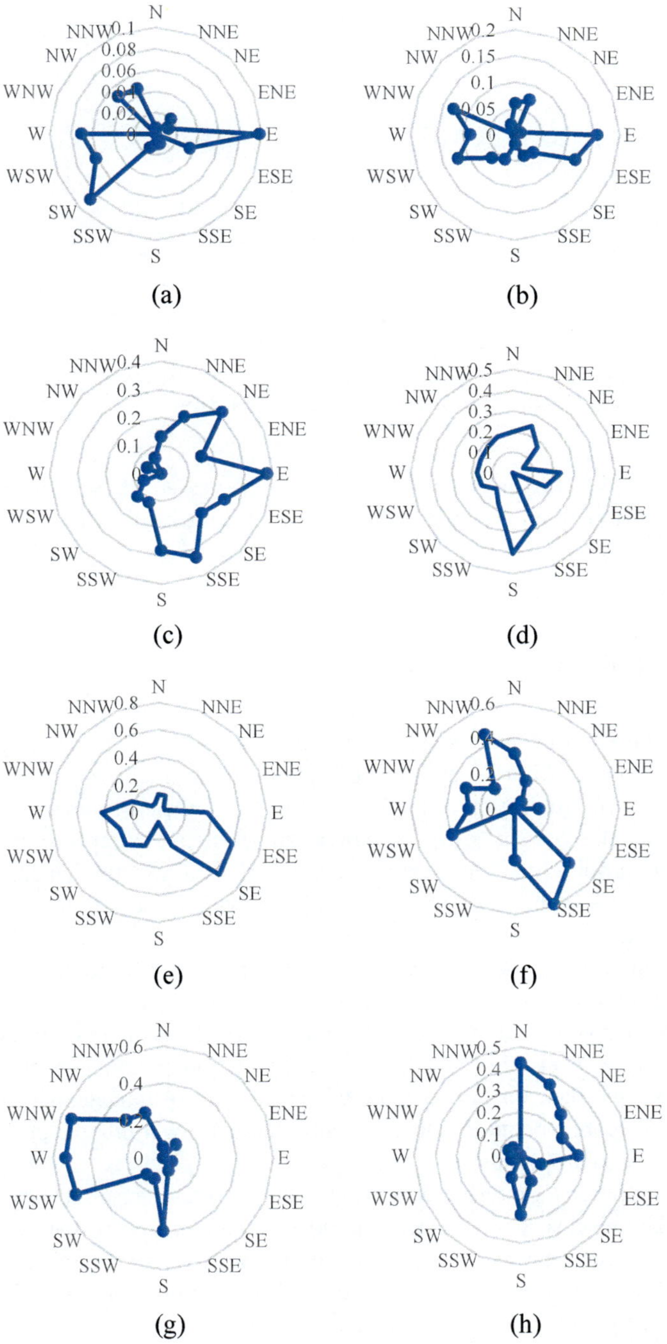

Fig. 8.12 Graphs of the distribution of sulfur dioxide concentrations according to the 16 directional wind rose in Kyiv (metro station Chernihivska) from September 2018 to April 2019: **a** spring 2018; **b** October 2018; **c** November 2018; **d** February 2018; **e** January 2019.; **f** February 2019.; **g** March 2019; **h** April 2019

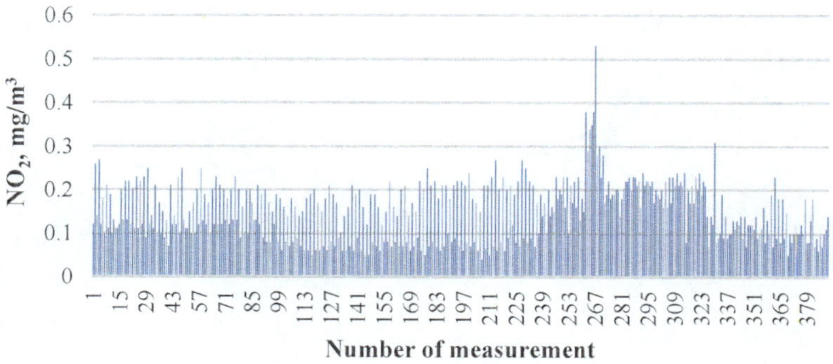

Fig. 8.13 Graph of the values of direct measurements of the nitrogen dioxide concentration in Kyiv (metro station Chernihivska) from September 2018 to April 2019

November 2018 to February 2019 according to the range of S-SE directions, which indicates the functioning of the energy object (Darnitskaya TPP) in the indicated period.

8.5 Air Pollution Monitoring Network

Improving monitoring efficiency is ensured by creating an extensive monitoring network. Recent technological developments in the miniaturization of electronics and wireless communication technologies have led to the emergence of many environmental wireless sensor networks, which can qualitatively improve the process of monitoring air pollution. Such systems facilitate the study of fundamental meteorological processes in the environment, and also provide information on various vital parameters necessary for safe human life.

The main tasks of atmospheric air monitoring systems include: monitoring compliance with air quality standards; identification of threats to natural ecosystems; identification of pollution sources and assessment of their contributions to the overall pollution level; assessing the effects of pollution on human health; obtaining objective source data on the basis of which air quality management is ensured; informing the public about the state of atmospheric air; warning the population of a sharp increase in air pollution [58–62].

The air pollution monitoring system based on a wireless sensor network includes a large number of sensors and their nodes and a communication system, which allows to quickly transfer the received data to the network server. Figure 8.15 shows a functional diagram of the interconnections of measuring modules in a monitoring system for air pollution based on a wireless network.

Figure 8.16 shows the functional diagram of the measuring module of the system as an element of the wireless network [63].

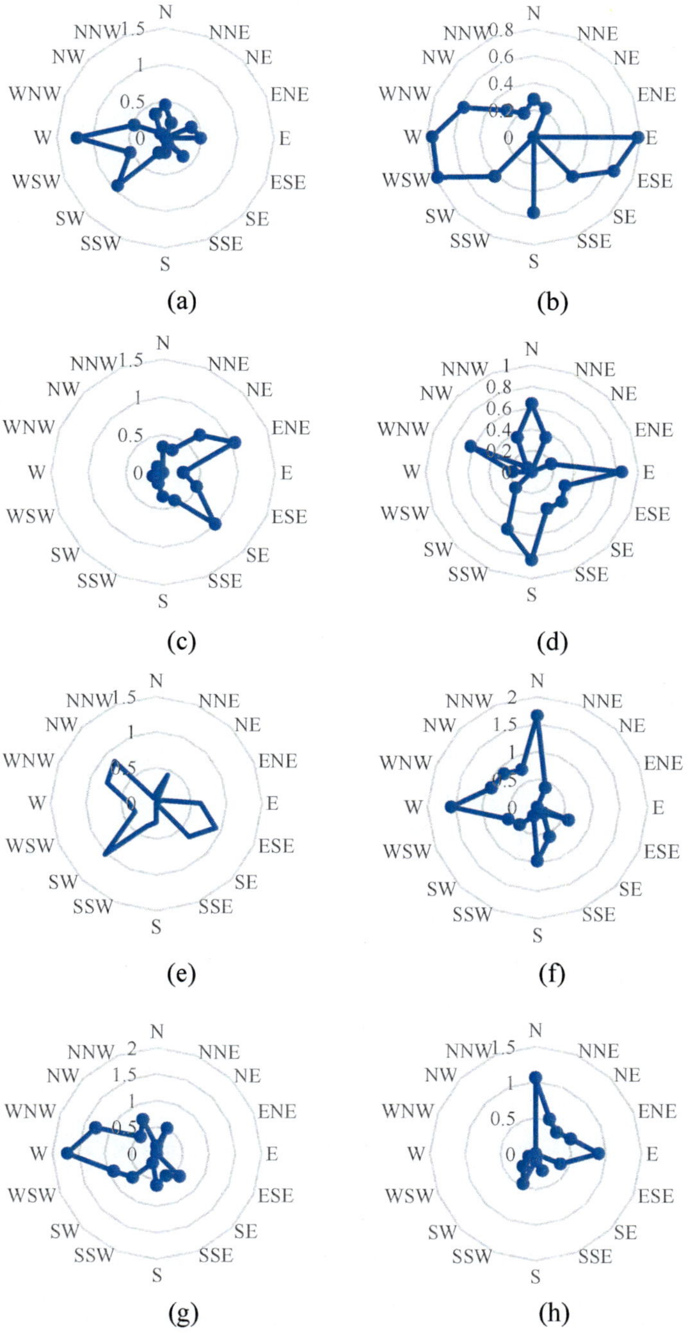

Fig. 8.14 Graphs of the distribution of mitrogen dioxide concentrations according to the 16 directional wind rose in Kyiv (metro station Chernihivska) from September 2018 to April 2019: **a** spring 2018; **b** October 2018; **c** November 2018; **d** February 2018; **e** January 2019; **f** February 2019; **g** March 2019; **h** April 2019

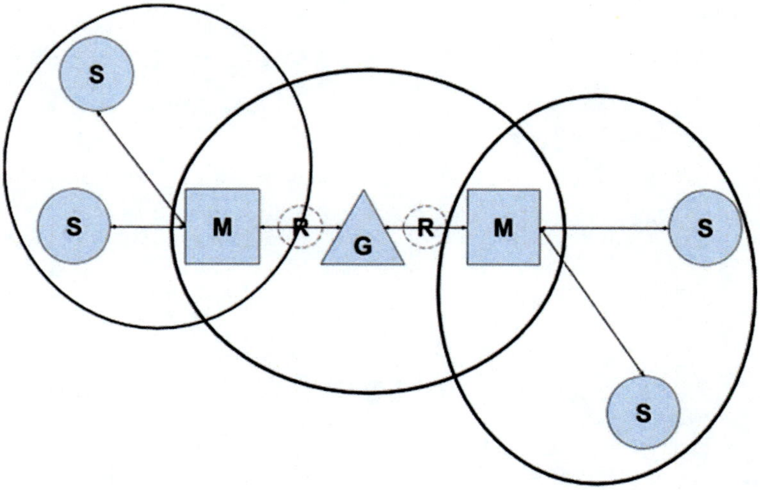

Fig. 8.15 Functional diagram of the relationship of the measuring modules of the air pollution monitoring system (S—slave module; M—master module (all modules in radius G); G—GSM receiver (main module with GSM transmitter), R—receiver)

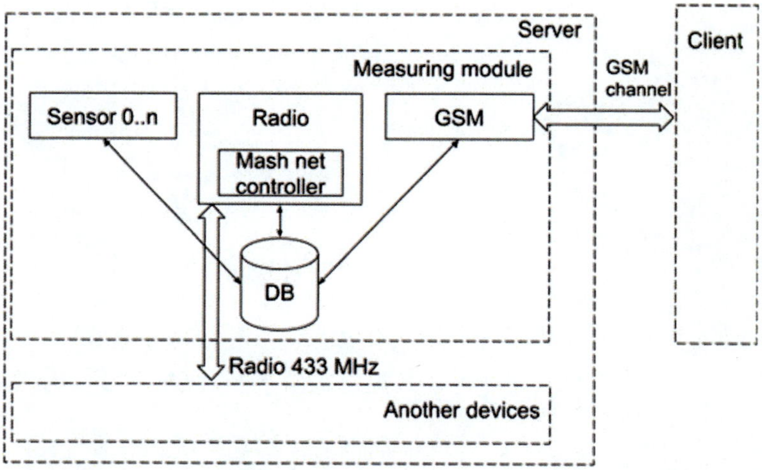

Fig. 8.16 Functional diagram of the measuring module of the system as a mesh element (DB—database)

Figure 8.17 shows an image of measuring modules developed at the Institute of Engineering Thermophysics of the National Academy of Sciences of Ukraine. These modules are elements of the Eco-city pollution monitoring system (eco-city.org.ua), which is currently the largest air pollution monitoring network in Ukraine.

Figure 8.18 shows the locations of the developed measuring modules of the monitoring system. Measuring modules are already operating in Kyiv, Dnipro, Lviv,

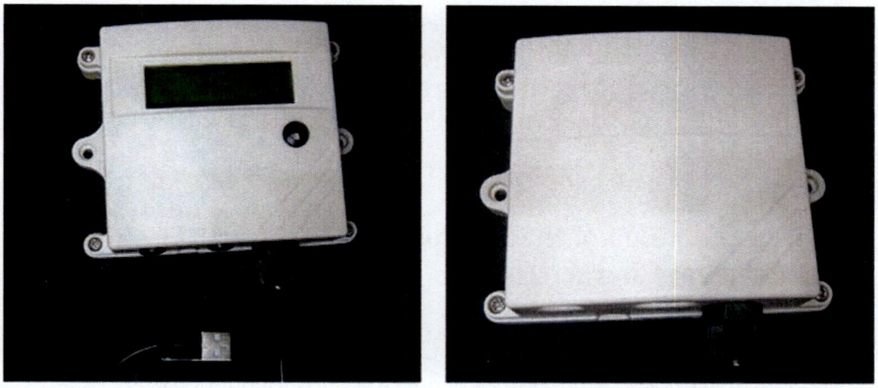

(a) (b)

Fig. 8.17 Photos of the developed measuring modules of the air pollution monitoring network: **a** with data indication; **b** without data indication

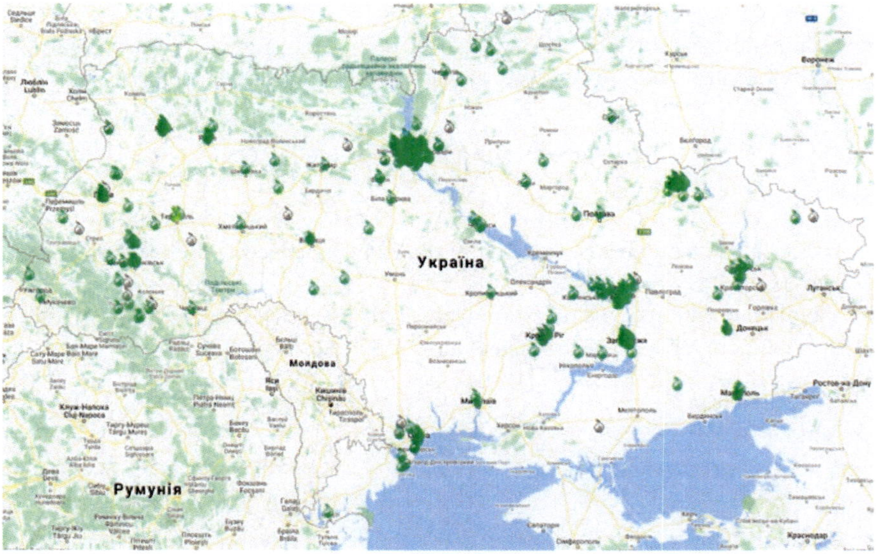

Fig. 8.18 Cartographic image of the locations of the developed measuring modules of the air pollution monitoring system in Ukraine

Kharkiv, Odesa, Zaporizhia, Kryvyi Rih, Poltava, Ivano-Frankivsk, Ternopil and others.

In contrast to published data of the Geophysical Observatory, the developed network data is updated every minute.

The following sensors may be included in the composition of the developed measuring module: PM_1, $PM_{2.5}$, PM_{10}, CO, CO_2, NH_3, NO_x, CH_2O, VOC (volatile organic compounds) and others. The types of sensors used in the developed monitoring system are recommended by The World Air Quality Project. Some parameters can be predictable [64]. The obtained data are also displayed on the global air quality aggregators: waqi.info and aqicn.org. Such system also can based on UAVs [65, 66].

8.6 Determination of the Air Pollution Measures Through Statistical Characteristics

Monitoring of pollutants concentrations at stationary posts can be considered as a set of random variables (single values of atmospheric air pollution).

To assess the level of air pollution over a certain period of time, it is advisable to apply statistical characteristics, also act as certain measures in environmental studies [67–69].

The following is a list of statistical characteristics (measures) that can be applied for a systematic assessment of air pollution at specified locations:

- arithmetic mean of the concentration of pollutants [mg/m^3, μg/m^3, ppm, ppb] (daily average, monthly average, annual average, long-term average):

$$\overline{q}_c = \sum_{i=1}^{n} q_i / n,$$

  where q_i is the number of single concentrations obtained with the corresponding period;
- standard deviation of the measurement results σ (allows to estimate the scatter of the results relative to the average value):

$$\sigma_i = \sqrt{\sum_{i=1}^{n} (q_i - \overline{q})^2 / (n-1)},$$

- coefficient of variation V (allows to assess the degree of variability of the pollutants concentration):

$$V = \sigma / \overline{q};$$

- maximum concentration value of pollutants q_m (the maximum concentration value from the arithmetic mean value);
- maximum concentration of the pollutant with a given probability P of its excess (allows to evaluate the probability of occurrence of concentrations exceeding a given level):

$$q_M^p = \overline{q} \cdot e^Z \sqrt{\ln(1+V^2)} \Big/ \sqrt{1+V^2},$$

where $\overline{q}$—average concentration, at P = 0.1% – Z = 3.08, at P = 1% – Z = 2.33, at P = 5% – Z = 1.65;

- background concentration level C_f (concentration of pollutants generated by all selected sources except the source under study);
- maximum surface concentration of pollutants C_m (calculated value of the concentration in the surface layer of atmospheric air created by individual emission sources):

$$C_m = \frac{AMFmn\eta}{H^2 \sqrt[3]{V_1 \Delta T}};$$

- air pollution indices:
- coefficient for expressing the concentration of pollutant in units of MPC (a);
- repeatability of the concentrations of pollutant in air above a given level:

$$g = 100 \cdot m/n; \quad g_1 = 100\, m_1/n,$$

where n—number of observations for the period under review; m, m_1, m_2—number of cases of excess of one-time concentrations at a stationary post or at all city level posts (a_i MPC), a = 1, 5, 10, 50, respectively;

- atmospheric pollution index with certain pollutants (API)—a quantitative characteristic of the level of atmospheric pollution with pollutants, taking into account differences in the rate of increase of the level of harmful substances, reduced to the level of harmful sulfur dioxide with increasing excess MPC:

$$IAP = (\bar{q}/MPC)^{C_i},$$

where C_i—a constant, takes values of 1.7; 1.3; 1, 0.9 for, respectively, 1, 2, 3, 4 hazard classes of substances, which allows to specify the degree of harmfulness of the i-th substance to the degree of harmfulness of sulfur dioxide;

- complex atmospheric pollution index (CAPI)—a quantitative characteristic of atmospheric pollution, is formed by substances that are present in the atmosphere of a particular agglomeration:

$$CAPI = \sum_{i=1}^{n} API_i,$$

where n—number of considered pollutants (may include all pollutants present in the atmosphere of agglomeration, or only priority substances that determine the state of the atmosphere).

CAPI is used to compare the degree of air pollution in different cities and regions.

The calculation of the API is based on the assumption that at the MPC level all pollutants are characterized by the same effect on humans. With a further increase in concentration, the degree of their harmfulness grows at a different rate, which depends on the hazard class of the substance.

Thus, the development of monitoring of air pollution is an urgent scientific task. Wireless sensor networks have become an important component of environmental science, as scientists and specialists have the opportunity to monitor the parameters of air pollution from anywhere online. The development of environmental research methods, sensors and wireless sensor networks allows to create a new scientific base, conceptually expands the possibilities of solving scientific and practical problems.

References

1. Karpinski, M., Pohrebennyk, V., Bernatska, N., Ganczarczyk, J., Shevchenko, O.: Simulation of artificial neural networks for assessing the ecological state of surface water. In: 18th International Multidisciplinary Scientific Geo-Conference SGEM 2018, pp. 693–700. https://doi.org/10.5593/sgem2018/2.1/S07.088
2. Mitryasova, O., Pohrebennyk, V.: Hydrochemical indicators of water system analysis as factors of the environmental quality state. In: Królczyk, G., Wzorek, M., Król, A., Kochan, O., Su, J., Kacprzyk, J. (eds.) Sustainable Production: Novel Trends in Energy, Environment and Material Systems. Studies in Systems, Decision and Control, vol. 198, pp. 91–104. Springer, Cham (2020). https://doi.org/10.1007/978-3-030-11274-5_7
3. Pohrebennyk, V., Petryk, A.: The degree of pollution with heavy metals of fallow soils in rural administrative units of psary and ploki in Poland. In: 17th International Multidisciplinary Scientific GeoConference SGEM, pp. 967–974 (2017). https://doi.org/10.5593/sgem2017/52/S20.123
4. Pohrebennyk, V., Karpinski, M., Dzhumelia, E., Klos-Witkowska, A., Falat, P.: Water bodies pollution of the mining and chemical enterprise. In: 18th International Multidisciplinary Scientific GeoConference SGEM, pp. 1035–1042 (2018). https://doi.org/10.5593/sgem2018/5.2/S20.133
5. Przystupa, K., Vasylkivsyi, I., Ishenko, V., Pohrebennyk, V., Kochan, O., Su, J.: Assessing air pollution from nuclear power plants. In: 2019 12th International Conference on Measurement, pp. 232–235. Slovakia (2019). https://doi.org/10.23919/MEASUREMENT47340.2019.8780088
6. Isaienko, V., Zaporozhets, A., Babikova, K., Gulevets, D., Savchenko, S: Review of methods and means of monitoring the air pollution. In: Proceedings of the National Aviation University, vol. 80, No. 3, pp. 61–70 (2019). https://doi.org/10.18372/2306-1472.80.14275
7. Popov, O., Iatsyshyn, A., Kovach, V., Artemchuk, V., Kameneva, I., Taraduda, D., Sobyna, V., Sokolov, D., Dement, M., Yatsyshy, T.: Risk assessment for the population of Kyiv, Ukraine as a result of atmospheric air pollution. J. Health Pollut. **10**(25), 200303 (2020). https://doi.org/10.5696/2156-9614-10.25.200303
8. Iatsyshyn, A., Iatsyshyn, A., Artemchuk, V., Kameneva, I., Kovach, V., Popov, O.: Software tools for tasks of sustainable development of environmental problems: peculiarities of programming and implementation in the specialists' preparation. In: E3S Web of Conference, vol. 166, 01001 (2020). https://doi.org/10.1051/e3sconf/202016601001
9. Iatsyshy, A., Artemchuk, V., Zaporozhets, A., Popov, O., Kovach, V.: Mathematical approaches for determining the level of impact of ash-slag dumps of energy facilities on the environment. In: Systems, Decision and Control in Energy I, Studies in Systems, Decision and Control, vol. 298. Springer, Cham. https://doi.org/10.1007/978-3-030-48583-2_1

10. Zaporozhets, A.: Analysis of control system of fuel combustion in boilers with oxygen sensor. Periodica Polytechnica Mech. Eng. **64**(4), 241–248 (2019). https://doi.org/10.3311/PPme.12572
11. Zaporozhets, A.O.: Research of the process of fuel combustion in boilers. In: Control of Fuel Combustion in Boilers. Studies in Systems, Decision and Control, vol. 287, pp. 36–60. Springer, Cham. https://doi.org/10.1007/978-3-030-46299-4_2
12. Zaporozhets, A.O.: Experimental research of a computer system for the control of the fuel combustion process. In: Control of Fuel Combustion in Boilers. Studies in Systems, Decision and Control, vol. 287, pp. 89–123. Springer, Cham. https://doi.org/10.1007/978-3-030-46299-4_4
13. Zaporozhets, A.: Development of software for fuel combustion control system based on frequency regulator. In: CEUR Workshop Proceedings, vol. 2387, pp. 223–230 (2019). https://ceur-ws.org/Vol-2387/20190223.pdf
14. Baralis, E., Cerquitelli, T., Chiusano, S., Garza, P., Kavoosifar, M.R.: Analyzing air pollution on the urban environment. In: 2016 39th International Convention on Information and Communication Technology, Electronics and Microelectronics (MIPRO), pp. 1464–1469. Croatia (2016). https://doi.org/10.1109/MIPRO.2016.7522370
15. Garrido, M.A., Font, R., Conesa, J.A.: Pollutant emissions during the pyrolysis and combustion of flexible polyurethane foam. Waste Manage. **52**, 138–146 (2016). https://doi.org/10.1016/j.wasman.2016.04.007
16. Wielgosinski, G., Lechtanska, P., Namiecinska, O.: Emission of some pollutants from biomass combustion in comparison to hard coal combustion. J. Energy Inst. **90**(5), 787–796 (2017). https://doi.org/10.1016/j.joei.2016.06.005
17. Zaporozhets A.: Methods and means for the control of the fuel combustion process. In: Control of Fuel Combustion in Boilers. Studies in Systems, Decision and Control. Springer, Cham, pp. 1–33 (2020). https://doi.org/10.1007/978-3-030-46299-4_1
18. Zaporozhets A.: Hardware and software implementation of modules of the system of the fuel combustion control process. In: Control of Fuel Combustion in Boilers. Studies in Systems, Decision and Control. Springer, Cham, pp. 61–87 (2020). https://doi.org/10.1007/978-3-030-46299-4_3
19. Kyrylenko Y., Kameneva I., Popov O., Iatsyshyn A., Artemchuk V., Kovach V.: Source term modelling for event with liquid radioactive materials spill. In: Babak V., Isaienko V., Zaporozhets A. (eds.) Systems, Decision and Control in Energy I. Studies in Systems, Decision and Control. Springer, Cham, pp. 261–279 (2020). https://doi.org/10.1007/978-3-030-48583-2_17
20. Ivanitckii, M.S., Sultanov, M.M., Trukhanov, V.M.: Analysis of the influence of operating modes of heat generating plants on the energy and environmental safety of thermal power plants. In: 2020 International Youth Conference on Radio Electronics, Electrical and Power Engineering (REEPE), pp. 1–5. Russia (2020). https://doi.org/10.1109/REEPE49198.2020.9059205
21. Szkarowski, A., Janta-Lipinska, S.: Improving of boiler efficiency by controlling the harmful substances concentration in the combustion products. Struct. Environ. **10**(4), 367–375 (2018). https://doi.org/10.30540/sae-2018-033
22. Zaporozhets, A., Babak, V., Isaienko, V., Babikova, K.: Analysis of the air pollution monitoring system in Ukraine. In: Systems, Decision and Control in Energy I. Studies in Systems, Decision and Control, vol. 298. Springer, Cham (2020). https://doi.org/10.1007/978-3-030-48583-2_6
23. Sukhinov, A.I., Khachunts, D.S., Chistyakov, A.E.: A mathematical model of pollutant propagation in near-ground atmospheric layer of a coastal region and its software implementation. Comput. Math. Math. Phys. **55**, 1216–1231 (2015). https://doi.org/10.1134/S096554251507012X
24. Xiaojun, C., Xianpeng, L., Peng, X.: IOT-based air pollution monitoring and forecasting system. In: 2015 International Conference on Computer and Computational Sciences (ICCCS), pp. 257–260. Noida, India (2015). https://doi.org/10.1109/ICCACS.2015.7361361

25. Kabeel, A.E., Elkelawy, M., El-Din, H.A., El-Banna, A.M., Sathyamurthy, R., Prakash, N.: Computational study of different turbulence models for air impingement jet into main air cross stream. Int. J. Fluid Mech. Res. **46**(5), 459–475 (2019). https://doi.org/10.1615/InterJFluidMechRes.2019026713
26. Zaporozhets, A.O.: Analysis of means for monitoring air pollution in the environment. Sci. Technol. **3**(35), 242–252 (2017). https://doi.org/10.18372/2310-5461.35.11844
27. Peng, L., Zhao, X., Tao, Y., Mi, S., Huang, J., Zhang, Q.: The effects of air pollution and meteorological factors on measles cases in Lanzhou China. Environ. Sci. Pollut. Res. **27**, 13524–13533 (2020). https://doi.org/10.1007/s11356-020-07903-4
28. Cichowicz, R., Wielgosinski, G., Fetter, W.: Dispersion of atmospheric air pollution in summer and winter season. Environ. Monit. Assess. **189**, 605 (2017). https://doi.org/10.1007/s10661-017-6319-2
29. Trinh, T.T., Trinh, R.R., Le, T.T., Nguyen, T.D.H., Tu, B.M.: Temperature inversion and air pollution relationship, and its effects on human health in Hanoi City Vietnam. Environ. Geochem. Health **41**, 929–937 (2019). https://doi.org/10.1007/s10653-018-0190-0
30. Yao, W., Zhang, C., Wang, X., Sheng, J., Zhu, Y., Zhang, S.: The research of new daily diffuse solar radiation models modified by air quality index (AQI) in the region with heavy fog and haze. Energy Convers. Manage. **139**, 140–150 (2017). https://doi.org/10.1016/j.enconman.2017.02.041
31. Zhao, Q., Yao, W., Zhang, C., Wang, X., Wang, Y.: Study on the influence of fog and haze on solar radiation based on scattering-weakening effect. Renew. Energy **134**, 178–185 (2019). https://doi.org/10.1016/j.renene.2018.11.027
32. Bai, Y., Li, Y., Wang, X., Xie, J., Li, C.: Air pollutants concentrations forecasting using back propagation neural network based on wavelet decomposition with meteorological conditions. Atmos. Pollut. Res. **7**(3), 557–566 (2016). https://doi.org/10.1016/j.apr.2016.01.004
33. Oprea, M., Ianache, C., Mihalache, S.F., Dragomir, E.G., Dunea, D., Iordache, S., Savu, T.: On the development of an intelligent system for particulate matter air pollution monitoring, analysis and forecasting in urban regions. In: 2015 19th International Conference on System Theory, Control and Computing (ICSTCC), pp. 711–716. Romania (2015). https://doi.org/10.1109/ICSTCC.2015.7321377
34. Conti, G.O., Heibati, B., Kloog, I., Fiore, M., Ferrante, M.: A review of AirQ models and their applications for forecasting the air pollution health outcomes. Environ. Sci. Pollut. Res. **24**, 6426–6445 (2017). https://doi.org/10.1007/s11356-016-8180-1
35. Ding, W., Zhang, J., Leung, Y.: Prediction of air pollutant concentration based on sparse response back-propagation training feedforward neural networks. Environ. Sci. Pollut. Res. **23**, 19481–19494 (2016). https://doi.org/10.1007/s11356-016-7149-4
36. Pawul, M., Sliwka, M.: Application of artificial neural networks for prediction of air pollution levels in environmental monitoring. J. Ecol. Eng. **17**(4), 190–196 (2016). https://doi.org/10.12911/22998993/64828
37. Tayebi, A., Sheraki, Y., Heydari, M.H.: A meshless method for solving two-dimensional variable-order time fractional advection–diffusion equation. J. Comput. Phys. **340**, 655–669 (2017). https://doi.org/10.1016/j.jcp.2017.03.061
38. Zaporozhets, A.O., Khaidurov, V.V.: Mathematical models of inverse problems for finding the main characteristics of air pollution sources. Water Air Soil Pollut. **231**, 563 (2020). https://doi.org/10.1007/s11270-020-04933-z
39. Bougoudis, I., Demertzis, K., Iliadis, L.: HISYCOL a hybrid computational intelligence system for combined machine learning: the case of air pollution modeling in Athens. Neural Comput. Appl. **27**, 1191–1206 (2016). https://doi.org/10.1007/s00521-015-1927-7
40. Adams, M.D., Kanaroglou, P.S.: Mapping real-time air pollution health risk for environmental management: combining mobile and stationary air pollution monitoring with neural network models. J. Environ. Manage. **168**, 133–141. https://doi.org/10.1016/j.jenvman.2015.12.012
41. Yi, W.Y., Lo, K.M., Mak, T., Leung, K.S., Leung, Y., Meng, M.L.: A survey of wireless sensor network based air pollution monitoring systems. Sensors **15**(12), 31392–31427 (2015). https://doi.org/10.3390/s151229859

42. Zaporozhets, A.O.: Correlation analysis between the components of energy balance and pollutant emissions. Water Air Soil Pollut. **232**, 114 (2021). https://doi.org/10.1007/s11270-021-05048-9
43. Marjovi, A., Afrire, A., Martinoli, A.: High resolution air pollution maps in urban environments using mobile sensor networks. In: 2015 International Conference on Distributed Computing in Sensor Systems, pp. 11–20. Brazil (2015). https://doi.org/10.1109/DCOSS.2015.32
44. Jamil, M.S., Jamil, M.A., Mazhar, A., Ikram, A., Ahmed, A., Munawar, U.: Smart environment monitoring system by employing wireless sensor networks on vehicles for pollution free smart cities. Proc. Eng. **107**, 480–484 (2015). https://doi.org/10.1016/j.proeng.2015.06.106
45. Morelli, X., Rieux, C., Cyrys, J., Forsberg, B., Slama, R.: Air pollution, health and social deprivation: a fine-scale risk assessment. Environ. Res. **147**, 59–70 (2016). https://doi.org/10.1016/j.envres.2016.01.030
46. Kuchansky, A., Biloshchytskyi, A., Andrashko, Y., Vatskel, V., Biloshchytska, S., Danchenko, O., Vatskel, I.: Combined models for forecasting the air pollution level in infocommunication systems for the environment state monitoring. In: 2018 IEEE 4th International Symposium on Wireless Systems within the International Conferences on Intelligent Data Acquisition and Advanced Computing Systems (IDAACS-SWS), pp. 125–130. Ukraine (2018). https://doi.org/10.1109/IDAACS-SWS.2018.8525608
47. Zaporozhets, A., Eremenko, V., Redko, O.: Metrological assessment of the indirect method of measuring the concentration of oxygen in the air. In: 2019 IEEE 8th International Conference on Advanced Optoelectronics and Lasers (CAOL), pp. 640–643. Sozopol, Bulgaria (2019). https://doi.org/10.1109/CAOL46282.2019.9019506.
48. Zaporozhets, A.O., Redko, O.O., Babak, V.P., Eremenko, V.S., Mokiychuk, V.M.: Method of indirect measurement of oxygen concentration in the air. Naukovyi Visnyk Natsionalnoho Hirnychoho Universytetu **5**, 105–114 (2018). https://doi.org/10.29202/nvngu/2018-5/14
49. Babak, V.P., Babak, S.V., Myslovych, M.V., Zaporozhets, A.O., Zvaritch, V.M.: Methods and models for information data analysis. In: Diagnostic Systems for Energy Equipments. Studies in Systems, Decision and Control, vol. 281, pp. 23–70. Springer, Cham (2020). https://doi.org/10.1007/978-3-030-44443-3_2
50. Belov, A.A., Proskuryakov, A.Yu.: Structures and algorithms of the automated ecological monitoring system of industrial enterprises. In: 2019 International Multi-Conference on Industrial Engineering and Modern Technologies (FarEastCon), pp. 1–5. Russia (2019). https://doi.org/10.1109/FarEastCon.2019.8934926
51. Kropotov, Y.A., Kolpakov, A.A., Belov, A.A., Proskuryakov, A.Yu.: Conditional local approximation in the representation of a discrete process by interpolation polynomials. In: 2019 International Siberian Conference on Control and Communications (SIBCON), pp. 1–4. Russia (2019). https://doi.org/10.1109/SIBCON.2019.8729569
52. Babak, V., Eremenko, V., Zaporozhets, A.: Research of diagnostic parameters of composite materials using Johnson distribution. Int. J. Comput. **18**(4), 483–494 (2019)
53. Eremenko, V., Zaporozhets, A., Babak, V., Isaienko, V., Babikova, K.: Using hilbert transform in diagnostic of composite materials by impedance method. Period. Polytech. Electr. Eng. Comput. Sci. **64**(4), 334–342 (2020). https://doi.org/10.3311/PPee.15066
54. Carta, J.A., Ramirez, P., Bueno, C.: A joint probability density function of wind speed and direction for wind energy analysis. Energy Convers. Manage. **49**(6), 1309–1320 (2008). https://doi.org/10.1016/j.enconman.2008.01.010
55. Kumar, P., Fennel, P., Britter, R.: Effect of wind direction and speed on the dispersion of nucleation and accumulation mode particles in an urban street canyon. Sci. Total Environ. **402**(1), 82–94 (2008). https://doi.org/10.1016/j.scitotenv.2008.04.032
56. Ozay, C., Celikates, M.S.: Statistical analysis of wind speed using two-parameter Weibull distribution in Alaçatı region. Energy Convers. Manage. **121**, 49–54 (2016). https://doi.org/10.1016/j.enconman.2016.05.026
57. Han, Q., Hao, Z., Hu, T., Chu, F.: Non-parametric models for joint probabilistic distributions of wind speed and direction data. Renew. Energy **126**, 1032–1042 (2018). https://doi.org/10.1016/j.renene.2018.04.026

58. McKercher, G.R., Vanos, J.K.: Low-cost mobile air pollution monitoring in urban environments: a pilot study in Lubbock. Texas. Environ. Technol. **39**(12), 1505–1514 (2018). https://doi.org/10.1080/09593330.2017.1332106
59. Rai, A.C., Kumar, P., Pilla, F., Skouloudis, A.N., Sabatino, S.D., Ratti, C., Yasar, A., Rickerby, D.: End-user perspective of low-cost sensors for outdoor air pollution monitoring. Sci. Total Environ. **607–608**, 691–705 (2017). https://doi.org/10.1016/j.scitotenv.2017.06.266
60. Shahbazi, H., Karimi, S., Hosseini, V., Yazgi, D., Torbatian, S.: A novel regression imputation framework for Tehran air pollution monitoring network using outputs from WRF and CAMx models. Atmos. Environ. **187**, 24–33 (2018). https://doi.org/10.1016/j.atmosenv.2018.05.055
61. Kizel, F., Etzion, Y., Shafran-Nathan, R., Levy, I., Fishbain, B., Bartonova, A., Broday, D.M.: Node-to-node field calibration of wireless distributed air pollution sensor network. Environ. Pollut. **233**, 900–909 (2018). https://doi.org/10.1016/j.envpol.2017.09.042
62. Boubrima, A., Matigot, F., Bechkit, W., Rivano, H., Ruas, A.: Optimal deployment of wireless sensor networks for air pollution monitoring. In: 2015 24th International Conference on Computer Communication and Networks (ICCCN), pp. 1–7. USA (2015). https://doi.org/10.1109/ICCCN.2015.7288443
63. Babak, V.P., Babak, S.V., Myslovych, M.V., Zaporozhets, A.O., Zvaritch, V.M.: Technical provision of diagnostic systems. in: diagnostic systems for energy equipments. In: Studies in Systems, Decision and Control, vol 281, pp. 91–133. Springer, Cham (2020). https://doi.org/10.1007/978-3-030-44443-3_4
64. Babak, V.P., Mokiychuk, V.M., Zaporozhets, A.A., Redko, A.A.: Improving the efficiency of fuel combustion with regard to the uncertainty of measuring oxygen concentration. Eastern-Euro. J. Enterprise Technol. **6**(8, 84), 54–59 (2016). https://doi.org/10.15587/1729-4061.2016.85408
65. Zaporozhets, A., Kovtun, S., Dekusha, O.: System for monitoring the technical state of heating networks based on UAVs. In: Shakhovska, N., Medykovskyy, M. (eds.) Advances in Intelligent Systems and Computing IV. CCSIT 2019. Advances in Intelligent Systems and Computing, vol. 1080, pp. 935–950. Springer, Cham (2020). https://doi.org/10.1007/978-3-030-33695-0_61
66. Zaporozhets, A.: Review of quadrocopters for energy and ecological monitoring. In: Babak V., Isaienko V., Zaporozhets A. (eds.) Systems, Decision and Control in Energy I, Studies in Systems, Decision and Control, vol. 298. Springer, Cham. https://doi.org/10.1007/978-3-030-48583-2_2
67. Stafoggia, M., Breitner, S., Hampel, R., Basagana, X.: Statistical approaches to address multipollutant mixtures and multiple exposures: the state of the science. Current Environ. Health Reports **4**, 481–490 (2017). https://doi.org/10.1007/s40572-017-0162-z
68. Xiao, C., Chang, M., Guo, P., Gu, M., Li, Y.: Analysis of air quality characteristics of Beijing–Tianjin–Hebei and its surrounding air pollution transport channel cities in China. J. Environ. Sci. **87**, 213–217 (2020). https://doi.org/10.1016/j.jes.2019.05.024
69. Luo, H., Han, Y., Lu, C., Yang, J., Wu, Y.: Characteristics of surface solar radiation under different air pollution conditions over Nanjing, China: observation and simulation. Adv. Atmos. Sci. **36**, 1047–1059 (2019). https://doi.org/10.1007/s00376-019-9010-4